Preparing for Algebra by Building the Concepts

Preliminary Edition

Martha Haehl

Maple Woods Community College

Kansas City, Missouri

UPPER SADDLE RIVER NJ 07458

Library of Congress Cataloging-in-Publication Data

Haehl, Martha.
Preparing for algebra by building the concepts / Martha Haehl.
p. cm.
Includes index.
ISBN 0-13-608878-3
1. Mathematics. I. Title.
QA 135.5.H32 1997
513'.1--dc21 97-42050
CIP

Acquisitions Editor: Karin Wagner
Editorial Assistant/Supplements Editor: Kate Marks
Editor-in-Chief: Jerome Grant
Editorial Director: Tim Bozik
Editorial/Production Supervision: Barbara Mack
Senior Managing Editor: Linda Mihatov Behrens
Executive Managing Editor: Kathleen Schiaparelli
Assistant Vice President of Production and Manufacturing: David W. Riccardi
Manufacturing Buyer: Alan Fischer
Manufacturing Manager: Trudy Pisciotti
Marketing Manager: Jolene Howard
Director of Marketing: John Tweeddale
Marketing Assistant: Jennifer Pan
Art Director: Jayne Conte
Creative Director: Paula Maylahn
Cover Designer: Wendy Allig Judy
Cover Illustration: Sarah Marquis

Simon & Schuster/A Viacom Company
Upper Saddle River, New Jersey 07458

Printed in the United States of America

10 9 8 7 6 5 4 3 2 1

ISBN 0-13-608878-3

Prentice-Hall International (UK) Limited, *London*
Prentice-Hall of Australia Pty. Limited, *Sydney*
Prentice-Hall Canada Inc., *Toronto*
Prentice-Hall Hispanoamericana, S. A., *Mexico City*
Prentice-Hall of India Private Limited, *New Delhi*
Prentice-Hall of Japan, Inc., *Tokyo*
Simon & Schuster Asia Pte. Ltd., *Singapore*
Editora Prentice-Hall do Brasil, Ltda., *Rio de Janeiro*

Preparing for Algebra
by Building the Concepts

Martha Haehl

Table of Contents

Chapter 4 Fractions

Chapter 5 Ratios, Proportions, Unit Conversions and Rates

Chapter 6 Percentages

Chapter 7 Statistical Information, Pie and Bar Charts, and Other Graphical Data

Chapter 8 Floor Plans and Other Summary Exercises

Preface

To the Instructor:

Memorizing arithmetic and algebraic processes out of the context of real-life does not necessarily lead to the understanding of numbers, numeric concepts, and mathematical concepts necessary for success in future mathematics courses. Nor does it necessarily prepare the student to apply the skills to a real-life situation. Prior to calculators, proficiency in pencil-and-paper arithmetic skills necessarily preceded higher mathematical thinking skills and using the skills in real-life applications. With calculators we can address algorithmic skills, mathematical concepts, and applications simultaneously. Calculators can even be used as a tool to help understand numeric concepts and skills. The materials in this book are designed for students to:

- use hands-on materials and discovery exercises to learn "pencil-and-paper" arithmetic and certain algebraic manipulation skills;
- solve multi-task, real-life problems with the aid of a calculator--the "pretty number" is no longer necessary; and
- work with hands-on materials as well as calculators to discover and reinforce mathematical rules and concepts.

The first materials were written as supplements to a traditionally taught basic math class. In the floor plan exercises, students add, subtract, multiply, divide whole numbers and fractions, make unit conversions, work with linear as well as square units, and prorate bills according to percentage of floor space. Students found these exercises difficult. Many students could not work with the numbers involved (with or without a calculator) or spot a mistake that yielded an unreasonable answer. Students also had difficulty analyzing many tasks in a single exercise. I began the task of writing materials to build the skills and mathematical thinking students need in order to solve such multi-task problems. For someone who was attracted to mathematics for its intrinsic beauty and the absence of term papers, writing with a pragmatic approach has been quite a challenge.

The predominant philosophies behind writing the book are:

- Every adult brings with her/him a wealth of experience and knowledge. We best set up the learning environment when we build on their experiences.
- Most adult students who take a developmental mathematics class have developed a fear and/or dislike of mathematics.
- Generally, adults who take a developmental mathematics class are capable of thinking mathematically and succeeding in future mathematics classes.
- A student who does not learn and retain mathematical processes previously taught in abstraction can benefit from learning the algorithms using manipulatives.

Special features include

1. Hands-on activities to introduce concepts and help build a number sense
2. Pencil-and-paper arithmetic skills taught along with efficient use of calculator
3. Use of calculator for tedious computations
4. Use of calculator to discover mathematical patterns and rules
5. Emphasis on estimation and reasonableness of answers
6. Emphasis on vocabulary and mathematical symbolism
7. Integration of topics--geometric, algebraic, and numeric

8. Real-life, multi-task projects and exercises
9. Materials needed identified at the beginning of each chapter as well as at the beginning of each exercise or activity
10. Vocabulary, processes, and formulas boxed-in for quick reference
11. Vocabulary, processes, and formulas summarized at the end of each chapter with page references to the more in-depth coverage
12. A summary exam at the end of each chapter
13. A summary chapter that includes floor plan and other summary exercises and a summary exam

Integrating technology with pencil-and-paper skills and integrating numeric, algebraic and geometric topics has presented particular challenges in the structure and order of the book. As a result, the chapters tend to be long and arranged differently than the more traditional text. In using the money model for arithmetic, for example, arithmetic of whole numbers is not separated from numbers with places after the decimal point.

Students who do not know the times tables or the names of numbers can work additional problems in Appendix I and Appendix II. Additional tips for doing string operations on a calculator to find a decimal approximation to an arithmetic expression involving fractions are in Appendix III.

Acknowledgments:

It is hard to know where to begin with acknowledgments, so I will just jump in. Publishing any mathematics text book at this time is risky business in that the market is a rapidly moving target. Prentice Hall took a chance on this book and has given it considerable attention. Prentice Hall runs a very professional operation with support in editing, proofreading, marketing, setting up class tests and reviews. Thanks to Karin Wagner, Cindy Kilborn, Melissa Acuna, and many others who have worked on the production of this book.

Deena Cloud, the developmental editor, played a major role in transforming an unruly manuscript into a cohesive text book. Her talents are far reaching. In a very short period of time, she contributed substantially to concise, clear writing as well as correct mathematics. Since this project did not start out to be a text book five years ago, and integration of topics makes chapter organization difficult, Deena performed magic on the structure of the book.

Fran Endicott Armstrong, Ph. D., mathematics instructor at St. Louis Community College at Meramec campus, has considerable experience in developing and using hands-on activities for math instruction. She not only was influential in the creation of the materials, but carefully read the manuscript in its final stages and helped with rewording and revision.

The following class-testers have been invaluable in their suggestions and encouragement. Debra Swedberg and Laura Bracken in particular have provided substantial feedback as well as encouragement. Below is a list of the class-testers:

Debra Swedberg, Casper College, Casper, Wyoming
Laura Bracken, Lewis-Clark State College, Lewiston, Idaho
Bridget Gold, Longview Community College, Kansas City, Mo.
Sara Woodward, Maple Woods Community College, Kansas City, Mo.
William Capotosto, Maple Woods Community College, Kansas City, Mo.

The comments by reviewers at several stages helped mold the project. They not only caught mistakes but made great recommendations for improvement. Reviewers:
Elizabeth Chu, Suffolk Community College
Irene Doo, Austin Community College
Rebecca Easley, Rose State College
Terry Y. Fung, Kean College of New Jersey
Sue King, Kansas Newman College
Frank A. Rivera, Glendale Community College
Debra M. Swedberg, Casper College
Lana Taylor, Siena Heights College

Many friends helped with proofreading and suggested projects. Sarah Marquis, friend and artist, provided art work for the cover design, suggested some project ideas, scanned in a protractor design, and helped on several occasions with the tedious task of checking page references and page numbers. Jennifer Tacy did an excellent job helping proofread the final camera-ready copy of the book. Lynn Snyder, mathematician and computer programmer, proofread some of the earlier manuscripts and offered many suggestions and ideas.

Several people early in the process read the materials and made suggestions as well as gave me ideas for exercises. Karin McAdams, GED instructor, tried many of the early materials and suggested the ruler exercises for division of fractions. Dr. Sue Sundberg, Central Missouri State University, gave me the idea of using red and white beans for signed numbers and for the percent thermometer image. Beverlye Brown, English Instructor, helped me incorporate writing into mathematics classes. Many friends told me how they use mathematics at home or work. Their experiences were the basis for some of the projects in the book.

Thanks to all of the math instructors who have given me ideas in person, with their presentations at conferences, or with comments on the MATHEDD listserve. Thanks to Brian Smith for seeing that the MATHEDD discussion group continues to exist and for the many people who put together conferences and workshops so that we can learn from each other.

Many friends and colleagues at my school and district, Maple Woods Community College (one of the Metropolitan Community Colleges in Kansas City, Missouri), have continually encouraged me through this very long process. It would have been difficult to write the book without the support from colleagues--faculty, staff, and administrators.

Such a project would have been difficult without the interest and support of many family members. A particular thanks goes to my sister, Ruth Engle, and to my mother, Glennis Harmonson, who read the manuscript and said "This looks like it would be fun. It makes me want to teach it" -- a sentiment she had not expressed since retiring from teaching school.

Martha Haehl, email haehl@kcmetro.cc.mo.us
Maple Woods Community College, Kansas City, Missouri
B.S., Mathematics, Wayland Baptist University
M.A., Mathematics, University of Kansas

To the Student:

"Mathematics is the study of patterns" (Lynn Steen). A solid understanding of numbers, arithmetic, and arithmetic concepts builds the foundation for the study of mathematics. Throughout the book, concepts and skills are introduced with hands-on group activities designed to understand the patterns of arithmetic and algebra. Even if you can add numbers, for example, the hands-on activities are designed so that you learn why the process works. As a topic of study, mathematics is too big to memorize your way through to success. To remember the skills needed in subsequent mathematics classes as well as to apply mathematics to personal and professional decision making, you need to learn the concepts behind the mathematical procedures you learn.

Working with other students can help you learn mathematics. Listening to an instructor who "knows the answers" already may give you false impressions of mathematics--that being successful at mathematics means you know how to analyze a problem at the first look and that deriving a correct answer is a straight path, done efficiently and correctly the first time. Sometimes finding a direct path to a correct answer comes after several false starts in the wrong direction. Working both inside and outside of class with at least one other person helps you to explore a problem and check answers. When working mathematics by yourself, if you are stuck you often stay stuck. When working with someone else, you can help each other analyze the problem differently and then move on. Many adults do not have the time to form study groups; however, here are some ways even for a busy person to work with someone else outside of class:

1. Form a "distance" study group by exchanging phone numbers or email addresses.
2. Help a grade school student, middle school, or high school student with the mathematics concepts you are learning. They can benefit from the extra attention and mathematics practice. You can benefit from the practice and different insights someone younger may have.
3. Communicate with coworkers about mathematics. Many of them will learn from you as well as help you out when you are stuck.

Many activities require materials. You will need a scientific or graphics calculator. I recommend the type that allows you to input an entire arithmetic expression before computing the answer. If subsequent mathematics classes at your school require a graphics calculator, you might consider purchasing (or renting, if your school offers that option) the type of calculator that you can use later. Most of the other materials are common house-hold items (ruler or measuring tape, red and white beans, empty cereal boxes, for example) or can be made from paper or cardboard. The activities are designed around inexpensive materials so that schools as well as students can afford to use the materials at school as well as at home.

Most real-life problems deal with tedious computations that are best done with a calculator. It is easy, however, to press the wrong key or sequence of keys on the calculator. Coupled with efficient calculator use, you need to understand the concepts and processes of arithmetic and estimation in order to determine whether the answers derived using calculators are reasonable. Calculators can also be a tool in learning a number sense. While learning pencil-and-paper arithmetic, use a calculator to check your accuracy.

When doing computations on the calculator, use your estimation skills to determine if the answer is reasonable.

Without a calculator, you should:

1. Know the times tables. If you are weak on these, do exercises in Appendix II.
2. Understand our "base 10 system" of arithmetic; in other words, understand place value and know the result of multiplying or dividing any number by 10, 100, 1000, 0.1, 0.01, and so on.
3. Be able to add, subtract, multiply and divide positive and negative integers.
4. Be able to add, subtract, multiply, and divide numbers with decimal points.
5. Be able to add, subtract, multiply, and divide simple fractions.
6. Estimate answers to "messy" arithmetic expressions.
7. Solve simple algebraic equations and proportions.
8. Solve simple problems involving ratios and percents.

With a calculator, you should be able to:

1. Analyze and solve real-life problems.
2. Use the calculator efficiently to evaluate decimal approximations to numeric expressions.
3. Solve algebraic equations or simplify algebraic expressions involving "messy" arithmetic.

Learning mathematics involves doing the work. People do not become a great basketball player by watching the professionals play ball. Neither do they become good at mathematics merely by watching a professional (generally the instructor) do math. Be active in your own learning. Participate in the discovery exercises to understand the concepts rather than just memorize a skill; then practice and review the skills and concepts outside of class. When you do not understand the concept, get help from the instructor, a tutor or a classmate.

Martha Haehl

Chapter 1 Arithmetic of Whole and Decimal Numbers, the Money Model

Key Topics

- Addition, subtraction, multiplication, and division of whole and decimal numbers
- Approximating arithmetic
- Place value, rounding-off and significant digits.
- "$<, \leq, >, \geq$" and number comparisons.
- Mathematics in the news
- Averages
- Combining like terms

Materials Needed for the Exercises (one set per group)

- Fake money--at least 40 each of $1-bills, $10-bills, $100-bills, $1000-bills, $10,000-bills, and dimes and pennies (red and white beans may be substituted for the coins). Do not include any denomination of bill or coin not listed above.
- Library materials (encyclopedia, local and world maps, *World Almanac*)
- Car and odometer (optional)
- Scientific or graphics calculator
- Bag of individually wrapped candy
- Metric ruler or meter stick
- 40 small cardboard squares (about 1/2 in. x 1/2 in.), 40 large cardboard squares (about 1 in. x 1 in.), 40 toothpicks
- 50 red beans and 50 white beans

Section 1-1 *Introduction--What Is a Number?*

The difficulty in understanding numbers and arithmetic comes when we do not attach numbers to real-life situations. Numbers in and of themselves are abstract concepts. Ask a 3-year-old how old she is and she may hold a certain number of fingers in the air as an answer to the question, but she has no idea what 3 years means. This certainly proves true on her fourth birthday when she has to relearn the response to the question of how old she is. The same child, however, may have some working understanding of the number 3, because on request she can pick up 3 crayons, 3 blocks, 3 cookies, or she can stomp her foot 3 times.

At any age, we understand numbers when we associate them with real things, which we call **units**, that have meaning in our lives. Units may be apples, inches, dollars, gallons or years. We only compare numbers for relative size when the units are comparable. For example, we know that 10 inches is longer than 8 inches, because we are comparing inches to inches. If we know enough about how the metric system (meters, centimeters, etc.) compares with the English system (feet, inches, etc.) we can compare the sizes of 10 meters and 9 yards to determine which is longer. Although we are using different units, we can compare the units since they both measure length. To compare 10 meters with 9 yards, however, we would need to know how 1 meter compares to 1 yard.

The following exercise is designed to observe in what context it makes sense to compare the numbers 7 and 9.

Exercise 1--Which Is Bigger, 7 or 9? (Group)

Purpose of the Exercise

- To understand the abstract nature of numbers

1. Discuss the following questions:
   a. Which is bigger, 7 apples or 9 oranges?
   b. Which is bigger, 7 McDonald's hamburgers or 9 White Castle hamburgers?
   c. Which is bigger, 7 "blue spot" softballs or 9 "blue spot" softballs?
   d. Which is bigger, 7 softballs or 9 baseballs?
   e. Which is bigger, 7 feet or 9 inches?
   f. Which is bigger, 7 kilometers or 9 kilometers?
   g. Which is bigger, 7 kilometers or 9 miles?
   h. Which is bigger, 7 miles or 9 kilometers?
   i. Which is bigger, 7 jumps or 9 bicycles?
   j. Which is bigger, 7 carrot cakes or 9 lemon meringue pies?
   k. Which is bigger, 7 circles (radius = 3 in.) or 9 squares ($\text{side} = \frac{1}{3}$ ft.)?
   l. Which is bigger, 7 years or 9 countries?
   m. Which is bigger, 7 or 9?

2. List the parts of Question 1 for which you were able to determine which is bigger, 7 or 9.

____

3. List the parts of Question 1 for which you could determine which is bigger if you had more information.

____

4. List the parts of Question 1 where no answer is possible.

____

5. Explain what is involved in deciding which is bigger, 7 or 9. Why were some parts of question 1 impossible to answer?

____

____

____

In solving math problems, we very often use letters as a shorthand notation to represent unknown numbers. In the following exercise we will use letters to represent objects, unknown numbers, or unknown units. When you are comparing "pure" numbers to each other for size, it is implied that the numbers count or measure something comparable.

Exercise 2--Using Numbers and Letters (Group)

Suppose that X represents a marble of a certain size, while Y represents a birthday candle of a certain size. L represents still another kind of object of a certain size.

1. For each of the following, circle "T" if the statement is true, "F" if the statement is false, and "N" if there is not enough information to answer the question.

a. 7 X's	are smaller than	9 Y's	T	F	N
b. 7 Y's	are bigger than	9 Y's	T	F	N
c. 7 L's	are bigger than	9 L's	T	F	N
d. 7 L's	are smaller than	9 X's	T	F	N
e. 7 L's	are smaller than	9 L's	T	F	N

2. For each of the above questions marked "N," explain why you could not determine if the statement was true or false.

____

____

____

3. Suppose that X, Y, and L each represent a different positive number. "7X" means 7 groups of X or, 7 times the number that X represents. In each of the following, circle "T" if the statement is true, "F" if the statement is false and "N" if there is not enough information to answer the question.

a. 7X	is smaller than	9Y	T	F	N
b. 7Y	is bigger than	9Y	T	F	N
c. 7L	is bigger than	9L	T	F	N
d. 7L	is smaller than	9X	T	F	N
e. 7L	is smaller than	9L	T	F	N

4. For each of the above questions marked "N," explain why you could not determine whether the statement was true or false.

___

___

In summary, when we say that 11 is bigger than 5, it is assumed that we are comparing 11 and 5 of the same size and type of object. In mathematics, when letters such as X and Y are used to represent numbers, each time you see an X it represents the same number as every other X in the same problem and each time you see a Y in a particular problem, it represents the same number as every other Y in that problem.

In Section 1-2 we will create our own unit of measurement, as people did before measurements were standardized. Don't worry; we won't have to use a slate, quill, and knots in a rough rope to make measurements--you will have paper, cardboard straight edges, and pencils.

Section 1-2: *Place Value*

Before studying arithmetic, we need to develop a solid concept of numbers and digits within numbers and the importance of the "place" or position of digits in a number. First, here is some terminology.

> **Vocabulary**:
> There are 10 **digits**. They are 0, 1, 2, 3, 4, 5, 6, 7, 8, and 9. A number may have one or more digits in it.

Our numeric and arithmetic system is based on 10 digits instead of 7 or 9 or some other number of digits because we have 10 fingers--including thumbs of course. Throughout time people have used their fingers to help them count. Our fingers and toes are even called digits.

Example 1:
The number 35 is a two-digit number. The digits are 3 and 5.
The number 7095 is a four-digit number. The digits are 7, 0, 9, and 5.
The number 7.089 has the four digits, 7, 0, 8, and 9.

The position of the digits in a number determines the value of the number. The different positions have different "place values." The arithmetic in most of this chapter will be done with a money model. To gain a solid understanding of place value and the importance of the number "10" to numbers and arithmetic, and why arithmetic works as it does, we will only use the pennies, dimes, $1-bills, $10-bills, $100-bills, $1000-bills and $10,000-bills in the money exchanges.

Exercise 1--Place Value and the Money Model (Group)

Purpose of the Exercise
- To use money to understand what is meant by place value

Materials Needed
- Fake money--Each group needs at least 40 each of $1-bills, $10-bills, $100-bills, $1000-bills, $10000-bills and dimes and pennies. <u>Do not include</u> any denomination of bill or coin not listed.

1. Take 22 $1000-bills, no $100-bills, 31 $10-bills, 25 $1-bills, and no dimes or pennies. In the blank below, record how much money you have.

Amount of Money	$10,000's	$1000's	$100's	$10's	$1's	dimes	pennies
__________	____	22	0	31	25	0	0

2. Choose one person to be the banker, to make change as necessary and make your stack of money more efficient in the following manner:
   - Since there are no pennies or dimes, start with the dollars and trade in 20 of the 25 $1-bills for 2 $10-bills. Add the 2 $10-bills to the 31 $10-bills you already have.
   - Continue this type of money exchange until you have the fewest possible bills.
   - After all of the exchanges are finished, how much money do you have? Record your answer in the blank below.
   - How many of each kind of bill do you have in the pile? Record your answers in the blanks below.
   - If you have a different amount of money than when you started, repeat the steps above to find your mistake.

Amount of Money	$10,000's	$1000's	$100's	$10's	$1's	dimes	pennies
________________	_____	_____	_____	____	____	_____	_____

You should not have more than 9 of any one type of bill. Check to make sure this is true. If not, keep trading until you have the fewest possible bills. The digits of the number in the amount of money blank should be the same as the numbers put in the blanks as the count of each kind of bill.

3. Take 15 $1000-bills, 40 $100-bills, 24 $10-bills, 2 $1-bills, 35 dimes, and 29 pennies. In the blank below, rec^rd the number of each coin and bill you have and how much money you have.

Amount of Money	$10,000's	$1000's	$100's	$10's	$1's	dimes	pennies
________________	_____	_____	_____	____	____	_____	_____

4. Repeat part 2 above, starting with the pennies this time to exchange money to minimize the number of coins and bills. Record the money amount and the number of bills/coins after making the money exchanges.

Amount of Money	$10,000's	$1000's	$100's	$10's	$1's	dimes	pennies
________________	_____	_____	_____	____	____	_____	_____

5. When writing money in dollars and cents format, where do you put the decimal place? (between the count of which bills/coins?) ______________________________

6. What is the largest of the 10 digits? ______________________________

7. What would be the largest number you could make with the digits 0, 1, 3, 4, 7, 9 (with no decimal point)?

   ____

8. What would be the smallest number you could make with the digits 0, 1, 3, 4, 7, 9 (with no decimal point) using each digit exactly once?

   ____

9. What would be the smallest 6-digit number with 2 digits after the decimal point you could make with the digits 0, 1, 3, 4, 7, 9?

   ____

10. What would be the largest 6-digit number with 2 digits after the decimal point you could make with the digits 0, 1, 3, 4, 7, 9?

   ____

11. If you had the digits 0, 1, 3, 4, 7, 9 to make a 6-digit number, what would be the smallest number you could make with the 6 digits? You choose the number of digits after the decimal point.

   ____

Referring to the money model, the place value of a position is the denomination of the bill(s) in that position. For example, in the number $567, the digit 5 is in the hundreds place because there are 5 $100-bills, 6 is in the tens place because there are 6 $10-bills, and 7 is in the ones place because it represents 7 $1-bills. Answer the following questions about place value.

12. What are the digits that make up the number, $2536.78? __________ Which of those is the largest digit in the number? ______________

13. In the number \$2536.78, which digit has the highest place value (i.e., accounts for the most money?) ________________

14. In the number \$0.06, which digit has the highest place value? ________________

15. In the number \$3907.76, what is the place value of the digit 0? ________________

16. In the number \$591.82, what is the place value of the digit 8? ________________

17. In the number \$98,976.58, what digit is in the ten-thousands place? ________________

18. In the number \$98,976.58, what digit is in the tens place? ________________

19. In the number \$98,976.58, what digit is in the tenths place? ________________

20. In the number \$98,976.58, what digit is in the ones place? ________________

21. In the number \$98,976.58, what is the place value of the digit 7? ________________

22. What is the largest 4-digit number? ________________

23. What is the smallest 4-digit number (without a decimal point)? ________________

24. What is the largest 4-digit number that can be made with the digits 4, 4, 2, and 9 each used exactly once? ________________

25. What is the smallest number that can be made with the digits 7, 0, and 3 using each digit exactly once? You may include a decimal point. ________________

When you first took the money from the bank, you could not just write out the count of the bills and coins to represent the amount of money. After you traded in excess bills/coins to get the fewest possible bills and coins, the digit in a particular position coincided with the number of bills/coins of the corresponding denomination. When writing a number, each position or place has a **place value**. The digit in the "thousands place," for example, represents the number of thousands (in this case of dollars), the digit in the tenths place represents the number of dimes (tenths of a dollar). The digit in the hundredths place represents the number of pennies, since it takes one hundred pennies to make a dollar. The paragraphs below define and explain place value.

The **ones place**, the position where you put the count of \$1-bills, is the first place to the left of the decimal point if there is one. If there is no decimal point, as in a whole number, the ones place is the last digit reading from left to right. The place value is 1, because the digit in that place is the count of how many "ones" are in the "ones" column.

The other place values are relative to the "ones place value."

The **tens place** (or 10s place) is the position of \$10-bills and is one position left of the ones place. The place value is 10. It is called the 10s place because its value is 10 times that of the ones place. The digit in the tens column is the count of how many "tens" are in that column.

The **hundreds place** (or 100s or 10^2-place) is two positions to the left of the ones place. Its value is 100 times that of the ones place. The digit in the hundreds place is the count of how many hundreds are in the number.

The **thousands place** (or 1000s or 10^3-place) is three positions to the left of the ones place and its place value is 1000 times that of the ones place.

The **ten-thousands** place (or 10,000s or 10^4 place) is four positions to the left of the ones place, and it has a place value of 10,000.

When there is a decimal in the number, the first position to the right of the decimal (the dimes-place in the money model) is called the **tenths place** (or 10th's or 10^{-1} place) since a dime is one tenth (1/10) of a dollar.

The pennies place then is the second position to the right of the decimal and is called the **hundredths** place (or 100th's or 10^{-2}-place) since 1 penny is 1/100 of a dollar.

The difference between the **hundreds place** and the **hundredths place**:
The hundreds place is the third place left of the decimal. The hundredths place is the second place right of the decimal. There is a huge difference in the value of the positions. In the money model, the hundreds place has a value of $100 while the hundredths place has a value of pennies.

The difference between the **tens place** and the **tenths place**:
The tens place is the second place to the left of the decimal. The tenths place is the first place right of the decimal. The tens place has a value of 10 (as in $10) and the tenths place has a value of 1/10 (or a dime.)

Similar distinctions are made between the thousands place and thousandths place, millions place and millionths place, and so on.

- The third place to the right of the decimal is called the **thousandths place** (or 10^{-3}-place.)
- A digit in the fourth place to the right of the decimal would be in the **ten-thousandths** place because the place value is 1/10,000.

Historical note: In the 1950s, the United States mint made red and green plastic coins. The red coin was called a "mill" and the green coin, "green mill" was a 5-mill coin. The red mill was 1/1000 of a dollar and in writing money, the mill showed up in the third place to the right of the decimal. The green mill was worth 5 red mills.

When "milli" is used as a prefix in a word, it means a thousandth. For example, a **milli**meter is 1/1000 of a meter. A **milli**gram is 1/1000 of a gram.

Example 2: Place value

In the number 15, 1 is in the tens place and 5 is in the ones place since 15 represents one $10-bill and five $1-bills in the money model.

In the number 375, 3 is in the hundreds place, 7 is in the tens place, and 5 is in the ones place.

In the number 45,309, 4 is in the ten-thousands place, 5 in the thousands place, 3 in the hundreds place, 0 in the tens place, and 9 in the ones place. (In the money model, $45,309 is equivalent to 4 $10,000-bills, 5 $1000-bills, 3 $100-bills, no $10-bills, and 9 $1-bills.)

In the number 756,928, 7 is in the hundred-thousands place, 5 in the ten-thousands place, 6 in the thousands place, 9 in the hundreds place, 2 in the tens place, and 8 in the ones place.

In the number 7859.89, 7 is in the thousands place, 8 in the hundreds place, 5 in the tens place, 9 in the ones place, 8 in the tenths place, and 9 in the hundredths place.

In the number 0.0256, 0 is in the ones place, 0 in the tenths place, 2 in the hundredths place, 5 in the thousandths place, and 6 in the ten-thousandths place.

Numbers written with a superscript, like 3^2 (read "3 to the second power" of "3 squared") or 4^5 (read "4 to the fifth power), have a **base** and a **power or exponent.** In the number 3^2, 3 is the base and the superscript, 2, is the exponent (power). In the number 4^5, the base is 4 and the power (exponent) is 5. The power or exponent determines how many times the base is multiplied by itself.

$3^2 = 3 \times 3 = 9$	The base 3 occurs twice as a factor. This means you write 3 twice with a multiplication sign.
$4^5 = 4 \times 4 \times 4 \times 4 \times 4 = 1024$	The base 4 occurs five times as a factor. This means you write 4 five times with multiplication signs between them.

When the exponent is 1, the base is written only once, so there is no multiplication performed.

$3^1 = 3$
$4^1 = 4$
$7^1 = 7$

When the exponents are negative, the definition changes. We will not generally cover exponents that are negative or zero except when working with a base of 10 and those will be defined as shown in the box and examples below.

Numbers to show place value

10^2 means $\mathbf{10 \bullet 10 = 100}$

10^3 means $\mathbf{10 \bullet 10 \bullet 10 = 1000}$

10^{-1} means $\mathbf{1/10^1}$ **or 0.1**

10^{-2} means $\mathbf{1/10^2}$ **or 1/100 or 0.01**.

Notice that in the decimal forms of 10^2, 10^3, 10^{-1}, and 10^{-2} above, "1" is one of the digits and the rest of the digits are zero.

Example 3: When the exponent is positive, the exponent (or power) is the number that tells how many zeros appear to the right of the digit 1.

$10^1 = \underline{1}0$	The digit 1 is in the tens place. The power is 1 and there is 1 zero after the digit 1.
$10^2 = \underline{1}00$	The digit 1 is in the hundreds place. The power is 2 and there are 2 zeros after the digit, 1.
$10^3 = \underline{1}000$	The digit 1 is in the thousands place. The power is 3 and there are 3 zeros after the digit 1.
$10^4 = \underline{1}0000$	The digit 1 is in the ten-thousands place. The power is 4 and there are 4 zeros after the digit 1.
$10^5 = \underline{1}00{,}000$	The digit 1 is in the one hundred thousands place. The power is 5 and there are 5 zeros after the digit 1.

When the exponent is negative, the exponent determines the position of the digit 1 to the right of the decimal place.

$10^{-1} = 0.\underline{1}$	The exponent, -1, places the digit 1 in the first position right of the decimal point.
$10^{-2} = 0.0\underline{1}$	The exponent, -2, places the digit 1 in the second position right of the decimal point.
$10^{-3} = 0.00\underline{1}$	The exponent, -3, places the digit 1 in the third position right of the decimal point.

Above, we defined $10^1 = 10$ and $10^{-1} = 0.1$ and we did not show a notation for 10^0. We define $10^0 = 1$ modeling after the positive powers--write the digit 1, followed by no zeros.

As we know, numbers get very large, or extremely small. The following diagram shows the names of the place values of numbers up to a trillion and down to one-trillionth and where we would place the commas in a number.

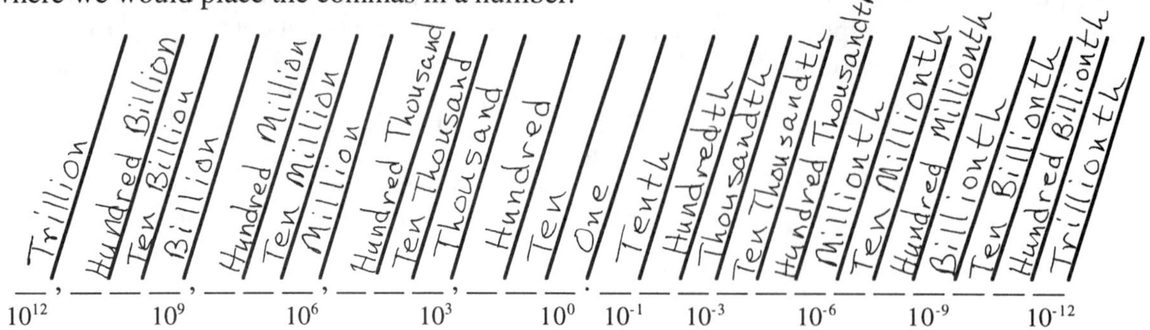

Exercise 2--Largest Digit and Highest Place Value

Purpose of the Exercise

- To practice the names of the place values in decimal numbers

In each of the following, write the name of the place for each digit. What is the largest digit and the smallest digit and which digit has the highest and which digit has the lowest place value? The first one is done as an example.

1. 137.985 1 is in the hundreds place 3 is in the tens place 7 is in the ones place
   9 is in the tenths place 8 is in the hundredths place 5 is in the thousandths place
   Largest digit 9 Digit with highest place value 1 is in the hundreds place
   Smallest digit 1 Digit with the least place value 5 is in the thousandths place
2. 0.2875 ______________________________
   ____
   Largest digit ______ Digit with highest place value ______________________________
   Smallest digit ______ Digit with the least place value ______________________________
3. 97,876.503 ______________________________
   ____
   Largest digit ______ Digit with highest place value ______________________________
   Smallest digit ______ Digit with the least place value ______________________________
4. 5,000,770 ______________________________
   ____
   Largest digit ______ Digit with highest place value ______________________________
   Smallest digit ______ Digit with the least place value ______________________________

5. 0.00306 ____

____

Largest digit ______ Digit with highest place value ______________________

Smallest digit ______ Digit with the least place value ______________________

6. 2.0467 ____

____

Largest digit ______ Digit with highest place value ______________________

Smallest digit ______ Digit with the least place value ______________________

7. 30,468.23 ____

____

Largest digit ______ Digit with highest place value ______________________

Smallest digit ______ Digit with the least place value ______________________

Which Is Bigger? 121 or 99.990000000?

As we learned earlier with our comparisons of 7 and 9 and apples to oranges, when we ask the question "Which is bigger, 7 or 9?" without attaching the numbers to real items, it is assumed that we are comparing "like" items, so when we ask the question, "Which is bigger, 121 or 99.99000000?", we will assume these numbers measure or count similar things. In our money model, 121 would represent $121. The 99.990000000 is a little more difficult to express in dollars and cents until we realize that all of the places to the right of the penny's (hundredths) place are zeros. Since there are no nonzero digits to the right of the hundredths place, we can shorten the number and represent it in dollars and cents as $99.99

When thinking of our money model, it is obvious that $121 is a larger amount of money than $99.99 so that 121 is greater than 99.990000000. The 1 in the hundreds place of the number 121 "out-ranks" the 9's in places of lesser value in 99.99.

To Compare Two Numbers

Find the highest place value in both numbers. If these are different, then the number with the higher place value is the larger number. For example, 121 is greater than 99.990000000 because the highest place in the 121 is the hundreds place, while the highest place in 99.990000000 is the tens place.

When the two numbers have the same highest place value, then the number with the larger digit in the highest place value is the larger number. For example, 576 is larger than 411 because 5 hundreds is larger than 4 hundreds.

Example 4: Determine which is bigger, 7894.67 or 7893.876.

The highest place value in both numbers is the thousands place. In both numbers, the digit in the thousands place is 7, so we cannot yet determine which number is larger.

The next highest place is the hundreds place. In both numbers the digit in the hundreds place is 8, so we cannot yet determine which number is larger.

The next highest place is the tens place. In both numbers the digit in the tens place is 9, so we cannot yet determine which number is larger.

The next highest place is the ones place. The digit in the tens place of the first number is 4 and in the second number it is 3. Since 4 is greater than 3, we conclude that

7894.67 is greater than 7893.876

Mathematical symbols

The symbol, “>“ is read “**is greater than**.”

The symbol “≥” is read “**is greater than or equal to.**”

The symbol, “<“ is read “**is less than**.”

The symbol, “≤” is read “**is less than or equal to.**”

An easy way to remember the direction of the sign is the following: the **smaller number** is always written **on the small end** of the sign and the **larger number** is **written on the large end** of the sign. In other words the sign always points towards the smaller number.

Exercise 3--The "<, ≤, >, ≥" Symbols and Comparing Numbers

Purpose of the Exercise

- To get accustomed to symbolism of "<, ≤, >, ≥"
- To compare decimal numbers

Write the following mathematical statements in English sentences. The first two are done as examples.

1. $5 < 7$ stated in words is: ___7 is more than 5___
2. $121.32 > 99.99x$ stated in words is: ___121.32 is greater than or equal to 99.99x___
3. $290 > 85$ stated in words is: ______________________
4. $0.256 < 0.267$ stated in words is: ______________________
5. $1.95x \leq 2$ stated in words is: ______________________
6. $899 > 763.87364$ stated in words is: ______________________
7. $2{,}345y \geq 435$ stated in words is: ______________________

In each of the following pairs of numbers, determine which number is larger and put the correct sign, "<" or ">", between the numbers.

8. 7 _____ 6.2
   Which is larger? ______________

9. 365 _____ 9.9687
   Which is larger? ______________

10. 7893.435 _______ 7893.492
    Which is larger? ______________

11. 356.79 _____ 356.798
    Which is larger? ______________

Section 1-3 *Rounding Off and Significant Digits*

According to the Web site (http://www.brillig.com/debt_clock/) as of June 10, 1997 at 12:51:53 p.m., the national debt was $5,359,122,646,128.76. The basic size of the number is hard enough to comprehend, but what do all those digits really mean? Since the debt is constantly changing as bills, payroll and other expenses are paid, taxes or other debts are collected, and interest on loans is added to the debt, it is impossible to know at any one time exactly what the national debt is, so although the Web site shows a number "accurate to the penny," the site cannot possibly be that accurate. At best we can expect an accuracy "to the nearest trillion or 100-billion dollars" or to "2 or 3 significant digits." Even then, the accuracy of the number depends on how the managers of the site get and update the data. For clarity, we might say that the national debt is $5,359,000,000,000 or even state that it is $5.4 trillion.

In this section, we will learn how to "round off" such a huge number as the national debt as well as a very small number in a way that we might better comprehend its meaning. We will not tackle the question of whether the data are accurate--that would be an excellent study in a political science class. To learn to interpret very large numbers and very small numbers, we will learn what is meant by "significant digits" as well as how and why to round a number to a particular place or to a certain number of significant digits.

Terminology

When we say "round a number to one significant digit," or to two significant digits, and so on, we determine the one, two, three significant digits in the following manner:

If a number has nonzero digits left of the decimal point:

- 1 significant digit is the first digit in the number.
- 2 significant digits are the first two digits.
- 3 significant digits are the first three digits.
- and so on

If the only digit to the left of the decimal place is a zero or if there are no digits to the left of the decimal point, then the first significant digit is the first nonzero digit.

In a number the first nonzero digit (reading from left to right) and all of the digits following are significant. In fact, placement of zeros after the decimal point indicate accuracy to the decimal place indicated. For example, the number 4078.000 is accurate to the one-thousandth place. All of the digits, including the zeros, are significant.

Five cents written in decimal notation is $0.05. In this situation, it is customary to write the 0 in the ones place to help us notice the decimal point. The 0 in the tenths place is necessary so that the significant digit, 5, is in the correct place and represents pennies and not dimes.

Criteria for Determining the Number of Significant Digits

In usage, there are several different criteria for deciding where to round off a number, and when in a problem to round off numbers.

1. A very important consideration is **What kind of accuracy is needed in a particular situation**?

   If a carpenter is to cut a piece of trim and she carefully calculates that the measurement should be 7.2395828 inches, her calculations are no doubt more accurate than her cutting tools are capable of handling. One, two, three, or at most four decimal places would be sufficient. (1/16 of an inch in decimal notation is 0.0625 inches.)

When measuring amperes, one ampere can kill you, so the number of amperes in an electronic circuit (such as in live speaker wires) are in measurements like 0.003 amps or 3 milliamps. (Remember, the prefix "milli" means "one thousandth." 3 thousandths of an amp is 3 milliamps.) Rounding off to two decimals places would miss the current in the circuit.

The number accurate to 10 significant digits is 3.141592654. The accuracy needed varies with the application. In a "high-tech" situation, you may need many decimal places of accuracy. When buying lace to sew around a circular table cloth, 3.14 or 22/7 would work just fine.

If the information you have is only accurate to a certain number of decimal places, the final answer will not be more accurate than your original data. The final answer cannot have more significant digits than the number with the fewest significant digits.

2. Another consideration is the difficulty in comprehending many nonzero digits in a number.

   The national debt of $5,359,122,646,128.76, according to the Web Site is difficult to comprehend because of the size of the number as well as the number of digits. It is more comprehensible if rounded and expressed as $\$5.4 \times 10^{12}$ (or $5.4 trillion.) At least the size of the number is not cluttered by the many digits.

   If you won the sweepstakes and were to receive an annual check of $287,459.87, you might refer to that amount of money as $287,000 or as $287,500 because either of the second descriptions is an easier number to comprehend.

3. When doing computations with calculators, we often come up with decimal approximation of the answers. If we round off too severely, too early, the final answer may have a large error as a result of several "round-off" errors in the entire process. In general, when using a calculator to do a series of computations, learn how to do string computations on the calculator because the calculator typically rounds numbers to eight or nine digits internally. Also learn how to store answers on your calculator. When you get the final answer on the calculator, round off as appropriate to the current situation.

Example 1: In each of the following numbers, underline the significant digits indicated or the digit in the place indicated.

7089.067	1 significant digit is underlined.
7089.067	2 significant digits are underlined.
87.903	4 significant digits are underlined.
87.903	The tenth's place is underlined.
87.903	The first decimal place is underlined.
0.1789	Three significant digits are underlined. In this case, three significant digits are the same as three decimal places.
0.00589	Three significant digits are underlined.
1.00589	Four significant digits are underlined. The last underlined digit is in the thousandths place or third decimal place.

Now we will learn to round numbers off to a certain number of significant digits or to a particular decimal place.

> **Process to Round Off a Number to a Particular Place or a Certain Number of Significant Digits,**
> Locate the place indicated or the place of the last of the significant digits and take note of the digit to the right of that place.
> 1. If the digit on the right of the located position is less than 5, (i.e., 0, 1, 2, 3, or 4), then convert all the digits to the right of the position to 0. We call this rounding a number **down.**
>   If the digit on the right is 5 or greater (i.e., 5, 6, 7, 8, or 9), then change the digit in the place of round off to one digit higher and convert all of the digits to the right to zero. We call this rounding a number **up.**
> 2. If rounding to a place left of the decimal point, delete all digits to the right of the decimal point.
>   If rounding to a place right of the decimal point, delete all digits to the right of the rounded place.

Example 2: Round 27,824.9706 off to the nearest **hundred**.
Locate the hundreds place: 27,<u>8</u>24.9706. The digit 8 is in the hundreds position.
The digit to the right of 8 is 2, so convert all of the numbers to the right of the hundreds place to 0. We have **rounded** the number **down**.
The rounded off number is 27,800.

Example 3: Round off 27,824.9706 to the nearest **thousandth**.
Locate the thousandths place 27,824.97<u>0</u>6. The digit, 0, is in the thousandth position.
The digit to the right is 6, so **round** the digit 0 **up** to a 1. All digits to the right become 0. The rounded off number is 27,824.971.

Example 4: Round off 27,824.9706 to **one decimal place**.
The first decimal place is the tenth's place 27,824.<u>9</u>506.
The digit to the right is 5, so convert the 9 to one larger digit (**round** 9 **up**) and convert all of the numbers on the right to 0. Notice this presents a problem because 9 is the largest digit. The next largest number is 10, so changing the 9 to a 10 means we have to "carry" the 1 to the ones position.
The rounded off number is 27,825.0.
Although 27,825.0 = 27,825, we keep the zero to the right of the decimal place to indicate that the accuracy is to one decimal place. If we wrote the answer as 27,825, we would be suggesting that the accuracy is only to the ones place. This also indicates that the digit 0 is one of the significant digits.

Exercise 1--Rounding Off and Significant Digits

Purpose of the Exercise

- To practice rounding off decimal numbers

In each of the following exercises, underline the significant digits, or decimal place as indicated. In the third and fourth columns record the digit following the digit(s) underlined and state whether the number will be rounded up or down. Write the rounded answer in the last column. The first six are worked as examples.

The Number to Round Off	Underline the Digit(s)	The Next Digit	Round Up/Down?	Answer
1. 108.935 Round to 3 significant digits	108.935	9	Up, since 9 > 5	109
2. 0.0167362 Round to 3 significant digits	0.0167362	3	Down, since 3 < 5	0.0167
3. 17.0983 Round to 1 decimal place	17.0583	5	Up, since 5 5	17.1
4. 17.0483 Round to 1 significant digit	17.0483	7	Up, since 7 5	20
5. 17.0483 Round to the nearest ten	17.0483	7	Up, since 7 5	20
6. 17.0483 Round to the nearest tenth	17.0483	4	Down, since 4 5	17.0
7. 3467.892 Round to the nearest 100				
8. 3467.892 Round to the nearest 100th				
9. 2089.0056 Round to 3 significant digits				
10. 45.0408 Round to the nearest 100th				
11. 45.0408 Round to the nearest 1000th				
12. 99.9978 Round to 1 significant digit				
13. 99.9978 Round to the nearest 10th				
14. 0.0008735 Round to 1 significant digit				
15. 0.0008735 Round to 2 significant digits				
16. 0.0006378 Round to the nearest 100th				
17. 0.0006378 Round to the nearest 10th				
18. 18.98 Round to the nearest ten				
19. 18.98 Round to 1 significant digit				
20. 0.0956 Round to 3 decimal places				
21. 0.0956 Round to 3 significant digits				
22. 205.736 Round to the nearest hundred				
23. 205.736 Round to the nearest hundredth				
24. 89.72 Round to 1 decimal place				
25. 89.72 Round to 1 significant digit				

The Number to Round Off	Underline the Digit(s)	The Next Digit	Round Up/Down?	Answer
26. 90,099.99 Round to the nearest hundred				
27. 90099.99 Round to the nearest tenth				
28. 307,564 Round to one significant digit				
29. 307,564 Round to 3 significant digits				
30. 0.00570834 Round to 3 significant digits				
31. 89.00345 Round to the nearest tenth				
32. 5.8074 Round to the nearest 10,000th				
33. 403.0942 Round to 2 decimal places				
34. 274.9847 Round to the nearest tenth				
35. 274.9847 Round to the nearest ten				
36. 0.00095 Round to the nearest ten-thousandth				
37. 0.0009355 Round to the nearest thousandth				
38. 123.578948 Round to the nearest hundred				
39. 123.578948 Round to the nearest hundredth				
40. 9,877,207,305 Round to the nearest million				
41. 9,877,207,305 Round to the nearest ten million				
42. 5,870,384.75 Round to 5 significant digits				
43. 5,870,384.75 Round to the nearest ten thousand				
44. 2.05789 Round to the nearest thousandth				
45. 2.05789 Round to the nearest one				
46. 2.05789 Round to the nearest ten				

Exercise 2--Maps, Distances, and Rounding Off
(Using Reference Materials outside of Class)

Purpose of the Exercise

- To understand why we choose to round off to a particular place

Materials Needed

- Reference materials--encyclopedia
- Local and world maps
- Car and odometer (optional)

Use maps, encyclopedias, and/or other reference materials, as well as your own measurements when appropriate to find the following distances or measurements. Put all distances in terms of miles. In some cases, you may have to decide on a particular route and add several distances. Write the distances in the blanks given. Name the cities and landmarks you chose for distances.

1. Distance from Earth to the Sun. ____________________

2. Distance from Earth to the Moon. ____________________

3. Radius of the Earth. ____________________

4. Distance from your town/city to Tokyo. ____________________

5. Distance from your town/city to another city in the same state. ____________

6. Distance from your town/city to another city at least four states away. __________

7. Distance from your house to some nearby landmark or building. _____________

8. Distance between two towns near you. ________________________________

9. Distance from your house or apartment to the nearest edge of the property your house or apartment is on. (Measure this as best you can in miles. Get help if you need to convert feet to miles--this will be covered later in this book.) __________

10. In reporting the distance from Earth to the Sun, to what place was the number rounded? _______

11. In reporting the distance from your city to Tokyo, to what place was the number rounded? _______

12. In reporting the distance from Earth to the Sun, how many significant digits did you keep (i.e., how many nonzero digits were kept in the number)?

13. In computing the distance from your living space to the nearest edge of the property, how many significant digits did you decide to keep?

14. In computing the distance from your living space to the nearest edge of the property, to what place did you decide to round off?

In this section and the next three sections, we will use the money model to understand arithmetic better. In this section, we will focus on addition and learn from the money model how and why the process we use for addition works. It is important in all of mathematics to understand the associated vocabulary, so let's start with that.

Vocabulary

Addition, subtraction, multiplication, and division are called **arithmetic operations.**

When two numbers are **added**, the answer is called the **sum**, or **total**.
When one number is **subtracted** from another, the answer is called the **difference.**
When two numbers are **multiplied**, the answer is called the **product**.
When one number is **divided** by another, the answer is called the **quotient**.

In the following discovery exercise, group members will take money out of the bank, combine the money into a common pool, and then make money exchanges that model our procedure for adding numbers.

Exercise 1--Money Addition (Group)

Purpose of the Exercise

- To use money to understand addition

Materials Needed

- Fake Money--at least 40 each of $1-bills, $10-bills, $100-bills, $1000-bills, $10,000-bills and dimes and pennies. Do not include any denomination of bill or coin not listed above.
- Calculator

1. Take the following amounts of money using the minimum number of bills/coins.

Member 1:	$1059	Member 2:	$3995.37
Member 3:	$5308.98	Member 4:	$7006.90

2. For each member, record the dollar amount of money and the number of bills/coins of each denomination.

MEMBER 1

Amount of Money	$10,000's	$1000's	$100's	$10's	$1's	dimes	pennies
____________	____	____	____	____	____	____	____

MEMBER 2

Amount of Money	$10,000's	$1000's	$100's	$10's	$1's	dimes	pennies
____________	____	____	____	____	____	____	____

MEMBER 3

Amount of Money	$10,000's	$1000's	$100's	$10's	$1's	dimes	pennies
____________	____	____	____	____	____	____	____

MEMBER 4

Amount of Money	$10,000's	$1000's	$100's	$10's	$1's	dimes	pennies
____________	____	____	____	____	____	____	____

3. Combine all of the money into one "pot." Count and record the number of each kind of bill/coin you have in the common pot. How much money do you have?

Amount of Money	$10,000's	$1000's	$100's	$10's	$1's	dimes	pennies
________	____	____	____	____	____	____	____

4. Minimize the number of bills/coins by going to the bank to trade excess pennies for dimes, and so forth. After the money exchanges, count the number of each kind of bill/coin you have in the common pot. How much money do you have?

Amount of Money	$10,000's	$1000's	$100's	$10's	$1's	dimes	pennies
________	____	____	____	____	____	____	____

5. Use a calculator to add the original amounts of money. Compare the answer on the calculator with the money amount you got through the money exchanges. If the amounts differ, check your money exchanges as well as the calculator computations.

6. Repeat parts 1-5 with the following money amounts:
   a. $275.35, $2356.90, and $760
   b. $39.50, $69.98, and $935.70
   c. $157, $1035.35, and $598.98
   d. $79, 85¢, and $4.35
   e. 92¢, 79¢ , and $7598.35

The Process of Addition

From our observations in Exercise 1, when adding whole or decimal numbers, we must add digits of equal place value. As we know, $7 + 3 cents is written as $7 + $0.03 and the answer is $7.03, not 10 dollars or 10 cents. When setting up an addition problem, it is usually helpful to write the numbers in columns that line up the ones place with the ones place, the tens place with the tens place, the tenths place with the tenths place, and so on. A sure way to accomplish this is to <u>line up the decimal points</u>. If the number of digits to the right of the decimal point is different in the two numbers, the numbers will not line up on the right. To make such an arrangement look better we can put zeros in columns on the right of the last digit after the decimal point to line up the right column.

Example 1: Add 7 to 0.03 (or $7 to 3¢).

Since there are two digits to the right of the decimal place in 0.03 but none in the number 7 we rewrite 7 as 7.00 so that the numbers line up on the right when the decimal points line up.

$$\begin{array}{r} 7.00 \\ +\ \underline{0.03} \\ 7.03 \end{array}$$

After the digits of like place value are lined, just add the digits down each column.

Process for Addition in Columns

When the problem is set up in columns, **start on the right and add all of the digits in that right column.**

1. If the sum of the digits in the right column is 9 or less, write that sum in the right column of the total.
2. If the sum of those digits is more than 9, then regroup as shown in the example below.
3. Continue this process, moving left through the columns, until the addition is completed.

Example 2: Add 28 and 46.

$$\begin{array}{r} 28 \\ +\,46 \\ \hline 4 \end{array}$$

6 + 8 = 14 in the ones column. Exchange 10 of the ones for 1 ten. This regroups the 14 ones to be 4 ones and 1 ten.

$$\begin{array}{r} 1 \\ 28 \\ +46 \\ \hline 74 \end{array}$$

Copy the 4 to the ones column of the total and write the 1 in the 10s column (above the top number.) Add the digits in the 10's column.

Example 3: Find the sum of 7895.36 and 988.6.

First set up the numbers in columns with the decimal points lined up; then write in zeros on the right to even out the right columns.

Add the digits in the right column, which in this case is the hundredths place.

$$\begin{array}{r} 7895.36 \\ +\quad 988.60 \\ \hline 6 \end{array}$$

There are no digits to regroup, so write the 6 in the hundredths place and then add the digits in the tenths column.

$$\begin{array}{r} 7895.36 \\ +\quad 988.60 \\ \hline .96 \end{array}$$

There are still no digits to regroup, so write the 9 in the tenths place and add the digits in the ones column.

$$\begin{array}{r} 1 \\ 7895.36 \\ +\quad 988.60 \\ \hline 3.96 \end{array}$$

The digits add to 13, so write the 3 in the ones place of the sum and regroup the 1 to the tens place. Add the tens column including the digit that was regrouped in the tens place.

$$\begin{array}{r} 11 \\ 7895.36 \\ +\quad 988.60 \\ \hline 83.96 \end{array}$$

The digits add to 18, so write the 8 in the ten's place of the sum and regroup the 1 to the hundreds place. Add the hundreds column.

$$\begin{array}{r} 111 \\ 7895.36 \\ +\quad 988.60 \\ \hline 883.96 \end{array}$$

The digits add to 18, so write the 8 in the hundreds place of the sum and regroup the 1 to the thousands place. Add the thousands column.

$$\begin{array}{r} 111 \\ 7895.36 \\ +\quad 988.60 \\ \hline 8883.96 \end{array}$$

The sum of the digits in the thousands column is 8, so there is no regrouping.

The **sum** of 7895.36 and 988.60 is 8883.96

Adding three, four, or more numbers follows the same process.

Example 4: Add 7850.87, 578, 0.5892, and 3.96.

First, write the numbers in columns (showing the hidden decimal point behind 578) with the decimal points lined up. Since there are four digits after the decimal point in one of the numbers, we write extra zeros at the end of the other two numbers to line up the right columns.

2	There was no regrouping in the last two decimal places.
7850.8700	In the hundredths place, the sum of the digits is 21, so the 2 was regrouped to the tenths place and 1 written in the total.
0.5892	In the tenths place, the digits added to 24, so regroup the 2 to
3.9600	the ones place and write the 4 in the total.
7855.4192	In the other columns, there is no regrouping.

Exercise 2--Adding Whole and Decimal Numbers

Purpose of the Exercise

- To practice adding whole and decimal numbers

Materials Needed

- Calculator

On a separate sheet of paper, set up the following groups of numbers in columns then add the numbers. Show your work and do not use a calculator. After you finish, use a calculator to check your work.

1. 245.67 + 17.9
2. 7893.56 + 3465.9
3. 1008 + 25.973 + 0.2
4. 0.0157 + 0.693 + 0.0008
5. 80000 + 579 + 0.8975
6. 9.17 + 3.56 + 8
7. 3647.458 + 0.00003 +19
8. 278 + 378.1 + 3.002
9. 4500 + 85 + 976
10. 279.35 + 96.007
11. 4000 + 300 + 50 + 9
12. 50 + 7 + 0.6 + 0.07
13. 298 + 967 + 0.29
14. 0.256 + 0.004 + 0.5
15. 6008 + 450 + 1.7
16. 0.45 + 0.06 + 0.9786
17. 89 + 0.987
18. 77 + 2033 + 96.2

Approximations

When adding numbers, we most often use a calculator to get the answer. When using a calculator, however, it is important to know when the answer is "**reasonable**." It is easy to miskey the problem when putting numbers into the calculator. It is particularly easy to overlook exact placement of decimal points.

There are several considerations in checking the **reasonableness** of answers. One of the most important in real life is **common sense**.

Example 5: Suppose you are computing the distance between you and a thunderhead and come up with an answer of 15 feet. The answer makes no sense; if that is the situation, you are dead with the first bolt of lightening. Maybe you made a mistake in your calculations, or maybe the unit is miles instead of feet.

Example 6: Suppose you computed the speed of a bicyclist to be 127 miles per hour. That is an unlikely, if not impossible, speed for a bicyclist to travel (unless he has just fallen off a cliff). This is a clue to check your computations.

The other aspect of checking for reasonableness is estimation of the arithmetic itself. Outlined below is a process for estimating the sum of two numbers.

To Estimate the Sum When Adding Numbers:

First: Round off each number to one significant digit.

Second: Add the rounded-off numbers.

When you are practicing estimating arithmetic, check your process by adding the original numbers on the calculator to see how close the answer is.

Example 7: Estimate the sum of 3896.76 and 3.008947586.

First: The numbers rounded to 1 significant digit are 4000 and 3.

Second: The sum of the rounded numbers is 4003, so 4003 is the estimate.

Using a calculator, 3896.76 + 3.00894786 = 3899.7689479.

Compare the answer, **3899.7689479** to the estimated answer, **4003**. Not bad! At least we are in the ballpark.

In addition to the process outlined above, always keep a lookout for the obvious. The first significant digit in the second number is in the ones place. In the first number, there are digits in the thousands, hundreds and tens places. The whole second number is insignificant compared to the first number. A better estimate might simply be the first number, or the first number rounded to the 10s or 100s place.

Example 8: Estimate the sum of 0.0784736, 0.0547345, and 0.961023.

First: The numbers rounded to 1 significant digit are 0.08, 0.05 and 1.

Second: The sum of the rounded numbers is 1.13; this is the estimate.

Using a calculator to check, 0.0784736 + 0.0547345 + 0.961023 = 1.0942311.

Compare the answer, **1.0942311** to the estimated answer, **1.13**. Not bad! At least we are in the ballpark.

Exercise 3--Estimating Sums

Purpose of the Exercise

- To practice estimating sums

Materials Needed

- Calculator

For each of the groups of number below, follow the process as outlined above to estimate the sum of the numbers. Use a calculator to add the numbers and compare your estimate to the calculator answer. If your answers are very far off, check your work or get help.

1. 356.7892, 37,928 and 38,472

   The numbers rounded ____________________

   ____

   Sum of the rounded numbers ______________

   Sum on the calculator ____________________

2. 0.15637, 0.55374, 0.6343495

   The numbers rounded ____________________

   ____

   The sum of the rounded numbers is __________

   Sum on the calculator ____________________

3. 475.68 and 585.93

   The numbers rounded ____________________

   ____

   Sum of the rounded numbers ______________

   Sum on the calculator ____________________

4. 2378.87 and 5,693

   The numbers rounded ____________________

   ____

   The sum of the rounded numbers is __________

   Sum on the calculator ____________________

5. 0.005348 and 0.009463

The numbers rounded ____________________

____

Sum of the rounded numbers ____________

Sum on the calculator ____________________

6. 4983.87 and 599.20

The numbers rounded ____________________

____

The sum of the rounded numbers is ________

Sum on the calculator ____________________

7. 17.93, 20, and 197.304

The numbers rounded ____________________

____

Sum of the rounded numbers ____________

Sum on the calculator ____________________

8. 0.000678, 0.003756, and 0.000009999

The numbers rounded ____________________

____

The sum of the rounded numbers is ________

Sum on the calculator ____________________

9. 5,799,987 and 8,394,457

The numbers rounded ____________________

____

Sum of the rounded numbers ____________

Sum on the calculator ____________________

10. 1,078,837,876.98 and 5,970

The numbers rounded ____________________

____

The sum of the rounded numbers is ________

Sum on the calculator ____________________

11. 18.375, 19.8, and 22

The numbers rounded ____________________

____

Sum of the rounded numbers ____________

Sum on the calculator ____________________

12. 4793, 6376, and 3849

The numbers rounded ____________________

____

The sum of the rounded numbers is ________

Sum on the calculator ____________________

13. 11, 968, and 8739

The numbers rounded ____________________

____

Sum of the rounded numbers ____________

Sum on the calculator ____________________

14. 0.0354, 0.5, and 8.9

The numbers rounded ____________________

____

The sum of the rounded numbers is ________

Sum on the calculator ____________________

15. 8967.87 and 378.99

The numbers rounded ____________________

____

Sum of the rounded numbers ____________

Sum on the calculator ____________________

16. 47,567.89 and 5788.95

The numbers rounded ____________________

____

The sum of the rounded numbers is ________

Sum on the calculator ____________________

17. 0.678, 0.987, and 0.003

The numbers rounded ____________________

____

Sum of the rounded numbers ______________

Sum on the calculator ____________________

18. 145.78 and 0.05

The numbers rounded ____________________

____

The sum of the rounded numbers is ___________

Sum on the calculator ____________________

19. Without a calculator, circle the best answer for each of the following addition problems. Use your money as needed to estimate the answer.

a. $84 + $0.75 + $19	$178	$93.75	$103.75
b. 84 + 0.75 + 19	178	93.75	103.75
c. $19.24 + $20 + $0.68	$39.92	$20.12	$107.24
d. $19.24 + $20 + $0.68	39.92	20.12	107.24
e. $2398.76 + $56,798 + $9.98	59,000	17,900	2976.72
f. $467.9 + $0.99 + $0.53	470	4830	48.31
g. $0.27 + $0.96 + $1	1	2.3	124
h. 14.9 + 0.09 + 8	166	15.26	20.99
i. 0.5 + 0.03 + 0.006	0.536	0.14	0.59
j. 567 + 30 + 9	1400	600	800
k. 50 + 8 + 0.5 + 0.07	70	130.57	58.57
l. 2 + 40 + 500 + 7000	18000	1800	7542
m. 9876 + 0.987 + 0.0993	9878	9876	9874

20. With a calculator, check your work above and see how close you got to a right answer.

Here are some observations about rounding to estimate sums:

When the last significant digit of each rounded off number is in the same place:

If all of the numbers were rounded up, the estimated sum is a high estimate.

If all of the numbers were rounded down, the estimated sum is a low estimate.

When the last significant digits of the rounded numbers are in a different places, the digits with the highest place value have the most impact on the estimate.

Section 1-5 *Subtracting Whole and Decimal Numbers*

In this section, we will work with money exchanges to understand the process of subtraction and then learn to estimate differences. Recall that when one number is subtracted from another, the answer is called the **difference** of the numbers. In general, when we refer to the difference between two numbers, we mean the larger number minus the smaller number.

Exercise 1--Money Subtraction (Group)

Purpose of the Exercise

- To use money to understand subtraction

Materials Needed

- Fake Money: pennies, dimes, $1-bills, $10-bills, $100-bills, $1000-bills, $10,000-bills

Divide the class into groups of three and have each group decide who is Member 1, Member 2, and Member 3. In each of the following, do the money exchanges as indicated with everyone in the group keeping track of the transactions. If there are only two members in the group, both members will serve as the banker for the other person.

1. Member 2 will be the banker.
   a. Member 1 takes $47,360.24 from the bank, getting the fewest possible bills and coins.
   b. Member 3 reminds Member 1 to pay him/her the $18,195.32 owed.
   c. Member 1 pays Member 3 the money, having the banker make change whenever it is necessary.
   d. After the money exchange, how much money does Member 1 have left over? Record the answer below.

$$\begin{array}{r} 47{,}360.24 \\ \underline{-\ 18{,}195.32} \end{array}$$

2. Member 3 is the banker.
   a. Member 2 takes $55,000.00 from the bank, getting the fewest possible bills and coins.
   b. Member 1 reminds Member 2 to pay him/her the $29,295.81 owed.
   c. Member 2 pays Member 1 the money, having the banker make change whenever it is necessary.
   d. After the payment, how much money does Member 1 have left? Record the answer below.

$$\begin{array}{r} 55{,}000.00 \\ \underline{-\ 29{,}295.81} \end{array}$$

3. Member 1 is the banker.
   a. Member 3 takes $29,253.89 from the bank, getting the fewest possible bills and coins.
   b. Member 2 reminds Member 3 to pay him/her the $38,105.00 owed.
   c. Member 3 pays the $29,253.89 to Member 2.
   d. Member 2 reminds Member 3 that he/she still owes money.
   e. Member 3 borrows the rest of the money from the bank and pays the rest of the debt.
   f. How much money did Member 3 have to borrow from the bank to pay the debt?
   g. Discuss how the following subtraction would be done.

$$\begin{array}{r} 29{,}253.89 \\ \underline{-\ 38{,}105.00} \end{array}$$

The Process of Subtraction

Subtraction of whole and decimal numbers has some similarities to addition in that the actual subtraction involves individual columns, and the subtraction process starts with the right column and moves left. The process differs, however, because the individual columns are subtracted, not added. If we are subtracting a larger digit from a smaller digit, we have to convert a digit from the adjacent column on the left to 10 in the current column. The digit in the column to the left, is then reduced by 1.

Example 1: Just as in the money exchanges, to subtract $9 from $37, we start by subtracting the ones from the ones.

```
          3 tens  and  7 ones
minus                  9 ones
```

Since we cannot subtract 9 from 7, we convert one of the 10s to 10 ones and regroup the 3 tens and 7 ones to be 2 tens and 17 ones. The problem restated is

```
          2 tens  and 17 ones
minus                  9 ones
          2 tens  and  8 ones
```

We show this more concisely in the following way:

```
 2
 3̸¹7    Regroup to change the 3 tens and 7 ones to 2 tens and 17 ones
- 9
 2 8
```

To check the answer, 28 + 9 = 37, so the answer is correct.

The only type of subtraction problem that will be discussed in this chapter is the one in which we subtract a smaller number from a larger number. As we know from experience and the money exercise, it is possible to subtract a larger number from a smaller number, but we end up "in the hole" and have to borrow money. This is how we end up with negative answers. We will wait until Chapter 2 to work extensively with negative numbers.

Example 2: Subtract $56 - $17.

Interpreted in bills we get

```
          5 tens  and  6 ones
minus     1 ten   and  7 ones
```

We cannot take 7 ones from 6 ones, so we exchange one of the tens for 10 ones and regroup so that we have 4 tens and 16 ones. The problem restated is

```
          4 tens  and 16 ones
minus     1 ten   and  7 ones
          3 tens  and  9 ones
```

We show this more concisely in the following way:

```
  4
  5̸¹6    Take 1 from the 5 in the 10s place to give 16 in the 1s place.
- 1 7    Change the 5 to a 4 in the 10s place. Subtract 7 from 16.
    9    Write the result in the 1s place of the answer.
  4
  5̸¹6    Subtract the 1 from the 4 in the 10s place.
- 1 7    Write the result in the 10s place of the answer.
  3 9      The difference of 56 and 17 is 39.
```

Check: 39 + 17 = 56, so the answer is correct.

Example 3: Subtract 988.6 from 7895.36.

First set up the problem in columns and subtract the digits in the right column, which in this case is the hundredths place. (Be sure to line up the decimal points and write zeros on the right as necessary.)

Work	Explanation
7895.36 - 988.60 6	Subtract 0 from 6 in the hundredths column. Write the result in the hundredths column of the answer.
4 7 8 9 ~~5~~.^{1}36 - 9 8 8. 60 .76	Subtract 6 from 3 in the tenths column. Since 6 is bigger than 3, regroup the 5 in the ones column to be 4 ones and 10 tenths. Change the 3 (in the tenths column) to a 13, and subtract.
8^{1}4 7 8 ~~9~~ ~~5~~.^{1}36 - 9 8 8. 60 6. 76	Subtract 8 from 4 in the ones column. Since 8 is bigger than 4, regroup the 9 in the tens column to 8 tens and 10 ones, Change the 4 (in the ones column) to a 14 and subtract.
8^{1}4 7 8 ~~9~~ ~~5~~.^{1}36 - 9 8 8 .60 0 6. 76	Subtract 8 from 8 in the tens column. Write the result in the tens column in the answer.
6 8^{1}4 ~~7~~^{1}8 ~~9~~ ~~5~~.^{1}36 - 9 8 8. 60 9 0 6. 76	Subtract 9 from 8 in the hundreds column. Regroup the 7 one thousands into 6 one thousands 10 hundreds. Add the 10 to the 8 in the hundreds column to get 18. Subtract.
6 8^{1}4 ~~7~~^{1}8 ~~9~~ ~~5~~.^{1}36 - 9 8 8. 60 6 9 0 6. 76	There is nothing in the thousands column to subtract from 6, so bring down the 6. The **difference** between 7895.36 and 988.60 is 6906.76.

Check: 6906.76 + 988.60 = 7895.36, so the answer is correct.

Many times we use words to describe arithmetic. It is important to recognize the various ways subtraction can be expressed in words.

29 - 15 can be expressed in the following ways:

"29 minus 15."

"Subtract 15 from 29."

"The difference between 29 and 15."

Exercise 2--Practicing Subtraction

Purpose of the Exercise

- To practice the process and vocabulary of subtraction

Without a calculator, subtract the following numbers as indicated. Show your work.

1.	2.	3.	4.
14.79 - 4.31	89.365 - 6.351	1873.98 - 17.39	0.6723 - 0.607

5. 82.79 - 79.36	6. 263.012 - 15.930	7. 8745.60 - 29.45	8. 7000.35 - 0.60
9. 375 - 78	10. 193.90 - 14.98	11. 187.00 - 99	12. 1097 - 12.67
13. 489 217	14. 145.09 - 139.00	15. 571.08 - 14.67	16. 300.00 - 0.09

On a separate sheet of paper, write each of the following problems and then subtract. Write the numbers in columns if needed.

17. 89 minus 0.35	18. 892.3 - 4.5	19. Subtract 87 from 320
20. 1970.40 - 9.80	21. 7000 minus 0.06	22. Subtract 30,009 from 40,000
23. 2000 - 999.9	24. 75 minus 26	25. Subtract 0.067 from 51
26. 135.98 minus 12.9	27. Subtract 99 from 102	28. Subtract 12.9 from 82.3
29. 100 minus 84	30. 823 minus 99	31. Subtract 451 from 567.92
32. 85 - 79	33. 85 minus 75	34. Subtract 367 from 400
35. 867.34 - 86.734	36. 89.23 minus 8.923	37. Subtract 0.67 from 12
38. 289 - 76	39. 210 minus 83	40. Subtract 124.56 from 200
41. 0.045 - 0.0045	42. 0.84 minus 0.095	43. Subtract 0.2756 from 10
44. 100 - 23.45	45. 27.04 minus 27.02	46. Subtract 89 from 134.5
47. 7456 - 864	48. 800 minus 732	49. Subtract 76.345 from 76.345
50. 456 - 256.78	51. 0.00345 minus 0.002	52. Subtract 2000 from 20000

Estimating Differences

As in addition, when the numbers are tedious to work with, most computations are done on a calculator. Estimated answers help safeguard against the ridiculous answers that we may get when the wrong buttons are pressed.

Process for Estimating Differences

Estimations of differences in subtraction problems are computed in a similar manner as estimating sums in addition.

First: Round the two numbers to one significant digit.

Second: Use rounded numbers to estimate the difference.

While learning to estimate answers, it is a good idea to check our process by calculating the exact answers on a calculator and compare them to the estimated answer. In real-life problems, we generally do the reverse--that is, calculate the answers with a calculator and then estimate the answers to see if our answer is reasonable.

Example 4: Estimate 75.9643 - 19.4 and then calculate the difference with a calculator and compare results.

The 1st number rounded is 80

The 2nd number rounded is 20

Estimated difference 80 - 20 = 60

Exact difference 56.5643

The estimated difference is 60 which is in the ballpark of the exact answer of 56.5643.

Exercise 3--Estimating Differences

Purpose of the Exercise

- To practice estimating differences

Materials Needed

- Calculator

Estimate the answers to the following subtraction problems and then use a calculator to calculate the answers and see how close the estimate is.

1. 789.987 - 453

   The numbers rounded ____________________

   ____

   Sum of the rounded numbers ____________

   Sum on the calculator ____________

2. 129.467 - 0.589

   The numbers rounded ____________________

   ____

   Sum of the rounded numbers ____________

   Sum on the calculator ____________

3. 289.015 - 89.99

   The numbers rounded ____________________

   ____

   Sum of the rounded numbers ____________

   Sum on the calculator ____________

4. 8091.0006 - 893.87

   The numbers rounded ____________________

   ____

   Sum of the rounded numbers ____________

   Sum on the calculator ____________

5. 28734.3942 - 5.0798

   The numbers rounded ____________________

   ____

   Sum of the rounded numbers ____________

   Sum on the calculator ____________

6. 0.04983 - 0.0057664

   The numbers rounded ____________________

   ____

   Sum of the rounded numbers ____________

   Sum on the calculator ____________

7. 28375.698 - 19345.67

   The numbers rounded ____________________

   ____

   Sum of the rounded numbers ____________

   Sum on the calculator ____________

8. 5678.93 - 9.890945

   The numbers rounded ____________________

   ____

   Sum of the rounded numbers ____________

   Sum on the calculator ____________

9. 4567 - 3569

   The numbers rounded ____________________

   ____

   Sum of the rounded numbers ____________

   Sum on the calculator ____________

10. 78927 - 9834.879

   The numbers rounded ____________________

   ____

   Sum of the rounded numbers ____________

   Sum on the calculator ____________

Section 1-6 *Multiplying Whole and Decimal Numbers*

Exercise 1--Money Multiplication (Group)

Purpose of the Exercise
- To understand how multiplication works

Materials Needed
- Fake Money
- Calculator

1. Each group member takes $6247.87 from the bank, using the fewest possible bills/coins. Combine all of the money into a pool. Record the number of members in the group and the total number of each denomination bill and coin in the money pool.

   Number of members in the group: ______________________

Amount of Money	$10,000's	$1000's	$100's	$10's	$1's	dimes	pennies
__________	_____	_____	_____	_____	_____	_____	_____

2. With one group member playing the banker, minimize the number of bills/coins from above. Record the amount of money and the number of each denomination bill/coin.

Amount of Money	$10,000's	$1000's	$100's	$10's	$1's	dimes	pennies
__________	_____	_____	_____	_____	_____	_____	_____

3. Write the multiplication problem that explains what you have done. ______________

   ____

4. Check your final money amount with a calculator to make sure no mistakes were made in the money exchanges. If the calculator answer does not match the amount of money after the exchange, repeat the exercise.

5. Repeat parts 1 - 4 for each of the money amounts below with each member taking from the bank:
   a. $2099.07 b. $387.59 c. $98.88 d. $7904.73

6. If you pay 6 people $4789.54 each, how much did you pay in total? ______________

7. If you owe 7 people $360.87 each, how much do you owe in total? ______________

8. If you make 12 monthly payments of $895.20 each, what is the total of your payments for the year? ___________

9. Calculate the following products:

   a. 4789.54 times 6 b. 360.87 times 7 c. 895.2 times 9

Multiplication by 10, 100, 1000 and 10,000. Do not use a calculator.

10. If you have ten quarters, how much money do you have? ________________

11. One quarter is $0.25. What is 10 times 0.25? ______________________

12. If you have ten dimes, how much money do you have? ____________________

13. One dime is $0.10. What is 10 times 0.10? ____________________

14. If you have 10 half-dollars, how much money do you have? ____________

15. A half-dollar is $0.50. What is 10 times 0.50? ____________________

16. If ten people each have $7.50, how much money will they have if they pool their money? ____________________

17. What is 10 times 7.50? ____________________

18. If you bought 10 items for $29.50 each, how much money did you spend altogether? ____________

19. What is 10 times 29.5? ____________________

20. In general, what is a shortcut for multiplying a number by 10? ____________ ____________________________________

21. If you have 100 quarters, how much money do you have? ____________

22. One quarter is $0.25. What is 100 times 0.25? ____________________

23. If you have 100 dimes, how much money do you have? ____________

24. One dime is $0.10. What is 100 times 0.10? ____________________

25. If you have 100 nickels, how much money do you have? ____________

26. One nickel is $0.05. What is 100 times 0.05? ____________________

27. If a hundred people each have $7.50, how much money will they have if they pool their money? ____________________

28. What is 100 times 7.50? ____________________

29. If you bought 100 items for $29.50 each, how much money did you spend altogether? ____________

30. What is 100 times 29.5? ____________________

31. In general, what is a shortcut for multiplying a number by 100? ____________ ____________________________________

32. Explain how you would multiply a number by 1000. ____________ ____________________________________

33. Explain how you would multiply a number by 10,000. ____________ ____________________________________

The Process of Multiplication

Multiplication of whole and decimal numbers can be explained with the money model. There are a couple of different processes. Multiplication is just repetitive addition, meaning that 3 times 87 is the same as 87 + 87 + 87. Before jumping into the process of multiplication, let's cover some of the ways we write a multiplication problem.

Example 1: 3 times 87 can be written several ways.

3• 87	Sometimes we use a "dot" to indicate multiplication. This can sometimes be hard to see or possibly confused with a decimal point.
(3)(87)	We can put the two numbers in side-by-side parentheses. This notation is very clear, but not always necessary. Variations would be 3(87) which is commonly used, or (3)87 which is correct but seldom used.
3 x 87	We can use the times sign, "x." In algebra, we seldom use this symbol because it is too easy to confuse with the letter "x," which is often used as a symbol for an unknown quantity.

There are several algorithms that work to compute multiplication. The most common one models the process we covered for addition and is a shorthand way of doing the repetitive addition.

Example 2: We can multiply 87 by 3 by setting up the problem as repetitive addition (below, left). Or we can set up the problem as multiplication (below, right). The addition is carried out just like any other addition process. In multiplication, we begin by multiplying 7 times 3, which is 21. We write the 1 in the ones column and convert 20 to two 10s. We write a "2" above the 8 in the tens column. We then multiply 3 times 8 and then add the 2 carried over. 3 x 8 + 2 = 24 + 2. This gives us 26 tens which we interpret as 2 100s and 6 10s and record the digits in their proper place in the answer.

```
   2                          2
  87                         87
+ 87                        x 3
+ 87                        ---
----                        261
 261
```

(87)(3) = 261

Multiplication by 10 and Powers of 10

In the money model, if 10 different people pay us $430.00, the total of the payments is the **product** of $430.00 and 10 which is $4300.00. The 3 in the tens place multiplied by 10 gives us 30 $10-bills, which is $300, and the 4 $100-bills multiplied by 10 gives 40 $100-bills which is $4000. Multiplication by 10 just shifts each digit one place to the left--multiplication by 10 has an end result of moving the decimal point one position to the right.

Similarly, multiplication by 100 (or 10^2) shifts each digit two place values to the left and moves the decimal point two places to the right.

Continuing in this pattern, multiplication by 1000 (or 10^3) moves the decimal point three places right, multiplication by 10,000 (or 10^4) moves the decimal point four places right, and so on.

We will now begin to find out how to multiply a number by a two- or three-digit number. First, we will learn how to multiply by multiples of 10, 100, and 1000. (These are numbers like 20, 30, 40...200, 300, 400...2000, 3000, 4000, etc.)

Example 3: Compute (792)(30).

We will break the computation into two parts. Since 30 = (3)(10), we will first multiply 792 by 3 and then that product by 10.

First: (792)(3) = 2376
Second:(2376)(10) = 23,760

To show this more concisely, we will write the entire problem in column form, multiply 792 by 3, but write a "0" in the ones place to indicate multiplication by 10.

```
    2        The 2 is "carried" after multiplying 9 by 3.
   792
 x  30
23,760
```

To multiply a number by 300, first multiply the number by 3, then that product by 100.

Example 4: Compute (792)(300).

First: (792)(3) = 2376

Second: (2376)(100) = 237,600

In column form,

```
     2       The 2 is "carried" after multiplying 9 by 3.
    792
 x  300
237,600
```

Example 5: Multiply 83 by 40 in column form.

```
   1     4 x 3 = 12, so 40 x 3 = 120.  Record 20 on the answer line and
  83     carry the 1 to add to 4 x 8.
x 40
3320
```

When we multiply by a number with two or more digits, we will do the computation in two steps as shown in the examples below.

Example 6: Multiply 83 by 46. The basic process is to

(1) multiply 83 by 6	83 x 6 = 498
(2) multiply 83 by 40	83 x 40 = 3320
(3) add the two products.	498 + 3320 = 3818

The sum of 498 and 3320 is 3818 so (83)(46) = 3818.

The more common way to write this calculation is in columns:

```
  83
x 46
 498     83 x 6 = 498
3320     83 x 4 = 332, so 83 x 40 = 3320
3818     Add the two numbers, 498 and 3320.
```

(83)(46) = 3818

Example 7: Multiply 135 by 372.

```
  135
x 372
  270     (135)(2) = 270
 9450     (135)(70) = 9450
40500     (135)(300) = 40,500
50220     Add the three lines above.
```

(135)(372) = 50,220

Multiplication by Numbers with Digits after the Decimal Place

As observed from the money model:

(84.71)(10,000) = 847,100
(84.71)(1000) = 84,710
(84.71)(100) = 8,471
(84.71)(10) = 847.1
(84.71)(1) = 84.71

From the numbers above, it appears that as the decimal place of 10,000, 1000, and so fourth, moves one place to the left in the number, the resulting products have decimal places moved one decimal place to the left. Following what seems to be an emerging pattern we get

(84.71)(10,000) = 847,100
(84.71)(1000) = 84,710
(84.71)(100) = 8,471
(84.71)(10) = 847.1
(84.71)(1) = 84.71
(84.71)(0.1) = 8.471
(84.71)(0.01) = 0.8471

Notice that when multiplying 84.71 by 0.01, the product has the same digits in the same order (8471) but the decimal point moved from in front of the 7 to in front of the 8, that is, the decimal point moved two places to the left. The multiplier, 0.01 has two decimal places.

When multiplying 84.71 by 0.001, the product has the same digits in the same order (8471) but the decimal point moved from in front of the 7 to in front of a 0 in front of the 8, that is, the decimal point moved three places to the left. The multiplier 0.001 has three decimal places.

Let's observe the rationale behind continuing the pattern.

As observed in the money model, 0.1 (or 0.10) has a 1 in the dimes place. A dime is 1/10 of a dollar, so multiplying 0.1 times 84.70 is like dividing 84.70 by 10.

If you have $84.70 that must be divided among 10 people, each person gets $8.47

Shortcuts for Multiplying by 0.1, 0.01, 0.001, and so on

To multiply a number by 0.1, move the decimal point one place to the left.
To multiply a number by 0.01, move the decimal point two places to the left.
To multiply a number by 0.001, move the decimal point three places to the left.
To multiply a number by 0.0...1, where the "dots" indicate an unknown number of zeros and the digit 1 is the only significant digit, the multiplication can be accomplished by moving the decimal point in the original number the correct number of decimal places to the left.

Example 8:

470.93 x 0.1 = 47.093
470.93 x 0.01 = 4.7093
470.93 x 0.001 = 0.47093
470.93 x 0.0001 = 0.047093
470.93 x 0.00001 = 0.0047093

Example 9: Multiply 82.3 and 0.2.

First:	Multiply the 82.3 by 2	$82.3 \cdot 2 = 164.6$
Second:	Multiply the answer by 0.1	$164.6 \cdot 0.1 = 16.46$
	The answer	$82.3 \cdot 0.2 = 16.46$

Let's observe a shorter process that gives the same answer. Notice that when we multiply 823 and 2, numbers with the same digits as above but without the decimal points, we get the same digits in the product as we did above.

```
 823
x  2
1646
```

To place the decimal point correctly, we count the number of digits to the right of the decimal place in the original numbers. We sometimes call this "the number of decimal places." In this example there is one decimal place in each number or a total of 2 decimal places. After multiplying the numbers without considering the placement of the decimal point, appropriately position the decimal point so that the product has 2 decimal places. Placing the decimal correctly is crucial to having the right answer. When we learn to approximate products, we will use estimates as a double check that the decimal is in the right position.

```
 82.3
x 0.2
16.46      The answer has 2 decimal places.
```

General Process for Multiplying Numbers with Decimal Points

(1) Multiply the numbers together without regard to the position of the decimal points.
(2) Count the number of decimal places in both numbers multiplied.
(3) Put the decimal point in the product so that the number of decimal places is the sum of the number of decimal places in the two original numbers. Count decimal places in the product starting from the right.

Example 10: Multiply 73.15 by 18.9.

Set up the problem in columns. There are a total of 3 decimal places in the two numbers.

```
   73.15
  x 18.9
   65835   7315 · 9   =  65835
  585200   7315 · 80  = 585200
  731500   7315 · 100 = 731500
1382.535   Add the three rows and put the decimal point 3 places from the right.
```

Estimating Products

In real life, most multiplication problems like the example above will be done on a calculator. Just as in addition, it is important to have an idea of the "size" of the answer before using the calculator. We will use the same kind of procedure for estimating answers in multiplication as in addition.

Procedure for Estimating Products

First: Round each of the factors to 1 or 2 significant digits.
Second: Multiply the rounded numbers.

While learning to estimate products, compute the exact product using a calculator to see how close your estimates are.

Example 11: Estimate the product of 73.15 and 18.9 (the problem from Example 10)

The first number rounded to 1 significant digit is 70.

The second number rounded to 1 significant digit is 20.

70 times 20 = 1400

From above, **(73.15)(18.9) = 1382.535** which rounds to 1383.

Example 12: Estimate 2575.73 times 0.312.

The first number rounded to 1 significant digit is 3000.

The second number rounded to 1 significant digit is 0.3.

(3000)(0.3) = 900.

The estimate is probably high because 3000 is considerably larger than 2575.73, but 0.312 is fairly close to 0.3. For comparison, compute the answer on the calculator.

(2575.73)(0.312) = 803.62776

The estimate is high, but it would at least catch decimal mistakes. For a better estimate, round 2575.73 to 2 significant digits instead of 1 significant digit.

In general, this type of estimation with multiplication will not be as accurate as the same kind of estimation process in addition. However, it is helpful. The process can be refined to more accuracy by rounding one or both of the numbers to be multiplied to 2 significant digits instead of 1 significant digit. As you gain experience with estimates, you will also develop other common sense observations.

Exercise 2--Estimating Products of Numbers

Purpose of Exercise

- To practice estimating products
- To practice multiplying

Materials Needed

- Calculator

In each of the following, round each number to one or two significant digits. Multiply the rounded numbers to estimate the products of the original numbers; then calculate the products on the calculator to compare results. The first one is done as an example.

1. (4896.30)(0.053)

   The 1st number rounded <u>5000</u>

   The 2nd number rounded <u>0.05</u>

   Estimation <u>(5000)(0.05) = 250</u>

   Calculator answer <u>259.5039 Not Bad!</u>

2. (0.0452)(0.3712)

   The 1st number rounded ____________

   The 2nd number rounded ____________

   Estimation ____________

   Calculator answer ____________

3. (129.78)(75.99)

   The 1st number rounded ____________

   The 2nd number rounded ____________

   Estimation ____________

   Calculator answer ____________

4. (99.57)(0.57)

   The 1st number rounded ____________

   The 2nd number rounded ____________

   Estimation ____________

   Calculator answer ____________

5. (0.5473)(0.0465)

The 1st number rounded ___________

The 2nd number rounded ___________

Estimation ___________

Calculator answer ___________

6. (407,893)(479)

The 1st number rounded ___________

The 2nd number rounded ___________

Estimation ___________

Calculator answer ___________

7. (17,387)(189)

The 1st number rounded ___________

The 2nd number rounded ___________

Estimation ___________

Calculator answer ___________

8. (2.579)(9.71)

The 1st number rounded ___________

The 2nd number rounded ___________

Estimation ___________

Calculator answer ___________

9. (4.7592)(2.1)

The 1st number rounded ___________

The 2nd number rounded ___________

Estimation ___________

Calculator answer ___________

10. (4.7592)(0.21)

The 1st number rounded ___________

The 2nd number rounded ___________

Estimation ___________

Calculator answer ___________

11. (47.592)(2.1)

The 1st number rounded ___________

The 2nd number rounded ___________

Estimation ___________

Calculator answer ___________

12. (47.592)(0.21)

The 1st number rounded ___________

The 2nd number rounded ___________

Estimation ___________

Calculator answer ___________

13. (475.92)(2.1)

The 1st number rounded ___________

The 2nd number rounded ___________

Estimation ___________

Calculator answer ___________

14. (475.92)(0.21)

The 1st number rounded ___________

The 2nd number rounded ___________

Estimation ___________

Calculator answer ___________

Without using a calculator, in each of the following decide which estimate of the product is the most reasonable. Check your estimate by calculating the products on the calculator.

15. $(752.96)(12) \approx$	8152	2512	29,273
16. $(4.375)(6.923) \approx$	512	81	31
17. $(76.93)(3.21) \approx$	240	2400	24
18. $(84.321)(0.15) \approx$	900	9	90
19. $(75.98)(92.387) \approx$	69000	690	6900
20. $(0.478)(0.037) \approx$	0.2	2	0.02
21. $(3456.78)(17) \approx$	6,000	59,000	80,000
22. $(14.3)(0.3) \approx$	42	6	4
23. $(0.05)(0.0189) \approx$	0.009	0.0009	0.10

Find the following products. Use a separate sheet of paper to set up problems as needed in columns. Products that involve powers of ten multipliers should be calculated mentally.

24. (756)(7)
25. 15 x 12
26. Multiply 178 and 13
27. Find the product of 893 and 6
28. (1.78)(1.3)
29. 89.3 x 6
30. 0.0897 x 100
31. 17 x 0.1
32. 146 x 100
33. (146)(99)
34. (25.8)(2.2)
35. (25.8)(0.22)
36. 98 x 0.8
37. 98 x 8
38. 98 x 80
39. 37.5 x 14
40. 375 x 14
41. 2987 x 132
42. (457)(632)
43. (4.57)(6.32)
44. 500 x 100
45. 30 x 1000
46. 197 x 0.01
47. 197 x 0.001
49. (99.9)(1000)
50. (9.99)(1000)

Section 1-7 *Dividing Whole and Decimal Numbers*

We have covered the processes of addition, subtraction, and multiplication. The remaining arithmetic operation is division. Division is really "divvying up." If $750 is to be divided among three people, we divide $750 by 3 to determine how much each person receives. As with addition, subtraction, and multiplication, we will use the money model to learn a process of division and understand what division means. Remember that the answer to a division problem is called the **quotient**.

Illustrated below are the various symbols of division and the vocabulary that we associate with division.

Symbols of Division

56,738 divided by 3 can be represented symbolically in the following ways:

$56{,}738/3 \qquad 56{,}738 \div 3 \qquad \dfrac{56{,}738}{3} \qquad \text{or} \qquad 3\overline{)56{,}738}$

"Divide 6 into 87" can be represented as illustrated below:

$87/6 \qquad 87 \div 6 \qquad \dfrac{87}{6} \qquad 6\overline{)87}$

Vocabulary

When dividing 56,738 by 3, the **divisor** is 3, the **dividend** is 56,738, and the answer to the division is the **quotient**.

When we divide 6 into 87, 87 is the **dividend** and 6 is the **divisor**. The answer to the division problem is the **quotient**.

The dividend, divisor, and quotient are shown below in their positions in the various representations of division.

dividend/divisor = quotient $\qquad$ dividend ÷ divisor = quotient

$\dfrac{\text{dividend}}{\text{divisor}} = \text{quotient} \qquad \begin{array}{r}\text{quotient}\\ \text{divisor}\overline{)\text{dividend}}\end{array}$

To do division on a calculator, first key in the dividend, then the divide by sign (÷), then the divisor, then press the "enter" or "=" key. The calculator will display the quotient or a decimal approximation to the quotient.

Example 1: Use a calculator to compute 56,738 ÷ 3.

Key strokes: [56738] [÷] [3] [=]

18912.67 (rounded to 2 decimal places) so the quotient is approximately 18,912.67.

Exercise 1--Money Division (Group)

Purpose of the Exercise

- To understand the concepts of division and reasonableness of answers

Materials Needed

- Fake money--dimes, pennies, $1-bills, $10-bills, $100-bills, $1000-bills, $10,000-bills

1. a. How many people are in your group? ______________

b. Take a $1000-bill from the bank. Change the $1000-bill into smaller bills and coins as needed to divide the money into equal stacks for each group member. **Do not use a calculator.**

How much money does each member get? ____________________

How much money (if any) is left over? ____________________

c. **Use a calculator** to divide 1000 by the number of people in the group. Compare the calculator answer to the answer in part b. The answer may differ slightly if the division does not come out exactly and there were some pennies left over when distributing the money.

2 a. How many people are in the class? ______________

b. Take a $1000-bill from the bank. Change the $1000-bill into smaller bills and coins as needed to divide the money into equal stacks for each class member. **Do not use a calculator.**

How much money does each member get? ____________________

How much money (if any) is left over? ____________________

c. **Use a calculator** to divide 1000 by the number of people in the class. Compare the calculator answer to the answer in part b.

3. a. If the money were evenly divided among all students at your school, would each student get more or less than when the $1000 was distributed among students in your group? ______________________________

b. If the $1000 were divided evenly among the immediate members of the family of the president of the United States, would each family member get more money, less money, or the same amount of money as when the $1000 was distributed among students in your group? ______________

4. a. How many people are in your group? ______________

b. Take $56,738 from the bank. Change the $56,738 into smaller bills and coins as needed to divide the money into equal stacks for each group member. **Do not use a calculator.**

How much money does each member get? ____________________

How much money (if any) is left over? ____________________

c. **Use a calculator** to divide 56,738 by the number of people in your group. Compare the answer to the answer in part b.

5. a. How many people are in the class? ______________

b. Take $56,738 from the bank. Change the $56,738 into smaller bills and coins as needed to divide the money into equal stacks for each class member. **Do not use a calculator.**

6. When you have a particular amount of money, will each person get more money if the money is divided evenly among 4 people or if the money is divided evenly among 5 people? ______________________________

7. Do the following money exchanges to observe what happens when dividing by 10 or 100.

a. Take \$250 dollars from the bank. Make money exchanges as needed to have 10 stacks of money with equal amounts.
How much money is in each stack? _______ $250 \div 10 =$ ___________

b. Divide \$75.30 into 10 equal stacks.
How much money is in each stack? _______ $75.3 \div 10 =$ ___________

c. Divide \$3789.40 into 10 equal stacks.
How much money is in each stack? _______ $3789.4 \div 10 =$ ___________

d. You have \$200 and want to divide it into 100 equal stacks.
How much money is in each stack? _______ $200 \div 100 =$ ___________

e. \$400 is divided among 100 people.
How much does each person get? _______ $400 \div 100 =$ ___________

f. \$50 is divided among 100 people.
How much does each person get? _______ $50 \div 100 =$ ___________

g. \$450 is divided among 100 people.
How much does each person get? _______ $450 \div 100 =$ ___________

8. Discuss the pattern that appears in Problem 7. Describe a shortcut for dividing a number by 10 or by 100. Generalize the shortcut for dividing by 1000, 10,000, and so forth. If you are not sure that your shortcut is correct, use your shortcut to do some division problems and check the quotients with your calculator.

9. Use your calculator to compute each of the following quotients and notice the connection between different divisions.

a. $28.95 \div 2.9 =$ _______ $2895 \div 290 =$ _______ $28{,}950 \div 2900 =$ _______

b. $48 \div 3 =$ _______ $480 \div 30 =$ _______ $4800 \div 300 =$ _______

c. $149 \div 0.08 =$ _______ $1490 \div 0.8 =$ _______ $14{,}900 \div 8 =$ _______

d. $14.9 \div 0.6 =$ _______ $149 \div 6 =$ _______ $1490 \div 60 =$ _______

10. From your observations in Problem 9, what impact does multiplying both the dividend and divisor by 10, 100, or 1000 have on the quotient?

11. In each of the following pairs, circle the division that has the largest quotient. In each problem, the letter represents a specific amount of money.

a. $x/27$ or $x/89$

b. $y \div 896$ or $y \div 966$

c. $377\overline{)A}$ or $298\overline{)A}$

d. $\dfrac{B}{4328.876}$ or $\dfrac{B}{5786.65}$

e.	$\frac{C}{0.1}$	or	$\frac{C}{0.01}$
f.	$0.5\overline{)D}$	or	$0.7\overline{)D}$
g.	$E \div 2.7$	or	$E \div 1.9$

Division and multiplication are reverse processes. After division is performed, we can check the division by multiplying the quotient times the divisor. If this product is the dividend, the answer is correct. $18 \div 9 = 2$ because $9 \bullet 2 = 18$.

Without a calculator, pick the best answer from the list by multiplying the choices for answers by the quotient. If the numbers are "messy," round off the quotient and possible answers before multiplying. After marking the best answer in all of the problems, use a calculator to do the division to see how accurate you were.

12. $5786.92 \div 17 \approx$	3400	340	34
13. $\frac{257}{389} \approx$	0.07	1.7	0.7
14. $0.98/0.4 \approx$	2.5	0.25	0.025
15. $84.0078/3.1 \approx$	300	15	27
16. $897/0.19 \approx$	4700	4.70	470
17. $8/0.2 \approx$	4	0.4	40
18. $18 \div 0.03 \approx$	6	60	600
19. $1.8 / 0.03 \approx$	6	60	600
20. $\frac{5798}{23} \approx$	25	250	2500
21. $\frac{5798}{2.3} \approx$	25	250	2500
22. $\frac{579.8}{23} \approx$	25	250	2500
23. $\frac{579.8}{2.3} \approx$	25	250	2500
24. $47.98/22.087 \approx$	0.7	2	32
25. $0.00089 \div 3 \approx$	0.03	0.0030	0.0003
26. $89 \div 0.0003 \approx$	300,000	30,000	3,000
27. $\frac{0.007}{7000} \approx$	0.000001	1	1,000,000

The Process of Division

There are several accurate pencil and paper ways to compute quotients. The best-known process is that of long division, but since many people have a hard time remembering how to do long division or understanding why it works, we will first cover a process that is closer to how we divvy up stacks of money. It works on the principle that you first distribute what is easy to distribute and then subtract the amount distributed. See how much is left over and continue to distribute until the money is gone.

Example 2: Divide $87 among 6 people (using iterative subtraction).

$$14\frac{3}{6} = 14\frac{1}{2}$$

	6) 87		
6 x $10	60	10	Everyone gets $10.
	27		This leaves $27 left to distribute.
6 x $4	24	4	Distribute $4 per person.
	3	14	This leaves $3 **remaining.**

So far, each person has received $10 + $4 for a total of $14. At this point, since the money left over is $3 and we cannot split that up without change, we can say that the quotient is 14 but we have a **remainder** of 3. If we want to divide the $3.00, we continue the process, but now the amounts distributed get into decimal amounts.

$$\frac{87}{6} = 14 \text{ with a remainder of } 3$$

We could further distribute the 3 if we broke it into parts and added that amount to the quotient (the amount already distributed). When writing the quotient plus the remainder divided by the divisor, we shorten the writing so that the "+" sign is omitted.

$$\frac{87}{6} = 14 + \frac{3}{6} = 14\frac{3}{6} = 14\frac{1}{2} \text{ or each person receives \$14.50}$$

We write the quotient above the divide box.

Example 3: Divide 2879 by 34.

Let's interpret this again in the money setting. We are dividing $2879 among 34 people. To do the first distribution, notice that 34 x $100 = $3400 so we would be distributing more than we have, so we have to distribute a number smaller than $100 to each person. 34 x $10 = $340, so we could distribute $10 to each person, but that doesn't distribute very much of the $2879, so lets guess somewhere in between. Guessing too low is not a problem because the excess will be distributed the next time around, but if the guess is too high, we have to adjust the guess-work with a pencil and keep a good eraser handy.

	84 R 23		
	34) 2879		
34 x $60 =	2040	60	Each of the 34 people get $60
	839		Subtract the amount distributed. There is $839 left.
34 x $10 =	340	10	Distribute another $10 per person.
	499		Subtract the $340. This leaves $499 to distribute.
34 x $10 =	340	10	Distribute another $10 per person.
	159		Subtract $340.
34 x $4 =	136	4	Distribute $4 to each person.
	23	84	This leaves $23 which cannot be distributed without breaking the $23 into change. This is the remainder.

To get the quotient, add the distributions. 60 + 10 + 10 + 4 = 84. Each person gets $84 and there is $23 remaining undistributed. The quotient = 84 and the remainder is 23.

$$\frac{2879}{34} = 84\frac{23}{34}$$

To check the answer when there is a remainder, multiply the quotient times the divisor and add the remainder. This should give the dividend.

Check: (84)(34) + 23 = 2856 + 23
= 2879 This is the dividend. The answer is correct.

The conventional **long division** algorithm is the multiplication and subtraction done more efficiently. The following example shows how to do long division.

Example 4: Divide $87 by 6.

First write the problem in the "box" format, making sure the dividend, 87, is inside the box and the divisor, 6, is outside the box.

14	**Quotient line.**
6) 87	6 divides into 8 <u>one</u> time. Put the digit, 1, in the 10s place of the quotient.
6	Multiply 1 times the divisor, 6, and write the answer under the 87.
27	Subtract. The first column, 8 - 6 is less than the divisor so continue.
24	6 will divide into 27 **four** times. Write **4** on the answer line.
3	Multiply **4** by the divisor, 6, to get **24**. Write 24 under the 27 .
	Subtract 24 from 27 to get **3** left over.

You might notice that this process is a variation of the subtraction process. In the first step, when we divided 6 into 8 and got 1, we put the "1" above the 8 on the answer line. That positioned the "1" in the 10s place, meaning we distributed $10 per person in the first round. We then subtracted the total amount, $60, from the $87.

The difference between this process and the subtraction method is that if the 8 - 6 in the first step had been more than the divisor, we would have had to erase what we had done and pick a larger amount to divide by. In the subtraction method you just pick up any extra amounts in the next step.

Long Division

Long division is a repetitive process. The same steps get done over and over until the process is done. The three basic steps are

Division Step:

If the first digit of the divisor is smaller than the first digit of the dividend, then decide how many times it will divide and write that number on the quotient line above the first digit of the dividend.

If the first digit of the divisor is larger than the first digit of the dividend, then divide into the first 2 digits of the divisor and write the quotient above the second digit of the dividend. (See Example 5 below.)

Multiplication Step: Multiply the <u>quotient</u> you got from the divide step <u>by the entire divisor</u>. Write the answer below.

Subtraction Step: Subtract the product from the multiply step from the line above it. If the difference is larger than the divisor, go back to the divide step and try a larger quotient.

Repeat the above three steps, until there are no numbers left to bring down and the remainder is smaller than the divisor. After the first time through the steps, however, divide into the last line, not into the dividend.

When there is a remainder, we can use long division to get a decimal quotient instead of a quotient and remainder or a mixed number quotient. Just write the decimal point in the dividend and write the desired number of zeros on the right of the decimal point.

Example 5: Using long division, divide 4367 by 53.

$$\begin{array}{r} 82\frac{21}{53} \\ 53\overline{)4367} \\ \underline{424} \\ 127 \\ \underline{106} \\ 21 \end{array}$$

Divide: 5 goes into 43 8 times. 8 x 53 = 424 so 8 will work
Multiply: 8 x 53 = 424
Subtract: 436 - 424 = 12 and bring down the 7.
Divide: 5 goes into 12 2 times. **Multiply:** 2 x 53 = 106
Subtract: 127 - 106 = 21

$$4367 \div 53 = 82\frac{21}{53}$$

Example 6: Divide 87 by 6 to get a decimal quotient.
To demonstrate, we will rework the above division:

$$\begin{array}{r} 14.5 \\ 6\overline{)87.0} \\ \underline{6} \\ 27 \\ \underline{24} \\ 3\,0 \\ \underline{3\,0} \\ 0 \end{array}$$

In this case, only one zero after the decimal was needed since the remainder was 0 in the next step. Check this answer on your calculator. Notice that your calculator gives the answer in decimal form, not in the fraction form that we got in Example 2.

To check the answer, multiply the divisor by the quotient. If the product is the dividend, the answer is correct.
Check: (14.5)(6) = 87. It checks.

In Problem 9 of Exercise 1, you noticed that if both the divisor and dividend are multiplied by 10, 100, 1000, and so on, the quotient is unchanged. A more general statement is true. Multiplying both the divisor and dividend by the same number leaves the quotient unchanged. When the divisor has a decimal place in it, we "get rid of the decimal point" in the divisor by multiplying by the appropriate power of 10 before doing the division process.

Example 7: Divide $0.007\overline{)278}$.

The divisor is 0.007, so we will multiply both the divisor and dividend by 1000 to get rid of the decimal in the divisor. We do this by moving the decimal point in the divisor right of the last digit and moving the decimal point in the dividend the same number of places to the right.

The problem restated is

$$7\overline{)278000}$$

Use your calculator to check that $0.007\overline{)278}$ and $7\overline{)278000}$ have the same quotient.

On your calculator, compute 278÷0.007. Quotient = ____________
On your calculator, compute 278000÷7 Quotient = ____________

Exercise 2--Practicing Division

Purpose of the Exercise

- To practice doing long division
- To practice identifying the divisor, quotient, dividend, and remainder

In each of the following, identify the divisor and dividend. Then compute the quotient by long division. Use your own paper and show the setup and work. Check your answer by multiplying the quotient times the divisor and adding the remainder. Record results in the blanks provided.

1. 67 divided by 9
divisor ________________
dividend ________________
quotient ________________
remainder ________________
check ________________

2. Divide 8 into 279
divisor ________________
dividend ________________
quotient ________________
remainder ________________
check ________________

2. 459 ÷ 0.9
divisor ________________
dividend ________________
quotient ________________
remainder ________________
check ________________

3. 18.5/7
divisor ________________
dividend ________________
quotient ________________
remainder ________________
check ________________

4. 0.92 ÷ 3
divisor ________________
dividend ________________
quotient ________________
remainder ________________
check ________________

5. 245 ÷ 22
divisor ________________
dividend ________________
quotient ________________
remainder ________________
check ________________

6. 245.79 ÷ 1.7
divisor ________________
dividend ________________
quotient ________________
remainder ________________
check ________________

7. 368 divided by 0.5
divisor ________________
dividend ________________
quotient ________________
remainder ________________
check ________________

Estimating Quotients by Rounding the Dividend and Divisor

As before, we may do tedious calculations on a calculator, but it is important to be able to estimate an answer and to recognize when an answer is reasonable. If we have computed a quotient either on a calculator or by hand, one way to check the reasonableness of the answer is to use the techniques from the last section and approximate the product of the divisor and quotient to see if the product is approximately the dividend.

Since division is for the most part a very tedious pencil and paper process but is quickly and efficiently done using a calculator, the rest of the discussion will focus on estimating the quotients. As in addition, subtraction, and multiplication, it is important to know when the answer is reasonable.

Estimating a Quotient

First: Round both the divisor and the dividend to 1 significant digit.

Second: If the first significant digit in the **divisor** comes **after** the decimal point, "move the decimal point" to the right of the first significant digit by multiplying both the divisor and dividend by the appropriate power of 10.

Third: Divide the rounded numbers to compute the estimated quotient.

Example 8: Estimate 782 divided by 0.094.

Round the dividend, 782, to 800 and round the divisor, 0.094, to 0.09.

In the rounded divisor, 9 is the **second** place after the decimal so multiply both the dividend and divisor by 100 to move the decimal place 2 places to the right.

The problem restated is 80,000 ÷ 9.

```
    8888
 9)80000
  72000
   8000
   7200
    800
    720
     80
     72
      8
```

The estimated quotient is 8888--or 8889 rounded since we obviously have a pattern going and the digit in the ones place will also be 8. On the calculator, 782/0.094 is 8319.1489362. At least this shows that the decimal point is in the right place. (The number of digits varies from calculator to calculator.)

Example 9: Estimate 7,820,000 divided by 940.

Round the dividend, 7,820,000, to 8,000,000.

Round the divisor, 940, to 900.

We can simplify problem by dividing both the divisor and dividend by 100.
The problem restated is 80000 ÷ 9. This is the same problem worked in Example 1.
The estimated quotient is 8777 (or 8778 since the remainder is 8).
On the calculator, 782/0/094 is 8319.1489362.

In division, we often round off a final quotient, even though a calculator will give us an answer with many digits. In most uses of mathematics, we only retain 3 or 4 significant digits.

Exercise 3--Division

Purpose of the Exercise

- To practice estimating quotients
- To practice the process of division

Materials Needed

- Calculator

On a separate sheet of paper without a calculator, estimate the following division problems by rounding the dividend and divisor and then computing an estimated quotient. Then use a calculator to compute the quotient and compare the estimated quotient to the quotient given by the calculator. The first one is done as an example.

1. $7986.98 \div 89$

The dividend rounded ______8000
The divisor rounded ______90
The adjusted problem ______8000÷90
The estimated answer ______87 (or 88)
The calculator answer 89.741348315

2. $80394.98 \div 21.82$

The dividend rounded ______________
The divisor rounded ______________
The adjusted problem ______________
The estimated answer ______________
The calculator answer ______________

3. $983.56 \div 0.054$

The dividend rounded ______________
The divisor rounded ______________
The adjusted problem ______________
The estimated answer ______________
The calculator answer ______________

4. $0.0967 \div 0.298$

The dividend rounded ______________
The divisor rounded ______________
The adjusted problem ______________
The estimated answer ______________
The calculator answer ______________

5. $29.6 \div 389$

The dividend rounded ______________
The divisor rounded ______________
The adjusted problem ______________
The estimated answer ______________
The calculator answer ______________

6. $2569 \div 0.05$

The dividend rounded ______________
The divisor rounded ______________
The adjusted problem ______________
The estimated answer ______________
The calculator answer ______________

7. $8976 \div 3456$

The dividend rounded ______________
The divisor rounded ______________
The adjusted problem ______________
The estimated answer ______________
The calculator answer ______________

8. $7856.9873 \div 47.2987$

The dividend rounded ______________
The divisor rounded ______________
The adjusted problem ______________
The estimated answer ______________
The calculator answer ______________

9. 14.568 ÷ 300.98

The dividend rounded ____________

The divisor rounded ____________

The adjusted problem ____________

The estimated answer ____________

The calculator answer ____________

10. 786.098 ÷ 9.99876

The dividend rounded ____________

The divisor rounded ____________

The adjusted problem ____________

The estimated answer ____________

The calculator answer ____________

11. 7895 ÷ 1057

The dividend rounded ____________

The divisor rounded ____________

The adjusted problem ____________

The estimated answer ____________

The calculator answer ____________

12. 7.895 ÷ 1.057

The dividend rounded ____________

The divisor rounded ____________

The adjusted problem ____________

The estimated answer ____________

The calculator answer ____________

13. 0.0078 ÷ 0.021

The dividend rounded ____________

The divisor rounded ____________

The adjusted problem ____________

The estimated answer ____________

The calculator answer ____________

14. 7.8 ÷ 21

The dividend rounded ____________

The divisor rounded ____________

The adjusted problem ____________

The estimated answer ____________

The calculator answer ____________

In Problems 15 and 16, verify by multiplying the quotient by the divisor that when the dividend is zero, the quotient is zero.

15. 0/7 = 0 Divisor = ________ Dividend = ________ Quotient = ________

Divisor x quotient = dividend? Show the multiplication to verify. ____________

16. 0/592.4 = 0 Divisor = ________ Dividend = ________ Quotient = ________

Divisor x quotient = dividend? Show the multiplication to verify. ____________

In each of the following see what happens when a number is divided by zero.

17. 8/0

Dividend = ____________ Divisor = ____________

8/0 = 0? ______ x ______ = ______ True? False?
Divisor quotient = Dividend

8/0 = 8? ______ x ______ = ______ True? False?
Divisor quotient = Dividend

8/0 = 1? ______ x ______ = ______ True? False?
Divisor quotient = Dividend

8/0 =1/8? ______ x ______ = ______ True? False?
Divisor quotient = Dividend

Are any of the possible quotients correct? If so, which one(s)? ________

18. 5.7/0

Dividend = ______________ Divisor = __________________

5.7/0 = 0? _______ x _______ = _________ True? False?
Divisor quotient = Dividend

5.7/0 = 5.7? _______ x _______ = _________ True? False?
Divisor quotient = Dividend

5.7/0 = 1? _______ x _______ = _________ True? False?
Divisor quotient = Dividend

5.7/0 = 0.157? _______ x _______ = _________ True? False?
Divisor quotient = Dividend

Are any of the possible quotients correct? If so, which one(s)? ___________

19. 0/0 =

Dividend = ______________ Divisor = __________________

0/0 = 0? _______ x _______ = _________ True? False?
Divisor quotient = Dividend

0/0 = 1? _______ x _______ = _________ True? False?
Divisor quotient = Dividend

0/0 = 7? _______ x _______ = _________ True? False?
Divisor quotient = Dividend

0/0 = 1/7? _______ x _______ = _________ True? False?
Divisor quotient = Dividend

Are any of the possible quotients correct? If so, which one(s)? ___________

Discussion questions:

20. Can you divide a nonzero number **by** zero? If so, what is the answer? ___________

____

____

____

21. Can you divide a nonzero number **into** zero? If so, what is the answer? ___________

____

____

____

22. What is wrong with the division, 0/0? How is it different from 7/0? _____________

____

____

____

Exercise 4--Fair Distribution of Candy and Inheritance (Group)

Purpose of the Exercise

- To understand the distribution of goods
- To understand the meaning of the quotient and remainder in a division problem

Materials Needed

- A bag of candy for each group Each group should have a different number of pieces of candy in the bag.
- Calculator (for the inheritance questions).

1. Count the number of pieces of candy in your group's bag and write the number (large enough to read) on the board. Each group is to devise a way to redistribute the candy to the class (without putting all of the candy in one bag) so that each student in the class gets the same number of pieces (the maximum amount possible) and the instructor gets the left over pieces. Record the amount of candy each group originally had and how many students are in each group:

   Group 1 ____ _____ Group 2 _____ _____ Group 3 ______ ______

   Group 4 ____ _____ Group 5 _____ _____ Group 6 ______ ______

   Group 7 ____ _____ Group 8 _____ _____ Group 9 ______ ______

   Total number of pieces of candy: ______ Total number of people in class: _______

2. Write your group's instructions for redistributing the candy. (For example, "Group 1 gives Group 2 three pieces of candy and Group 1 gives the instructor 2 pieces.")

3. Record the amount of candy each group has after redistribution:

   Group 1 _________ Group 2 ____________ Group 3 ______________

   Group 4 _________ Group 5 ____________ Group 6 ______________

   Group 7 _________ Group 8 ____________ Group 9 ______________

   How many pieces of candy does the instructor get? __________________

   How many pieces does each student in the class get? _________________

4. Each group will work one of the following problems and then report the problem and conclusions to the class.

GROUP 1: A friend of the family is in charge of distributing a $1,375,252 estate (after fees and probate court) evenly among 27 heirs. One of the heirs received a check for $49,273. Assuming the check for each heir was the same amount, did the family friend profit from this, and if so, by how much? Or did some of the checks bounce, and if so, how many and how much more money was needed to cover the costs?

GROUP 2: A friend of the family is in charge of distributing a $1,375,252 estate (after fees and probate court) evenly among 27 heirs. One of the heirs received a check for $52,321. Assuming the check for each heir was the same amount, did the family friend profit from this, and if so, by how much? Or did some of the checks bounce, and if so, how many and how much more money was needed to cover the costs?

GROUP 3: A friend of the family is in charge of distributing a $1,375,252 estate (after fees and probate court) evenly among 27 heirs. One of the heirs received a check for $51,123. Assuming the check for each heir was the same amount, did the family friend profit from this, and if so, by how much? Or did some of the checks bounce, and if so, how many and how much more money was needed to cover the costs?

GROUP 4: A friend of the family is in charge of distributing a $1,375,252 estate (after fees and probate court) evenly among 27 heirs. One of the heirs received a check for $50,215. Assuming the check for each heir was the same amount, did the family friend profit from this, and if so, by how much? Or did some of the checks bounce, and if so, how many and how much more money was needed to cover the costs?

GROUP 5: A friend of the family is in charge of distributing a $1,375,252 estate (after fees and probate court) evenly among 27 heirs. One of the heirs received a check for $42,793. Assuming the check for each heir was the same amount, did the family friend profit from this, and if so, by how much? Or did some of the checks bounce, and if so, how many and how much more money was needed to cover the costs?

Section 1-8 *Scientific Notation and Expanded Form of a Number*

Scientific notation is a way of writing very large numbers or very small numbers in a concise form, retaining the significant digits, but without writing so many zeros or all of the "insignificant" digits. For example, in scientific notation, the number 3,900,000,000,000 is written 3.9×10^{12} and the number, 0.00000987 is written 9.87×10^{-6}.

Recall from Section 1-6, that multiplying by 10, 100, 0.1, or 0.01 is performed easily by "moving the decimal point" the appropriate number of places. From Section 1-2, we defined the powers of 10 in the following manner:

$10^1 = \underline{1}0$
$10^2 = \underline{1}00$
$10^3 = \underline{1}000$
$10^4 = \underline{1}0000$
$10^5 = \underline{1}00{,}000$

In each problem above, the digit 1 is followed by a number of zeros that correspond with the power of 10. When the power is negative, the power determines how many places the digit 1 is positioned to the right of the decimal point. The pattern for negative powers continues.

$10^{-1} = 0.\underline{1}$
$10^{-2} = 0.0\underline{1}$
$10^{-3} = 0.00\underline{1}$

It is easy to verify by division that:

$10^{-1} = 1/10$
$10^{-2} = 1/100$ or $1/10^2$
$10^{-3} = 1/1000$ or $1/10^3$

Example 1: Multiplying a number by 10 to a power.

$4.56 \times 10^3 = (4.56)(1000) = 4560$ The decimal point is moved 3 places **right**.
$4.56 \times 10^{-3} = 4.56 \div 1000 = 0.00456$ The decimal point is moved 3 places **left**.
$27.8 \times 10^4 = (27.8)(10000) = 278{,}000$ The decimal point is moved 4 places **right**.
$27.8 \times 10\text{-}^4 = 27.8 \div 10000 = 0.00278$ The decimal point is moved 4 places **left**.

Vocabulary
Scientific notation is a system in which a number is written in the form of $N \times 10^n$ where N is a number 1 or larger, but less than 10, and n can be any integer.

In Example 1, 4560 written in scientific notation is 4.56×10^3 and 0.00456 written in scientific notation is 4.56×10^{-3}. However, 27.8×10^4 is not scientific notation for 278,000 because the number multiplied by 10 to a power must be between 1 and 10. 27.8 is not between 1 and 10

Example 2:

Write 5,739,987 in scientific notation.
5.739987×10^6
Write 0.00345 in scientific notation.
3.45×10^{-3}
Write 10,000 written in scientific notation.
1.0×10^4
The number 0.00001 written scientific notation is
1.0×10^{-5}

Exercise 1--Multiplication and Division by Powers of 10

Purpose of the Exercise

- To study the powers of 10 and multiplication and division by 10 to a power

Compute each of the following.

1. 34.1 x 10 = ________ 2. 34.1 x 100 = ________ 3. 34.1 x 1000 = ________

4. 8.23 x 10 = ________ 5. 8.23 x 100 = ________ 6. 8.23 x 1000 = ________

7. What happens to the decimal point in a number when the number is multiplied by 10?

____

8. What happens to the decimal point in a number when the number is multiplied by 100?

____

9. What happens to the decimal point in a number when the number is multiplied by 1000?

____

Write out each of the following.

10. 10^1 = ________ 11. 10^2 = ________ 12. 10^3 = ________

13. 10^6 = ________ 14. 34.1 x 10 = ________ 15. 34.1 x 10^2 = ________

16. 34.1 x 10^3 = ________ 17. 8.23 x 10 = ________ 18. 8.23 x 10^2 = ________

19. 8.23 x 10^3 = ________ 20. 174.6 x 10^4 = ________ 21. 1.067 x 10^2 = ________

22. 0.023 x 10^5 = ________ 23. 1.0 x 10^6 = ________ 24. 0.04 x 10^5 = ________

25. What happens to the decimal point in a number when the number is multiplied by 10 to a power?

____

Do the following divisions.

26. 14.3/10 = ________ 27. 14.3/100 = ________ 28. 14.3/1000 = ________

29. 14.3/10^1 = ________ 30. 14.3/10^2 = ________ 31. 14.3/10^3 = ________

32. What happens to the decimal point in a number when the number is divided by a 10 to a positive power? ____

Write out each of the following:

33. 10^2= 10 x 10 = ________ 34. 10^3 = 10 x 10 x 10 = ________

35. 10^4 = 10 x 10 x 10 x 10 = ________ 36. 10^5 = ________

37. 10^8 = ____________________ 38. 10^{12} = ____________________

Write out each of the following in decimal form.

39. $10^{-1} = 1/10$ = ____________________ 40. $10^{-2} = 1/10^2$ = ____________________

41. 10^{-5} = ____________________ 42. 10^{-4} = ____________________

43. 10^{-7} = ____________________ 44. 10^{-15} = ____________________

45. 10^{-10} = ____________________ 46. 10^{-12} = ____________________

Multiply each of the following.

47. 4×10 = ____________________ 48. 4.32×10 = ____________________

49. 51.73×10^2 = ____________________ 50. 1.923×10^{-1} = ____________________

51. 791.362×10^{-3} = ____________________ 52. 87.21×10^{-4} = ____________________

In each of the following, circle the correct answer, A or B.

53. **A.** $0.001239 = 1.239 \times 10^3$ **B.** $0.001239 = 1.239 \times 10^{-3}$

54. **A.** $2516.31 = 2.51631 \times 10^3$ **B.** $2516.31 = 2.51631 \times 10^{-3}$

55. **A.** $40135 = 4.0135 \times 10^4$ **B.** $40135 = 4.0135 \times 10^{-4}$

56. **A.** $0.0000318 = 3.18 \times 10^5$ **B.** $0.0000318 = 3.18 \times 10^{-5}$

Write the following numbers in scientific notation. Check your answer by multiplying to see if you get the original number. Also check that in your answer the number multiplied by 10 to a power is greater than or equal to 1 and less than 10.

57. 5,674,893 = ____________________ 58. 0.2456 = ____________________

59. 1762 = ____________________ 60. 28 = ____________________

61. 29.45 = ____________________ 62. 0.00345 = ____________________

63. 2,694.3241 = ____________________ 64. 4.35 = ____________________

65. 43.5 = ____________________ 66. 0.435 = ____________________

67. 39,238 = ____________________ 68. 0.0435 = ____________________

69. 125.67 = ____________________ 70. 12.567 = ____________________

71. 0.0000000567 = ____________________ 72. 1,389,000,000,000 = ____________________

73. 27 = ____________________ 74. 0.27 = ____________________

75. 100,000 = ____________________ 76. 0.00000001 = ____________________

Saying and Writing a Number in Words

When we say a number, we indicate the place values with the way we say the number. We are very familiar with naming 2- and 3-digit numbers. With larger numbers, we group digits in groups of three (called periods) with commas. We say the numbers in each period much as we say 3-digit numbers. Below are some examples of how the place values relate to the way we say a number.

Example 3: Say each of the following numbers in words.

4**378** is read "four thousand **three hundred seventy-eight**."
4 is in the thousands place. The next three numbers are read as a group of three.

17,000 is read "Seventeen thousand."

1,**457**,000,000 is read "one billion, **four hundred fifty-seven million**."
The 457 group of three has the last digit in the millions place (see below).

1	,	457	,	000	,	000
billions		millions		thousands		hundreds

1,**457**,000 is read "one million, **four hundred fifty-seven thousand**."
The 457 group of three has the last digit in the thousands place.

1**700** is read "one thousand, **seven hundred**." It is also sometimes called 17 hundred since 17 $100-bills is the same as one $1000-bill and 7 $100-bills.

3,507,872 is read "3 million, five hundred and seven thousand, eight hundred and seventy-two."

22.79 is read "twenty-two and seventy-nine hundredths."

When there are digits after the decimal point, wording can be confusing and misunderstood in communication. When in doubt, read the number out digit by digit, saying where the decimal point goes.

Exercise 2--Expanded Form of a Number and Scientific Notation

Purpose of the Exercise

- To practice saying or writing a number
- To practice putting a number in scientific notation

Write each of the following numbers in words.

1. 3786 Words__________________________________
2. 39,917 Words__________________________________
3. 1,078,703 Words__________________________________
4. 25,097,836 Words__________________________________
5. 16.08 Words__________________________________
6. 105.9 Words__________________________________

Round each of the following numbers to two significant digits and write in scientific notation.

7. 1,908 Scientific Notation ______________________________
8. 30,918,723 Scientific Notation ______________________________

9. 4,278.089 Scientific Notation ______________________

10. 0.005673 Scientific Notation ______________________

11. 0.005436 Scientific Notation ______________________

Number Search Projects

In the course of our daily lives we sometimes come across numbers that are either very large or very small. From the library, the Internet, or textbooks, look up the following information. Round the answers you find to 3 significant digits, and write the numbers in scientific notation. You may also need to look up some vocabulary words. Some of the numbers you may get from measuring, using a micrometer.

1. Check out the Web site for the national debt and record the most current reading. (http://www.brillig.com/debt_clock/)

2. Visit the Web site (http://www.odci.gov/cia/publications/95fact/index.html) and find how many miles of paved and unpaved roads there are in Egypt.

3. Look at several different computer manuals. How many commands are executed in one second? How long does it take for the computer to execute one command?

4. How many bytes of memory is a 1-gigabyte hard drive?

5. How much storage space is on your computer disk?

6. How long is a nanosecond? How long is a millisecond? How many nanoseconds make up a second? How many milliseconds make up a second?

7. About how thick is a fiber in a fiber optic cable? About how thick is the core of a fiber in a fiber optic cable?

8. What is the world's population?

9. How far is it from Earth to Venus?

10. How thick is a sheet of paper?

11. What is the world's population of ants?

Section 1-9 *Interpreting News Articles and Forms*

The concepts of arithmetic and procedures for doing arithmetic, whether they be pencil and paper or calculator computations, have little value to us unless we can apply the concepts and skills to help us make informed personal, business, and political decisions as well as to better understand the world around us. Quantitative information comes to us in many forms--in charts and graphs, in words, and on forms such as pay stubs, bills, or financial reports. In this section we will concentrate on reading and interpreting numeric information in several formats.

To interpret graphs, charts, and articles, among other things, we need to understand common phrases or words associated with arithmetic calculations. Here are some words and phrases and their meanings in mathematics.

Sum	The answer when numbers are added
Total	Same as "sum"
Cumulative Amounts	Same as "sum"
Difference	The answer when one number is subtracted from another, generally the larger number minus the smaller number
Product	The answer when numbers are multiplied
Quotient	The answer when one number is divided by another

To double something means to multiply by 2.
"2 more than ..." means to add 2.

"Judy is earning $2.00 more per hour than Sam." The key word is "more." Add 2.00 to Sam's wage to get Judy's or subtract 2.00 from Judy's to get Sam's.

"Judy is earning twice as much as Sam." The key phrase is "twice as much." Sam makes less than Judy, so multiply Sam's wage by 2 to get Judy's wage. (Or divide Judy's wage by 2 to get Sam's wage.)

"Judy's wage is double Sam's wage." This means the same as the phrase above.

"There were over four times as many methamphetamine labs seized in 1996 as in 1993." This means that if we multiply the number of labs seized in 1993 by 4, then more than that were seized in 1996, but probably not many more; otherwise we might have said "almost five times as many...."

"The difference between my salary and Jeff's salary is $12,879." Difference indicates that if we subtract the smaller salary from the larger salary, the answer is $12,879. We need to know who makes more money to know the order to set up the subtraction.

"Over the year, the account balance decreased (or declined or dropped) by $1,468." The account balance at the beginning of the year minus the account balance at the end of the year is 1468.

"Over the year, the account balance increased (or grew) by $1468." The account balance at the end of the year minus the account balance at the beginning of the year is 1468.

"The company profits tripled in the last six months." The profits in the latest month are three times the profits for the month six months ago.

"Everyone in the company got a $0.79 raise." Add $0.79 to everyone's hourly wage. (Notice the problem does not say that is an hourly increase, but $0.79 would be so insignificant as a monthly or annual increase, that it would hardly be worth recording, so we assume that the raise is per hour.)

"Everyone in the company got a 5% raise." Although we will not cover percents until Chapter 6, it is good to note that this means to multiply each person's current salary or wage by 0.05. This gives the amount of the raise. Once the amount of raise is calculated, the raise is added to the current salary or wage.

A deduction is subtracted from an amount.

Exercise 1--Reading and Interpreting Forms and Reports

Purpose of the Exercise

- To practice reading forms and reports
- To use arithmetic to help interpret the forms and reports

Materials Needed

- Calculator

1. Below is a part of a direct deposit notice form as used by the Metropolitan Community Colleges in Kansas City, Missouri.

	Total Hours	Regular Earnings	Premium Pay	Other Pay	Gross	Federal Tax
Current		3904.30	260.00	169.00	4333.30	336.75
Yr-to-Date					16644.20	1204.27

	State Tax	City Tax	F.I.C.A	Annuity	State Retirement	4038 Match
Current	152.00	36.58	62.94	200.00	483.04	100.00
Yr-to-Date	578.32	140.14	214.77	800.00	1859.80	400.00

	Medical	Life	Dental	Vision	United Way	Credit Union
Current		12.60			15.00	83.33

Net Pay
2951.06

To interpret the form, we first must understand that "Gross" means total, and in this case means "total pay" and is the sum of "Regular Earnings," "Premium Pay," and "Other Pay." After the gross pay is calculated, deductions are taken from the gross pay to compute the net pay--the amount deposited into the employee's account. The deductions are of two types. The required or mandatory deductions are the taxes, Social Security, and state retirement contributions. Other deductions are voluntary payments to an annuity, life insurance, United Way, and payments to the credit union. "F.I.C.A." is the social security contribution or tax.

"Yr-to-Date" means the total from all deposits for the year so far including this deposit.

The 4038 Match, an employee benefit paid by the employer, is not deducted from the gross pay. The employee contributed $200.00 to an annuity. If the employee chooses to contribute through payroll deduction to the annuity plan, the Metropolitan

Community Colleges will match half of that amount (up to $100) to the annuity. The $100 is not deducted from the employee's paycheck.

In each of the following, show the arithmetic setup, that is, what numbers you add, subtract, multiply, or divide along with the operation symbol, to answer the question. You may use a calculator to do the actual computations.

a. Check that the gross salary is computed correctly.

b. What are the gross wages so far this year including this paycheck?

c. What was the total gross salary for the year prior to this paycheck?

d. What is the total of the mandatory deductions?

e. What is the total of the voluntary deductions?

f. Check that the net pay was computed correctly.

g. If the annuity payments have been the same payment per month all year, how many months are accounted for in the "year-to-date" amounts?

h. In what month was this direct deposit made?

2.

Number of Legal Immigrants (in millions)

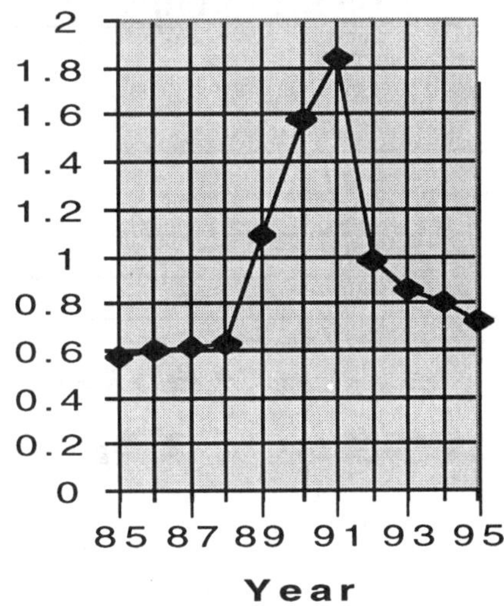

Source: Based on a chart in *USA Today* (September 30, 1996) based on Immigration and Naturalization Service Information.

Before answering questions about the chart, let's work on how to read information from the chart. The numbers on the left (0, 0.2, etc.) represent the numbers of legal immigrants stated in millions. 0.2 million immigrants means (0.2)(1,000,000) which is 200,000 immigrants. 1.3 million immigrants is (1.3)(1,000,000) or 1,300,000 immigrants. The numbers across the bottom are years (1985, 1987, etc.) The dates are labeled every other year, so the lines in between represent the years in between. Reading the graph, the dot on the 1990 vertical line crosses the horizontal line labeled "1.6." This means that in 1990 approximately 1.6 million legal immigrants entered the United States. In 1989, the dot is about halfway between 1 and 1.2 million. We estimate the number of legal immigrants in 1989 to be about 1.1 million people.

Answer the following questions based on the chart above. Show any related arithmetic setups. You may use your calculator for the computations.

a. Which year had the most legal immigrants during this period? About how many immigrants were there that year?
   year __________ number of immigrants ____________________

b. Were there more or fewer legal immigrants in 1995 than there were in 1989? How many more or fewer?
   more? or fewer? __________ How many more or fewer? _______________

c. About how many total legal immigrants were there in the years 1987 through 1990?
   ____

d. What was the total number of legal immigrants for the entire 11 years represented in the chart? ______________________________

e. In which year were there the fewest immigrants? __________________________

f. What is the difference between the most legal immigrants in one year and the fewest legal immigrants in one year?

3. Precipitation for Kansas City -- January 1996 through July 1996

Month	Total Precipitation	Normal
Jan	1.12	1.09
Feb	0.35	1.10
Mar	1.28	2.51
Apr	1.80	3.12
May	10.30	5.04
Jun	7.51	4.72
Jul	5.0	4.38

Precipitation in Kansas City--1996

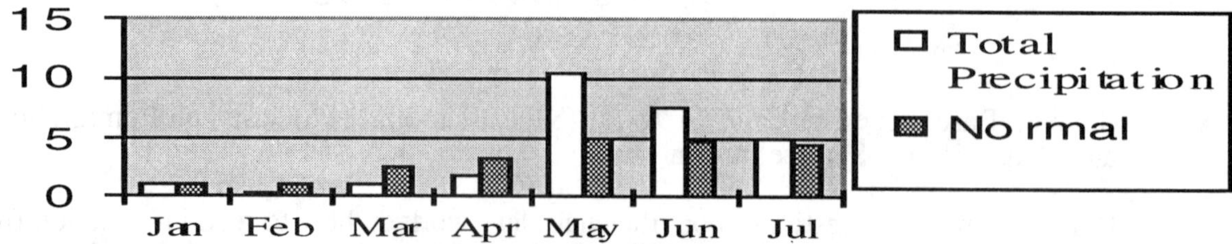

To read the above bar chart, notice that the unshaded bars represent the total actual precipitation for the month measured in inches, while the shaded bars represent the normal precipitation for that month. For example, in May, looking at the bar graph, the total precipitation was slightly over 10 inches (but not very close to 15 inches) and the normal precipitation for May was about 5 inches. To get more accurate information, we read from the table that May's total precipitation was 10.3 inches and May's normal precipitation is 5.04 inches. Subtracting 5.04 from 10.3, we see that in 1996 the precipitation in May was 5.26 inches more that the normal precipitation.

From the above table and bar chart, answer the following questions.

a. In what months was the total precipitation above normal? In those months, what was the difference between actual and normal precipitation?

____

____

b. What month had the most precipitation and how much precipitation occurred?

____

c. In what months was the total precipitation below normal? By how much? _______

____

d. What was the difference in total precipitation between the months with the highest and lowest amounts of precipitation?

____

4. The following is part of a statement sent out by *Merc Mortgage*.

Transactions Since Last Statement: Statement Date May 12, 1997

Description	Due Date	Date Paid	Total Payment	Principal	Interest	Escrow	Misc.
Mortgage Payment	05-97	05-08-97	387.80	190.48	139.83	57.49	

Current Principal Balance	Interest Paid Year to Date	Taxes Paid Year to Date	Current Escrow Balance
16,758.11	714.58	0.00	280.53

The current principal balance is the amount owed on this mortgage. The monthly payment above was broken down into principal, interest, and escrow. The principal is the part of the payment that goes toward paying off the amount owed. The home buyer pays a certain amount of money each month into an escrow account so that the mortgage company can pay the tax bills and insurance when they are due. The interest is based on a certain percent of the balance due and is how the bank makes money on the loan.

a. Verify that the total payment of $387.80 was computed correctly.

b. The current principal balance was computed after the payment was made. How much was the current principal balance before the payment?

c. How much interest had been paid for the year before this payment?

5. Airline near-collisions

Year	Planes within 100 feet	Planes within 101-500 feet	Total
1989	23	108	131
1990	12	79	91
1991	12	66	78
1992	7	52	59
1993	6	38	44
1994	4	45	49
1995	5	29	34

According to *USA Today*, (November 18, 1996), the Federal Aviation Administration, and Honeywell Inc., about 4500 U.S. airliners are now equipped with devices that warn pilots when other aircraft are near. The system (TCAS) also determines how fast planes are coming together, on what course, and at what altitude. It can track up to 30 planes at a time. The pilot is warned about another plane in the vicinity when the planes are about 45 seconds apart. When the planes are about 20 to 30 seconds apart, the pilot is warned to climb, descend, or stay at the same level. The sharp decline in the number of near midair collisions since 1989 has been attributed to these warning devices.

In the collision of a Saudi Airlines Boeing 747 and a Soviet-built Ilyushin-76 in India in November 1996, in which 349 people were killed, it is believed that the cargo plane did not have the TCAS warning system installed.

Smaller private planes without the warning system installed are most often involved in midair crashes. From January to November 1996, there were 16 midair collisions, killing 34 people, in the United States involving small aircraft.

Answer the following questions about the table above and the aforementioned article.

a. In 1992, how many planes came with 500 feet of each other? ______ How many planes came with 100 feet of each other? __________ How many planes came within 500 feet of each other but not within 100 feet? ____________

b. How many fewer near-collisions were there in 1995 than in 1989? ____________

c. How many fewer planes came within 100 feet of each other in 1995 than in 1989?

____

d. What year had the sharpest decrease from the previous year in number of planes that came within 500 feet of each other? ______________________

e. From January to November, 1996, how many midair collisions involving small aircraft were there in the United States? __________ What was the average number of deaths per collision? ______

f. Does the article or chart give us information about the number of midair collisions? Explain.

6. An electric bill may have the following numbers:

Total amount due May 1	\$57.32
After May 13	\$58.39

How much is the penalty if the payment is made after May 13? __________

Words to Mathematical Expressions, Mathematical Expressions to Words

In the questions above, you practiced interpreting words or phrases like "more than," "twice as much," "double," and "difference" appropriately as addition, multiplication, or subtraction. Now we will do more of the same, but express phrases algebraically with the use of "variables." We will talk more about variables later and will use them throughout the book in equations and formulas. For now, however, we will keep it simple.

Vocabulary
A **variable** is a letter that represents a numeric value.

A variable is so-named because its value may vary or change. For example, x may sometimes have a value of 10 and at another time have a value of 15.7. The following examples are to help you understand how variables are used.

Example 1: This year's budget was $4000 more than last year's budget.

If last year's budget was $143,650, how much is this year's budget?
Last year's budget + 4000 = this year's budget
143,650 + 4000 = 147,650
If last year's budget was $25,789, how much is this year's budget?
Last year's budget + 4000 = this year's budget
25,789 + 4000 = 29,789
If last year's budget is represented by the variable, B, how much is this year's budget?
Last year's budget + 4000 = this year's budget
B + 4000 = this year's budget

Example 2: In English there are many ways to say the same thing. If a cat, Trina, weighs 8.3 pounds and a dog, Bennie, weighs 10.3 pounds, which of the following statement(s) is true?

The dog is 2 pounds heavier than the cat.	<u>True</u>	False
The dog weighs 2 more pounds than the cat.	<u>True</u>	False
The dog weighs twice as much as the cat.	True	<u>False</u>
The cat is 2 pounds heavier than the dog.	True	<u>False</u>
The cat weighs 2 more pounds than the dog.	True	<u>False</u>
The cat weighs twice as much as the dog.	True	<u>False</u>
The cat weighs 2 pounds less than the dog.	<u>True</u>	False

Example 3: FooFoo the dog weighs 2 pounds more than Vanessa the cat.

Which weighs more, the dog or the cat? Ans: the dog
If FooFoo weighs 7.3 pounds, how much does Vanessa weigh?
Vanessa's weight = FooFoo's weight − 2 pounds
= 7.3 pounds − 2 pounds = 5.3 pounds
If Vanessa weighs 7.3 pounds, how much does FooFoo weigh?
FooFoo's weight = Vanessa's weight + 2 pounds
= 7.3 pounds + 2 pounds = 9.3 pounds
Would you add or subtract 2 pounds to Vanessa's weight to get FooFoo's weight?
FooFoo weighs 2 pounds more than Vanessa, so add 2 pounds to FooFoo's weight.

Would you add or subtract 2 pounds from FooFoo's weight to get Vanessa's weight?
FooFoo weighs more than Vanessa, so subtract 2 pounds from FooFoo's weight to get Vanessa's weight.

If the variable c represents the cat's weight, would the dog's weight be $c + 2$ or $c - 2$?
Vanessa's weight + 2 is FooFoo's weight.
$c + 2$

If the variable, d, represents the dog's weight, would the cat's weight be $d + 2$ or $d - 2$?
FooFoo's weight - 2 is Vanessa's weight.
$d - 2$

Example 4: Sue's new job pays double the salary of her old job. If Sue's old salary was $23,500 a year, what is her new salary?

New salary = 2(23,500) = $47,000 a year

If s is Sue's old salary, which of the following is her new salary?

$s + 2$ $\quad$ (2s) $\quad$ $s - 2$ $\quad$ $\frac{1}{2}s$

To double means to multiply by 2.

Example 5: Merv grew 9 inches between his 12th and 13th birthdays. If Merv was 5' 2" on his 12th birthday, how tall was he on his 13th birthday?

On his 13th birthday, Merv's height was 5'2" + 9" = 5'11."

If Merv's height on his 12th birthday was h (inches), his height on his 13th birthday was which of the following?

(h + 9) $\quad$ 9h $\quad$ $h - 9$

Add 9 inches to get his height on his 13th birthday.

Example 6: Jim makes $3 more than twice Matt's hourly wage.

If Matt's pay is $4.55 per hour, what is Jim's pay?
If Matt's pay is $7.32 per hour, what is Jim's pay?
If Matt's pay is $9.00 per hour, what is Jim's pay?

Let's analyze the statement, "Jim makes $3 more than twice Matt's hourly wage" and write it in mathematical symbolism. We will first paraphrase the sentence, and then move slowly toward a mathematical equation. After the sentence begins to look like an equation, we will use the variable J to represent Jim's hourly wage and M to represent Matt's hourly wage. Notice that Jim makes more money than Matt.

Jim's hourly wage is $3 more than twice Matt's hourly wage.
Jim's hourly wage = $3 more than twice Matt's hourly wage.
J = $3 more than twice M
J = 3 more than 2M
$J = 3 + 2M$

Notice that whatever wage we substitute for M (since M is a wage it is positive), the value of J is larger.

Exercise 1--Words to Algebra Expressions, Algebra Expressions to Words

Purpose of the Exercise

- To practice translating words to expressions using variables
- To practice interpreting expressions with variables to words

1. Joe started exercising and eating well and lost 33 pounds between March 1 and December 1.

   If Joe weighed 167 pounds on December 1, how much did he weigh March 1?

   If Joe weighed 192 pounds on March 1, how much did he weigh December 1?

   If Joe weighed w pounds on March 1, which of the following describes his weight on December 1?
   $w + 33$ $\quad$ $w - 33$ $\quad$ $33w$

2. A savings account earned $195.73 from June 1, 1996 to June 1, 1997. The earnings were left in the account.

   If the amount in the account on June 1, 1996 was $2350.75, how much was in the account June 1, 1997?

   If the amount in the account on June 1, 1997 was $2350.75, how much was in the account June 1, 1996?

   If A is the amount in the account on June 1, 1996, how much was in the account June 1, 1997?
   $A + 195.73$ $\quad$ $A - 195.73$

   If A is the amount in the account on June 1, 1997, how much was in the account June 1, 1996?
   $A + 195.73$ $\quad$ $A - 195.73$

3. The company's profits doubled this year.

   If last year's profits were $356,700, how much are this year's profits?
   If last year's profits were $7200, how much are this year's profits?
   If the company profits this year are $97,836, how much were last year's profits?
   If last year's profits are represented by the variable, P, which of the following represents this year's profits?
   $P + 2$ $\quad$ $P - 2$ $\quad$ $2P$

4. This year's profits were $27,300 more than last year's.

   If last year's profits were $356,700, how much are this year's profits?
   If last year's profits were $7200, how much are this year's profits?
   If the company profits this year are $97,836, how much were last year's profits?
   If last year's profits are represented by the variable, P, which of the following represents this year's profits?
   $P + 27,300$ $\quad$ $P - 27,300$ $\quad$ $23,700\ P$

5. This year the profits dropped $27,300 from last year.

   If last year's profits were $356,700, how much are this year's profits?
   If last year's profits were $29,200, how much are this year's profits?
   If the company profits this year are $97,836, how much were last year's profits?
   If last year's profits are represented by the variable, P, which of the following represents this year's profits?
   P + 27,300 P - 27,300 23,700 P

6. Jim makes double the hourly wage that Matt makes.

   If Matt's pay is $4.55 per hour, what is Jim's pay?
   If Matt makes M dollars per hour, what is Jim's pay?

Translate each of the following statements to a mathematical expression. In each case, state what variable you are using and say in words what the variable represents. Remember, a variable represents a number, so make sure the description is of something numeric, not just a name. You can use any letter you choose for the name of the variable. The first two are done as an examples.

7. 5 times Jane's bank balance.
   variable: x represents Jane's bank balance.
   expression: 5x

8. $7 less than twice Henry's hourly wage.
   variable: w represents Henry's hourly wage.
   expression: 2w - 7

9. Sally got a raise of $1.27 per hour. Express her new wage as an algebraic expression.
   variable: ____________________
   expression: ____________________

10. Terrence received a check in the mail and noticed that it was $157.89 less than it should have been. Write the expression that describes what the check should have been.
    variable: ____________________
    expression: ____________________

11. On June 1, 1900, the population of Peoplesville was 4534. Over the next year, 12 babies were born, a family of 5 and a family of 3 moved into town, a family of 6 moved out of town, and there were 9 funerals.

    a. On June 1, 1901, what was the population of Peoplesville? Show your calculations.

    b. Was there an increase or decrease in population that year? How much?

Federal Spending and the National Debt

Projections of future federal spending differ because of the unpredictability of some of the items. For example, no one can say exactly how much Medicare and Medicaid programs will cost the government in the future because the costs are determined by the price of health care, the general health of retired people, the general health of people needing public assistance, as well as the number of people receiving Medicare or Medicaid.

USA Today (May 2, 1997) made the following projections for federal spending, providing there is no deal reached to balance the federal budget. (Sources: Office of Management and Budget, Congressional Budget Office.) All numbers are given in billions of dollars, except the totals, which are given in trillions of dollars.

	1997	1998	1999	2000	2001	2002
Social Security	\$364.2	\$380.9	\$398.6	\$417.7	\$438.0	\$459.7
Medicare/ Medicaid	\$290.1	\$310.2	\$328.4	\$344.8	\$368.5	\$393.9
Defense	\$268.0	\$260.1	\$262.1	\$267.7	\$268.6	\$273.9
Other domestic programs	\$262.5	\$268.0	\$275.5	\$277.1	\$273.5	\$274.3
Net interest	\$247.4	\$249.9	\$251.8	\$248.2	\$245.0	\$238.8
Other	\$197.8	\$220.9	\$243.6	\$254.5	\$246.4	\$239.4
Total (in trillions)	\$1.63	\$1.69	\$1.76	\$1.81	\$1.84	\$1.88

When a number is given "in billions" it means that the actual number is the number times 1 billion.

Example 7: According to the chart above, Social Security spending for 2002 is projected to be about 459.7 billion dollars. Written out, this is

$\$459.7 \times 1{,}000{,}000{,}000 = \$459{,}700{,}000{,}000$

The total spending for 2002 is projected to be an estimated \$1.88 trillion.

$\$1.88 \times 1{,}000{,}000{,}000{,}000 = \$1{,}880{,}000{,}000{,}000$

Exercise 2--Federal Spending and the National Debt

Purpose of the Exercise

- To work with very large numbers and gain an understanding of their size
- To translate real data to algebraic symbols
- To translate words to algebraic symbols
- To better understand rounded-off numbers and significant digits

Materials Needed

- Calculator

1. Find amounts for each of the following in the chart and write the numbers in dollars (not in billions or trillions of dollars.) For example, \$262.1 billion dollars (projected defense budget, 1999) is written, 262,100,000,000.

   a. The projected amount that the federal government will spend on Medicare/Medicaid

      in the year 2002 ______________________

      How many significant digits are in the number? ______________

      To what place is the number rounded? ______________

b. The projected amount that the federal government will spend for defense in the year 2000 ____________________

How many significant digits are in the number? ____________________

To what place is the number rounded? ____________________

c. The projected total amount of federal spending in 1998? ____________________

How many significant digits are in the number? ____________________

To what place is the number rounded? ____________________

2. Compute the increases projected from 1997 to 2002 in each of the categories in the table.

a. Social Security ____________ b. Medicare/Medicaid ____________

c. Defense ____________ d. Other domestic programs ____________

e. Net interest ____________ f. Other ____________

3. Find the total amount of money spent in all six categories in 1999. ____________

What is the total shown in the chart for 1999? ____________________

Is the total given in the table correct? Explain.

4. If a variable, x, represents the amount spent on Social Security in 1997, circle all of the correct statements below.

a. For 1997, the projected amount to be spent on Social Security is $1.63 trillion - x.

b. The projected medicare/medicaid spending for 1997 is x - $74.1 billion.

c. For 1997, the projected amount to be spent on programs other than Social Security is $1.63 trillion - x.

In 1980, the national debt was about 1.1 trillion dollars, while in 1997, it was about 5.5 trillion dollars. (From *USA Today*, May 2, 1997. Sources: Office of Management and Budget, Congressional Budget Office.)

5. Circle T for true and F for false for each of the statements below. Let x represent the national debt in 1980.

T F a. The national debt in 1980 is about five times the national debt of 1997.

T F b. The national debt in 1980 = 5x.

T F c. The national debt in 1980 = 5 trillion + x.

T F d. The national debt in 1980 = 4.4x.

T F e. The national debt in 1980 = 4.4 trillion + x.

T F f. The national debt in 1997 is about five times the national debt of 1980.

T F g. The national debt in 1997 = 5x.

T F h. The national debt in 1997 = 5 trillion + x.

T F i. The national debt in 1997 = 4.4x.

T F j. The national debt in 1997 = 4.4 trillion + x.

6. If x represents a number, express each of the following in algebraic symbols.

   a. 12 more than the number ______________________

   b. 12 times the number ______________________

   c. 12 less than the number ______________________

   d. twice as much as the number ______________________

   e. double the number ______________________

   f. triple the number ______________________

   g. 3 more than the number ______________________

   h. 3 less than the number ______________________

   i. 3 less than twice the number ______________________

7. For x = 17, evaluate each of the following and determine which statement is correct.

   a. x + 3 = __________

      The answer is 3 more than 17.
      The answer is 3 times 17.
      The answer is 3 less than 17.
      17 is 3 less than the answer.

   b. 4.2x = __________

      The answer is 4.2 more than 17.
      The answer is 4.2 times 17.
      The answer is 4.2 less than 17.
      17 is 4.2 less than the answer.

   c. 82 + x = __________

      The answer is 82 more than 17.
      The answer is 82 times 17.
      The answer is 82 less than 17.
      17 is 82 less than the answer.

   d. 82x = __________

      The answer is 82 more than 17.
      The answer is 82 times 17.
      The answer is 82 less than 17.
      17 is 82 less than the answer.

Section 1-11 *Averages, Word Problems and Puzzles*

In everyday usage the word "average" can mean many things. When we say that a person has average food preferences, it means that most people have similar preferences. What that really means depends on what a person means by "most people" and how narrowly the food preferences are defined. If we say that a person is of average height, however, the number may be based on actual data from a sample of people. The averages with which we will work in this section are mathematical averages that are computed by the definition given in the box below.

> **Vocabulary**
> The **average** of several numbers is the sum of all of the numbers divided by how many numbers there are.
> The **range** of values of several numbers is
> (the largest number) - (the smallest number)

Example 1:

The average of 17, 19, 35, 67, and 93 is the sum of the five numbers divided by 5, or

average = (17 + 19 + 35 + 67 + 93)/5 = **46.2**

The **range** of the numbers is 93 - 17 = **76**

Example 2:

The average of 467, 29, 83, 156, 917, and 215 is the sum of the six numbers divided by 6, or

average = (467 + 29 + 83 + 156 + 917 + 215)/6 **311.17**

Notice that the average did not come out exact so we chose to round the answer to the nearest hundredth.

The **range** of numbers is 917 - 29 = **888**

Exercise 1--Averages (Group)

Purpose of the Exercise

- To compute averages from collected data

Materials Needed

- Calculator
- A metric ruler or a meter stick

1. In centimeters, measure the length of each group member's thumb. Record the lengths and compute the average thumb length. Compute the range of thumb lengths.

   Thumb lengths of each group member ______________________________

   Average ________________ Range ____________________

2. Have each group member walk the same path (for example from one wall to another.) Count the number of steps it takes to walk the path. Compute the average number of steps and the range of the number of steps.

   Number of steps it took each group member ___________________________

   Average ________________ Range ____________________

3. Measure each group member's height in centimeters. Compute the average height of the people in your group.

   Heights of each group member ________________________________

   Average ________________ Range ________________

4. Compute average and range of each of the following sets of numbers.

   a. 45, 87, and 90

      Average ________________ Range ________________

   b. 7.9, 14.7, 6.9, 14.3, 27.4

      Average ________________ Range ________________

   c. 2905, 479, 1099, 1493, 567, 3028, 1872

      Average ________________ Range ________________

   d. 5, 11, 19, 23, 29, 37, 39, 42, 47, 46, 52, 59, 69

      Average ________________ Range ________________

Exercise 2--Word Problems and Projects

Purpose of the Exercise

- To practice applying arithmetic concepts to real situations

Materials Needed

- A calculator

Work each of the following on a separate sheet of paper. Show the arithmetic set up and the answer to the questions. Use a calculator as needed and clearly label the answers.

1. Amelia had $378 cash until she and her children went shopping. Sarah bought shoes that cost $59.87. Ricky bought jeans and a T-shirt that came to $62.43. Amelia bought the baby, Josh, some diapers, sleepers, socks and a stuffed animal that came to $43.90. The athletic shoes and uniform for Brittany cost $135.87. Amelia spent $19.67 for gasoline and $22.67 for lunch.
   a. How much money did the shopping trip cost?
   b. How much money was left over after shopping?

2. The camp director, Ms. Haselworth, is to purchase souvenir gifts for each of the campers. The gifts come to $13.48 each including taxes and shipping. How much do the gifts for the summer cost if 487 campers come to the camp?

3. Father Strickland's church decides to have a fund-raiser. Half of the money made will go toward paying utility bills. The other half will be used to buy food for the food pantry (where families in need can pick up needed items). The utility bills that are due total $852.79. The fund-raiser cost the church $103.98 in advertising and printing costs. At the fund-raiser, $950.25 was collected from the garage sale and auction of donated items; $600.00 was collected from the sale of raffle tickets, and $210.00 was collected in donations.

   a. What was the total amount of money made at the fund-raiser ? What was the total after expenses?

b. Did the church make enough money (after expenses and after splitting the money with the food pantry) to pay the utility bills due? If not, how much were they short?
c. Was there money left over for future utility bills? If so, how much?
d. How much money did the church make to buy food for the food pantry?

4. Pat, Justin, Katherine, and Jeremy are planning a New Year's Eve Dance. The hall rents for $425, the decorations cost $145, the band charges $600, and the publicity costs will be $140.
   a. If 130 tickets are sold at $9.00 a ticket, how much would Pat, Justin, Katherine, and Jeremy each have to pay "out of pocket" to cover the loss?
   b. If 250 tickets are sold at $9.00 a ticket, how much would each organizer's share of the profit be?
   c. How many tickets would have to be sold to break even?

5. The surface area and volume of the five great lakes are as follows. (Source: *Water in Crisis, A Guide to the World's Fresh Water Resources*, Peter H. Gleick.)

	Superior	Huron	Michigan	Erie	Ontario
Area (km^2)	82,680	59,800	58,100	25,700	19,000
Volume (km^3)	11,600	3,580	4,680	545	1,710

   a. Compute the average surface area of the Great Lakes.
   b. Compute the average volume of the Great Lakes.
   c. Explain why it is possible for a lake with more surface area to have less volume than a lake with less surface area.

Projects

6. Plan a social function/school event. Estimate costs. Set a ticket price and determine how many people will have to come to recover costs.

7. Compute the total you have spent for rent/mortgage and utilities for the past year.
   a. What was your total cost for rent/mortgage and utilities for the year?
   b. What was your average monthly cost for rent/mortgage payment and utilities?
   c. What was your average daily cost for rent/mortgage payment and utilities?

8. For one month, keep track of the number of times you buy a meal at a restaurant or fast-food restaurant. Record the cost of the meals. Keep track of money spent for groceries and the number of meals made from the groceries.
   a. Compute the average cost per meal of food when eating out.
   b. Compute the average cost per meal made at home from the groceries.

Exercise 2--Three Steps Forward and Two Steps Back (Puzzle)

Sometimes we use the phrase that we are making progress "three steps forward and two steps back" meaning of course that we move forward, but have setbacks that take us back--not quite to where we started, but still backward. Mathematically speaking, let's analyze the literal statement. Suppose we are trying to climb a set of stairs and we start out by taking three steps up, but each time we take three steps upward, a wind then blows so hard that we are pushed two steps back down the stairs. In the following problems, determine how many total steps we would have to take to finally get to the top of the stairs. Assume that once you reach the top, the wind cannot blow you back down. Explain or show how you got each answer.

1. In the diagram below, how many steps are in the staircase? ________

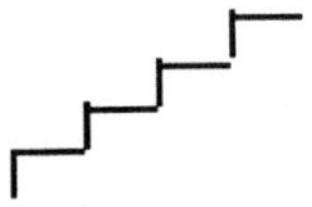

How many total steps (including backward and forward steps) do you have to take to reach the top step of the staircase?

How many of your steps are going up the staircase? ________ Down? ________

2. In the diagram below, how many steps are in the staircase? ________

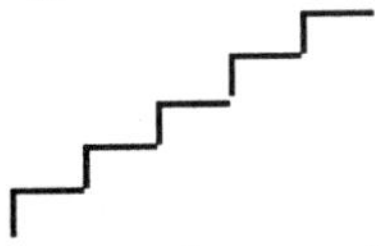

How many total steps (including backward and forward steps) do you have to take to reach the top step of the staircase?

How many of your steps are going up the staircase? ________ Down? ________

3. If there are 6 steps in the staircase, how many total steps does it take to reach the top?

   How many of your steps are going up the staircase? ________ Down? ________

4. If there are 7 steps in the staircase, how many total steps does it take to reach the top?

   How many of your steps are going up the staircase? ________ Down? ________

5. If there are 8 steps in the staircase, how many total steps does it take to reach the top?

   How many of your steps are going up the staircase? ________ Down? ________

6. If there are 9 steps in the staircase, how many total steps does it take to reach the top?

   How many of your steps are going up the staircase? ________ Down? ________

7. If there are 12 steps in the staircase, how many total steps does it take to reach the top?

   How many of your steps are going up the staircase? ________ Down? ________

8. If there are 15 steps in the staircase, how many total steps does it take to reach the top?

   How many of your steps are going up the staircase? ________ Down? ________

9. If there are 99 steps in the staircase, how many total steps does it take to reach the top?

   How many of your steps are going up the staircase? ________ Down? ________

10. If there are 100 steps in the staircase, how many total steps does it take to reach the top?

    How many of your steps are going up the staircase? ________ Down? ________

Section 1-12 *Terms, Like Terms and Coefficients*

In Section 1-1, we discussed and reinforced what we already know: "We do not compare or add apples to oranges," as the expression goes. In the money arithmetic model, we reinforced that concept even further. When adding money, we "line up" the decimal points before adding. Expressed more in layman's terms, we added pennies to pennies, dimes to dimes, $1-bills to $1-bills, $10-bills to $10-bills, and so on. In this section, we will work first with objects, then with variables, to see what adding "like terms" means in an algebraic setting.

Exercise 1--Like Terms (Group)

Purpose of the Exercise

- To work with objects to gain an understanding of what is meant by 5x and other such symbols

Materials Needed

- 420 small cardboard squares for each group

1. First member, take out 2 small squares.
   Second member, take out 5 small squares.
   Third member, take out 3 small squares.
   Fourth member, take out 4 small squares.

   a. Combine the squares. What is total number of squares the group has? ________

   b. Write an x on one side each square. How many x's does the group have? _______

   (xs means more than one x. It is the abbreviation we will use for "exes.")

   c. Write a 4 on the other side of each square. We are assigning the number 4 to x. How many 4's does the group have? _______________

   d. Add the 4's. Show the addition as well as the total. ______________________

   e. How can the addition in part d be written as a product? __________________

2. a. Erase the 4's on the squares. Write the number 7 on each square marked x. We are assigning the number 7 to the variable x.

   b. How many x's does the group have? ____________________________________

   c. How many 7's does the group have? ____________________________________

   d. Add the 7's. Show the addition as well as the total. ______________________

   e. How can the addition in part d be written as a product? __________________

3. a. Erase the 7's on the squares. Write the number 13 on each square marked x. We are assigning the number 13 to the variable x.

   b. How many x's does the group have? ____________________________________

   c. How many 13's does the group have? __________________________________

   d. Add the 13's. Show the addition as well as the total. ____________________

   e. How can the addition in part d be written as a product? __________________

In algebra, "5x" means literally 5 x's.

x x x

x x

When we say "evaluate 5x when x = 3," every one of those x's becomes a 3.

3 3 3 3 3

5 x's becomes five 3's, or 3 + 3 + 3 + 3 + 3 = 15. Another way to write five 3's is (5)(3) = 15.
To evaluate 5x when x = 3, we then multiply 5(3) = 15.

In the exercise set above, when the group members combined the cardboard squares they had 14 squares. Each one was labeled x, so they had 14 x's or 14x. As we see from above, when x = 4, 14x = (14)(4) = 56. Check your answer in Problem 1. Did you get 56 as an answer to part d? In a similar manner, the answer to Problem 2 d should be (14)(7) = 98 and the answer to Problem 3 d should be (14)(13) = 182.

Exercise 2--Like Terms (Group)

Purpose of the Exercise

- To work with objects to gain an understanding of what is meant by like terms

Materials Needed

- 40 small cardboard squares and 40 larger cardboard squares and 40 toothpicks for each group (1/2-in. squares and 1-in squares would be a good size--there must be space to write on the squares.)

1. Member 1: Take out 7 small squares, 3 large squares, and 6 toothpicks.
Member 2: Take out 3 small squares, 6 large squares, and 1 toothpick.
Member 3: Take out 8 large squares and 18 toothpicks.
Member 4: Take out 11 small squares and 8 toothpicks.

a. Combine all of the small and large squares and toothpicks and count the total of each.
Total number of sm. squares ________ Total number of lg. squares ________
Total number of toothpicks _______

b. On one line record how many of each type object Member 1 has, on the next line record how many Member 2 has, and so on. Line up the small squares with the small squares, the large squares with the large squares and the toothpicks with the toothpicks. Add the columns.

	Sm Squares	Lg. Squares	Toothpicks
Member 1:			
Member 2:			
Member 3:			
Member 4:	__________	__________	__________
Total:			

c. Write the letter "x" on each small square and "y" on each large square. Each toothpick represents the number 1. On the next page rewrite the lines in part b to give the number of x's, y's, and 1's. The first one is filled in. Include the x's and y's with your answer.

	x's	y's	Toothpicks
Member 1:	7x's	3y's	6
Member 2:			
Member 3:			
Member 4:	______	______	______
Total:			

d. Now we will assign a numeric value to each type of square. On each of the squares marked x, write the number 4 on the other side. On the back side of the squares marked y, write the number 17. On each line below, record the product that gives the sum of the 4's and sum of the 17's. The first is filled in as an example. Since Member 1 had 7 x's, Member 1 now has 7 4's, so the value is 7(4).

	x's (x = 4)	y's (y = 17)	Toothpicks
Member 1:	7(4)	3(17)	6
Member 2:			
Member 3:			
Member 4:			
Total:			= ________

e. Now we will assign a different numeric value to each type of square. Erase the numbers on the squares; then on each of the squares marked x, write the number 11 on the other side and on each of the squares marked y, write the number 9. Fill in the lines as you did in part d.

Member 1:

Member 2:

Member 3:

Member 4:

Total: = ________

2. Repeat Exercise 1 with different combinations of small squares, large squares, and toothpicks and with different values for x and y.

3. Mathematically, we will write 7 x's as 7x. In each of the following, determine what a number written directly in front of a variable means. We will again represent x's with small squares and y's with large squares .

a. [4] [4] [4] [4] [4] Represents 5 fours which is (5)(4) = 20.

[7] [7] [7] [7] [7] Represents 5 sevens which is (5)(7) = 35.

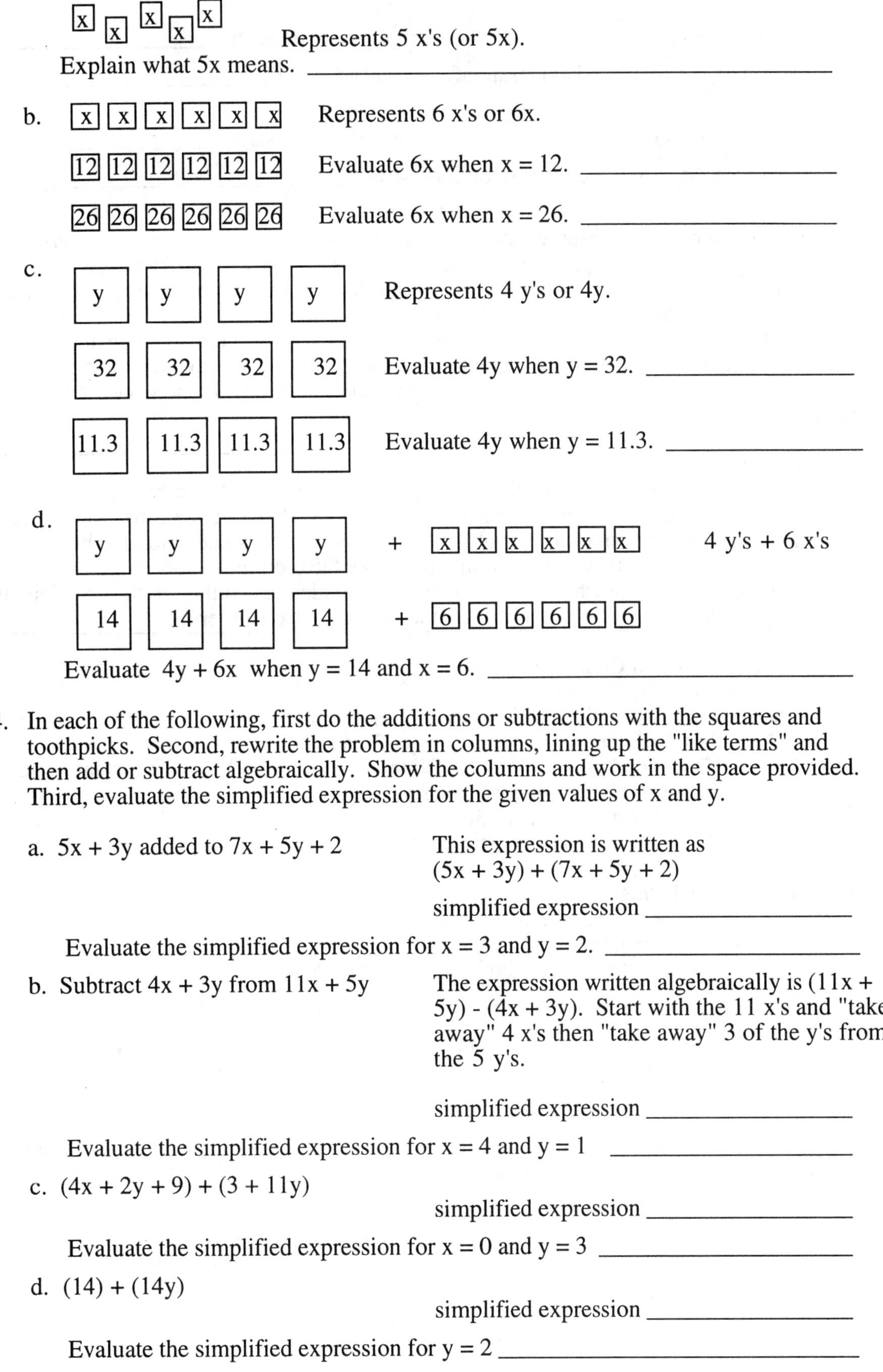

Represents 5 x's (or 5x).

Explain what 5x means. ____________________

b. Represents 6 x's or 6x.

Evaluate 6x when x = 12. ____________

Evaluate 6x when x = 26. ____________

c. Represents 4 y's or 4y.

Evaluate 4y when y = 32. ____________

Evaluate 4y when y = 11.3. ____________

d. 4 y's + 6 x's

Evaluate 4y + 6x when y = 14 and x = 6. ____________

4. In each of the following, first do the additions or subtractions with the squares and toothpicks. Second, rewrite the problem in columns, lining up the "like terms" and then add or subtract algebraically. Show the columns and work in the space provided. Third, evaluate the simplified expression for the given values of x and y.

a. 5x + 3y added to 7x + 5y + 2

This expression is written as (5x + 3y) + (7x + 5y + 2)

simplified expression ____________

Evaluate the simplified expression for x = 3 and y = 2. ____________

b. Subtract 4x + 3y from 11x + 5y

The expression written algebraically is (11x + 5y) - (4x + 3y). Start with the 11 x's and "take away" 4 x's then "take away" 3 of the y's from the 5 y's.

simplified expression ____________

Evaluate the simplified expression for x = 4 and y = 1 ____________

c. (4x + 2y + 9) + (3 + 11y)

simplified expression ____________

Evaluate the simplified expression for x = 0 and y = 3 ____________

d. (14) + (14y)

simplified expression ____________

Evaluate the simplified expression for y = 2 ____________

e. (11 + 5x) + (7y) simplified expression ______________

Evaluate the simplified expression for x = 0 and y = 0 ______________

f. (11 + 7x) - (7x) simplified expression ______________

Evaluate the simplified expression for x = 14 ______________

In the exercise above, you observed that 5 x's means x + x + x + x + x, which we write as 5x. Since multiplication is repetitive addition, 5x means "5 times x" and 7y means "7 times y." When we know a value for x, 5x is evaluated as 5 times the number that x represents. 5x + 7y means "5 times x plus 7 times y." Since x and y are different variables (they have different names) 5x and 7y are not "like terms" so we cannot add them algebraically. 5x + 7x, however means 5 x's plus 7 x's which is 12 x's. 5x and 7x are like terms so we can add them. We have been using some terminology so let's formally define what we mean algebraically by the terms.

Vocabulary

Terms are members of mathematical expressions that are added or subtracted.

A **coefficient** of a term is the number by which the variable part is multiplied. If there is no number in front of the variable part, the coefficient is understood to be 1.

A **constant term** is one with no variable part (like the toothpicks above).

Like terms are those in which the variable parts are identical, although the coefficients may be different. Two or more constant terms are also like terms.

Example 1: Terms and Coefficients.

3x - 7 has two terms, 3x and -7

The coefficient of 3x is 3, and -7 is a constant term.

4xy - 2t - 4 has three terms. They are: 4xy, -2t, and -4

The coefficient of 4xy is 4. The coefficient of -2t is -2, and -4 is a constant term.

(xy means to multiply the variables x and y.)

156 - 3m has two terms, 156 and -3m

156 is a constant term. The coefficient of -3m is -3.

Example 2: Like Terms.

3x - 7 - x + 18 has four terms; 3x, -7, -x, and 18.

3x and -x are like terms.

-7 and 18 are like terms since both are constant terms.

-147t has only one term so there are no like terms. The coefficient of t is -147.

14.3 - 21y + 14.7y - 23 has four terms; 14.3, -21y, 14.7y and -23.

14.3 and -23 are like terms.

-21y and 14.7y are like terms.

In Exercise 1, you added and subtracted algebraic expressions by adding or subtracting the x's to the x's and the y's to the y's. What you were doing is called "combining like terms." You observed, for example,

5x + 3x = 8x

[x] [x] [x] [x] [x] + [x] [x] [x] = [x] [x] [x] [x] [x] [x] [x] [x]

5x - 3x = 2x

[x] [x] [x] [x] [x] - [x] [x] [x] = [x] [x]

There was no way to add or subtract x's and ys without knowing the values of x and y. Because x and y are not like terms, we cannot add them.

5x + 3y

[x] [x] [x] [x] [x]   [y] [y] [y]

Example 3: Add $7x + 9y + 3$ to $4x + 8$

As you did with the squares, line up the problem in columns to add the x's to x's, y's to y's, and constant term to constant term.

$$\begin{array}{lr} & 7x + 9y + 3 \\ \text{Add} & \underline{4x \qquad\quad + 8} \\ & 11x + 9y + 11 \end{array}$$

Example 4: $(9y + 7x + 4) - (5x + y)$.

$$\begin{array}{r} 9y + 7x + 4 \\ \underline{-(\ y + 5x\ \ )} \\ 8y + 2x + 4 \end{array} \qquad \text{which is the same as} \qquad \begin{array}{r} 9y + 7x + 4 \\ \underline{-y - 5x\quad} \\ 8y + 2x + 4 \end{array}$$

Process of Combining Like Terms

To combine like terms just means to add or subtract the terms as indicated. First, identify like terms. We cannot combine unlike terms.

If the terms are constant, just add or subtract the numbers as indicated.

If the terms are not constant, add or subtract the coefficients, keeping the variable part the same.

Adding and Subtracting Algebraic Expressions:

If the expressions are added, add the like terms. If one expression is to be subtracted from the other, subtract like term from like term.

Exercise 3--Terms, Like Terms, and Coefficients

Purpose of the Exercise

- To practice identifying terms, like terms, and coefficients
- To practice combining like terms

State how many terms are in each of the following expressions, identify the like terms and their coefficients, and finally, combine all like terms to simplify the expressions. The first two are done as examples.

1. $12x - 7x + y + 9$

   number of terms 4

   Coef: 12 is the coefficient of 12x

   -7 is the coefficient of -7x

   1 is the coefficient of y

   9 is a constant term

   Like terms:

   12x and -7x are like terms

   There are no other like terms

   $12x - 7x + y + 9 = 5x + y + 9$

2. 234.6k

   number of terms 1

   Coef: 234.6 is the coefficient of 234.6k

   Like terms:

   None

   There are no like terms to combine

3. 89.3x - 17.9y - x + 7

number of terms ______________

Coef: ______________________

____

____

____

Like terms: __________________

____

____

4. 87a + b - a + 1

number of terms ______________

Coef: ______________________

____

____

____

Like terms: __________________

____

____

5. 17 + 24b - 17b -17

number of terms ______________

Coef: ______________________

____

____

____

Like terms: __________________

____

____

6. 12x + 3 - 6x + 9

number of terms ______________

Coef: ______________________

____

____

____

Like terms: __________________

____

____

7. 11 + 11t + 11s

number of terms ______________

Coef: ______________________

____

____

____

Like terms: __________________

____

____

8. 14x + 9 + 22x + 9y + 7 + 3y

number of terms ______________

Coef: ______________________

____

____

____

Like terms: __________________

____

____

9. 17 + 30x + y + 17y + 30

number of terms ______________

Coef: ______________________

____

____

____

Like terms: __________________

____

____

10. 25t - 11t + 17 + y - 11

number of terms ______________

Coef: ______________________

____

____

____

Like terms: __________________

____

____

Simplify each of the following expressions by adding or subtracting as indicated. Substitute the given values for x and y into the original expression as well as into the simplified expression to compare. If the original expression and the simplified expression evaluate differently, then check your work to find the mistake.

11. a. $(4x + 3y) + (7x + 9)$ = ________________

b. Let $x = 3$ and $y = 4$:

Evaluate $(4x + 3y) + (7x + 9)$ = ________________

Evaluate the answer from part a. ________________

c. Let $x = 1$ and $y = 2$:

Evaluate $(4x + 3y) + (7x + 9)$ = ________________

Evaluate the answer from part a. ________________

12. a. $(4x + 3y) - (x + 2y)$ = ________________

b. Let $x = 3$ and $y = 2$:

Evaluate $(4x + 3y) - (x + 2y)$ = ________________

Evaluate the answer from part a. ________________

c. Let $x = 2$ and $y = 1$:

Evaluate $(4x + 3y) - (x + 2y)$ = ________________

Evaluate the answer from part a. ________________

13. a. $(8 + 11x) - (2x + 3)$ = ________________

b. Let $x = 1$ and $y = 4$:

Evaluate $(8 + 11x) - (2x + 3)$ = ________________

Evaluate the answer from part a. ________________

c. Let $x = 5$ and $y = 3$:

Evaluate $(8 + 11x) - (2x + 3)$ = ________________

Evaluate the answer from part a. ________________

Exercise 3--Z-Hot Burger Cook Count

Purpose of the Exercise

- To practice adding and subtracting "like" objects in a real setting

In five minutes' time you, the cook at "Z-Hot Burger," get these orders shouted at you.

2 burgers, 3 French fries
1 grilled chicken, 1 onion ring
4 burgers, 2 onion rings, 1 French fries
4 burgers, 2 onion rings, 1 French fries
Cancel the 2 burgers, Make that 2 grilled chicken sandwiches instead
3 grilled chicken, 4 French fries, 1 fish sandwich
1 burger, 2 French fries
Cancel one of the grilled chickens
3 burgers, 2 fish sandwiches, 2 onion rings, 3 French fries
Cancel one of the French fries, make that an onion ring

Choose letters to represent the number of burgers, the number of orders of French fries, and so forth., and in the following, write the steps and compute how many of each food type to cook, The first one is done as an example.

1. Write the arithmetic steps and compute how many burgers to cook.

   2B + 4B + 4B - 2B + 1B + 3B

   The number of burgers cooked = 2 + 4 + 4 - 2 + 1 + 3 = 12 burgers

2. Write the arithmetic steps and compute how many chicken sandwiches to cook.

   ____

   Number of chicken sandwiches: ____________________

3. Write the arithmetic steps and compute how many fish sandwiches to cook.

   ____

   Number of fish sandwiches: ____________________

4. Write the arithmetic steps and compute how many orders of French fries to cook.

   ____

   Number of orders of French fries: ____________________

5. Write the arithmetic steps and compute the how many orders of onion rings to cook.

   ____

   Number of orders of onion rings: ____________________

Chapter 1 *Summary and Test*

Vocabulary and Symbols

- The ten **digits** are 0, 1, 2, 3, 4, 5, 6, 7, 8, and 9. (page 5)
- **Place value** is the value of a particular position in a number, such as the ones place, tens place, and so forth. (chart on page 10)
- 10^2, 10^3, and 10^{-3} are referred to as **powers of ten**. (page 9) The powers of ten are used in expressing numbers in scientific notation.
- **Symbols of inequalities,** (page 12)
  $<$ "less than" $>$ "is greater than"
  $\leq$ "less than or equal to" $\geq$ "is greater than or equal to"
- In a number, the first nonzero digit (reading from left to right) and all of the digits following are **significant digits**. (page 13)
- The four arithmetic **operations** are **addition, subtraction, multiplication, and division.** (page 19)
  When two numbers are **added**, the answer is called the **sum**.
  When one number is **subtracted** from another, the answer is called the **difference.**
  When two numbers are **multiplied**, the answer is called the **product.**
  When one number is **divided** by another, the answer is called the **quotient**.
- **Symbols of division**. (page 40)
- A number written in **scientific notation** is of the form, $N\text{x}10^n$, where N is a number between 1 and 10--it can be 1, but cannot be 10--and n is some integer. (page 54)
- To say a number in words, put the digits in groups of three starting at the decimal point. If there are digits on the left of the decimal point, start reading from the digit furthest from the decimal point. (page 57)
- Words and Phrases and their meaning in mathematics. (pages 59-60)
- A **variable** is a letter that represents a number. (page 65)
- The **average** of several numbers is the sum of all of the numbers divided by how many numbers there are. (page 72)
- The **range** of values of several numbers is:
  (the largest number) - (the smallest number) (page 72)
- **Terms** are members of mathematical expressions that are added or subtracted. (page 80)
- A **coefficient** of a term is the number by which the variable part of the term is multiplied. If there is no number in front of the variable part, the coefficient is 1. (page 80)
- A **constant term** is one with no variable part. (page 80)
- **Like terms** are those in which the variable parts are identical, although the coefficients may be different. Two or more constant terms are also like terms. (page 80)

Procedures

- When **rounding** a number to a particular place, round up if the next digit is 5 or greater and round down if the next digit is less than 5. (page 15)
- **To add**, write the numbers in columns, lining up the decimal points. Start by adding the right digits. Move left through the digits, carrying the extras. (page 20)
- **To subtract** one number from another, write the numbers in columns, lining up the decimal points. Start by subtracting the right-most digits. Move left through the digits, regrouping (borrowing) from the digit to the left as needed. (page 27)
- **To multiply**, write the numbers in columns. Multiply the top number by the right-most digit of the bottom number and then multiply by the second digit from the right (accounting for its place value). Add the resulting partial products. (pages 33 and 34)
- To **multiply numbers with decimal points**, follow the above process. Then count the total number of digits after the decimal places in the two number multiplied. Place the decimal point in the product left of that many digits counted from the right-most digit. (page 36)
- **Division Algorithms** (pages 44, 45)
- **To add or subtract like terms**: (page 81)
  If the terms are constant, just add or subtract the numbers as indicated.
  If the terms are not constant, add or subtract the coefficients, keeping the variable part the same.
- **Adding and Subtracting Algebraic Expressions**: (page 81)

  If the expressions are to be added, add the like terms.

  If one expression is to be subtracted from the other, subtract like term from like term.

Summary Test--Chapter 1

Part 1: Do Not Use a Calculator.

1. In each of the following arithmetic problems, circle the best answer from the choices given.

a. (978)(1.2786) ≈	9800	2100	1000	1300
b. 279.087÷352.98 ≈	0.25	0.79	1.34	2
c. 8753 + 76.98 + 0.987 + 603.8 ≈	9400	11200	15800	23500
d. 4678.97 - 98.999 - 0.7839 - 203.1 ≈	2500	4600	4400	3200

2. Do the arithmetic indicated in each of the following problems.

a. 78 + 9.83 b. 295.047 - 18

c. 245 ÷ 8 d. (86)(23)

e. (105.687)(100) f. 387.986 ÷ 1000

g. $156.98 \cdot 10^3$ h. 364 ÷ 10

i. 14 - 0.98 j. (1.34)(3)

3. Round off the following numbers as indicated.
   a. 8.90367 to the nearest hundredth.
   b. 1097.83 to one significant digit.
   c. 0.008063 to 2 significant digits.
   d. 3045.9876 to 1 decimal place.
   e. 3045.9876 to the nearest hundred.
   f. 17.9999 to the nearest hundredth.
4. Put the following numbers in scientific notation.
   a. 149,088.09
   b. 0.0000987
   c. 0.00000000067
   d. 67,900,000,000
5. Multiply each of the following numbers.
   a. 4.0897×10^{-5}
   b. 5.7×10^{5}
   c. 1.0×10^{3}
   d. 2.756×10^{-4}
6. Changing words to algebra.
   a. Jill is two inches taller than Amber. x = Amber's height.
      Write Jill's height as an algebraic expression in terms of Amber's height.

   b. Alan earns twice as much as Jody. J = Jody's salary.

7.

Year	Number of Methamphetamine Lab Seizures in the United States
1993	218
1994	263
1995	327
1996	881

Source: Drug Enforcement Administration.

Answer the following questions from the data given in the table above and illustrated in the bar chart.

a. How many methamphetamine labs were raided in 1995? ____________
   In 1996? ___________
b. How many more labs were raided in 1996 than in 1995? ____________
c. To get the number of labs raided in 1996, would you add 554 to the number of labs raided in 1995 or would you subtract 554 from the number of labs raided in 1995?
d. Which of the following statements are true and which are false? Show the computations to justify each answer.
   True False: The number of labs seized in 1996 is more than triple the number of labs seized in 1994.
   True False: The number of labs seized in 1996 is more than triple the number of labs seized in 1995.

8. Simplify each of the following expressions by combining like terms.
   a. $17 + 3x + 5y - x + 4y - 3$
   b. $4a + 5 + a$
   c. $3 + 7y + 1 + y$
   d. $17x - 17$

9. Evaluate each of the following expressions for $x = 3$ and $y = 7$.
   a. $4x + 3y$
   b. $5xy - 5$
   c. $5x + y$
   d. $17x + 2xy$

Part 2: Use a calculator as needed.

10. Perform the indicated operations. Write your answer correct to 3 decimal places.

    a. $\dfrac{23.45 - (7.5)(14.3)}{4 \div 16 \cdot 8}$

    b. $(11.5)(27.9) - (22) \div 17$

11. In the following pairs of numbers, circle the larger number.

    a. $\sqrt{8}$ or $89 \div 31$

    b. or 22/7

    c. $\sqrt{356}$ or (4)(4.7)

    d. 2 or 12/1.9

12. A company has a total of $4523 to give as bonuses for its employees, and each employee gets the same bonus. How much is each employee's bonus if there are 279 employees? Show the arithmetic setup of the problem and clearly state the answer.

13. A woman decides to give each of her nieces and nephews $125. She has 38 nieces and nephews. How much money will she need in order to give them the money?

Chapter 2 Signed Numbers
Bean Arithmetic, Checkbook Arithmetic, and the Number Line

Key Topics:

- Addition, subtraction, multiplication, and division of positive and negative numbers.
- The number line and comparison of two numbers, positive or negative, to determine which is bigger
- The absolute value of a number
- Rules of signs
- The order of operations and using variables
- Positive integer exponents
- Scientific notation
- Combining like terms
- Solving equations

Materials Needed for the Exercises

- 30 red beans and 30 white beans for each group
- 40 each per group: small white cardboard squares, small gray cardboard squares, large white cardboard squares, and large gray cardboard squares
- Calculator

Section 2-1 *Addition and Subtraction of Signed Numbers--The Bean Model*

In this section, there are several activities in which students use two colors of beans to understand positive and negative numbers. We will learn the rules of a game with the beans and use the rules to clear up misunderstandings about arithmetic with positive and negative numbers.

The first two rules of the game tell us how to add beans of the same or different colors.

Rule 1--Adding beans of the same color: To add beans of the same color, just add the way you normally would and include the color of beans in your answer.

Example 1: Add 3 red beans to 8 red beans.

We get 11 red beans.
Written symbolically, abbreviating red beans as RB, 3 RB + 8 RB = 11 RB

Example 2: Add 2 white beans and 3 white beans.

+ =

2 WB + 3 WB = 5 WB

Rule 2--Adding different colors of beans: When red beans are added to white beans, each red and white bean pair cancel out. The sum is the number and color of beans left over after all possible pairs of red and white beans are canceled.

Example 3: Add 3 white beans to 8 red beans.
Cancel red and white bean pairs until the remaining beans are one color.

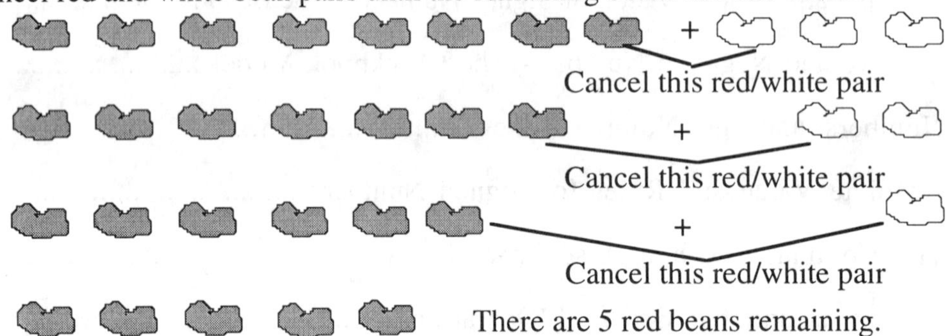

When we add 8 red beans to 3 white beans, the answer is 5 red beans.
Written symbolically, 3 WB + 8 RB = 5 RB

Example 4: Add 1 RB + 2 WB,

+ =

Cancel this red/white pair and we are left with one white bean.

1 RB + 2 WB = 1 WB

In this section are several games involving red and white beans. The purpose of the games is to learn a hands-on model for understanding arithmetic of positive and negative numbers. To play the games, each group will need a bag of about 30 red beans and 30 white beans.

Game 1--Adding Red and White Beans

For Problems 1-16, represent the addition with your beans and then use rules 1 and 2 to do the addition. **In the first column**, record the final result of the addition or subtraction after all white/red pairs are removed. Include the color of the beans in the answer, not just how many beans you ended up with. **In the second column**, state whether you actually added the two original numbers to get the answer or subtracted the two original numbers. **Leave the third column blank for now.**

	Total and Color	Add/Subtract?	
1. 7 red beans + 9 red beans			
2. 3 white beans + 2 red beans			
3. 5 white beans + 2 white beans			
4. 3 white beans + 2 white beans			
5. 5 red beans + 9 white beans			
6. 3 white beans + 8 red beans			
7. 5 red beans + 6 red beans			
8. 1 white bean + 6 red beans			
9. 1 white bean + 6 white beans			
10. 11 red beans + 7 white beans			
11. 11 red beans + 7 red beans			
12. 5 red beans + 6 white beans			
13. 5 white beans + 6 red beans			
14. 6 white beans + 5 red beans			
15. 8 red beans plus 8 white beans			
16. 8 red beans + 8 red beans			

17. In which problems above did you add the two original numbers of beans to get the answer?

____

18. In which problems above did you subtract the numbers of beans to get the answer?

____

19. Do each of the following additions. Use your beans to set up the model. "RB" is "red beans" abbreviated and "WB" is "white beans" abbreviated.

a. 7 RB + 7 RB = ____________ b. 7 RB + 7 WB = ______________

c. 1 WB + 1 WB = ___________ d. 1 WB + 1 RB = ______________

e. 3 WB + 3 RB = ___________ f. 3 RB + 3 WB = ______________

20. Describe what happens when equal numbers of beans of different colors are added.

____

21. What is the result when equal numbers of beans of the same color are added?

____

Now that we can add red and white beans, we need more rules to do bean subtraction. Some subtraction is obvious.

Example 5: 7 white beans − 2 white beans = 5 white beans.

Remove 2 white beans.

=

In this case, we just "take away" two of the white beans from the seven white beans, leaving us with five white beans. To check your answer, add the two white beans back to the five white beans. 2 WB + 5 WB = 7 WB. The answer is correct since the sum gives the total number of beans before subtraction.

The difficulty in subtraction comes when trying to subtract more beans of a particular color than are present. We cannot just "take away" beans that are not there, so we need to devise a way to get more beans in the problem without changing the value of the number of beans with which we are starting. As observed in Questions 19 and 20 above, when a certain number of white beans are added to the same number of red beans, all of the beans "cancel" so that there are no beans (or zero beans) left.

Example 6: Each of the following add up to zero. Notice that in each case, there are equal numbers of red and white beans.

5 WB + 5 RB = 0   15 RB + 15 WB = 0
11 RB + 11 WB = 0   9 WB + 9 RB = 0

When we start with a number of beans and add zero, the total number of beans does not change. This leads to the next rule of the "bean arithmetic" game.

Rule 3--Subtracting beans: Subtracting beans means to "take away" beans.
If there are enough beans of the right color to do the subtraction, just remove the beans that are to be subtracted. The difference is the number and color of beans left over.

Example 7: 5 red beans - 4 red beans.
Start with 5 red beans, take away 4 of the beans. This leaves 1 red bean.

Remove 4 red beans.

=

5 RB − 4 RB = 1 RB.

5 red beans - 4 red beans = 1 red bean. Check by adding: 1 RB + 4 RB = 5 RB.

Rule 4--Subtracting beans: Now suppose that we have to subtract a certain number of red beans, but we do not have that many red beans. Remember that pairs of red and white beans cancel out to zero; therefore, we can add as many red/white pairs as needed to do the subtraction. The answer is the number and color of beans left over.

Example 8: 4 red beans - 5 red beans.

You can't take away 5 red beans because there are not 5 red beans there. Add a red/white pair WHICH EQUALS ZERO so you do not change the value of what you have. Take away the 5 red beans. What remains?

Remove the 5 red beans.

4 RB

4 RB - 5 RB = 1 WB Check by adding: 1 WB + 5 RB = 4 RB

Example 9: Subtract 6 white beans from 4 red beans (4 red beans - 6 white beans).

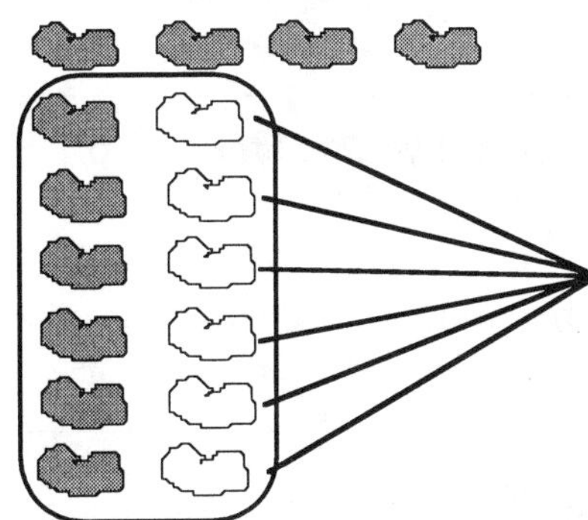

Add six red/white pairs (shown in the box) which equal zero.

Remove the 6 white beans. There are 10 red beans remaining.

4 RB - 6 WB = 10 RB Check by adding: 10 RB + 6 WB = 4 RB

GAME 2--Subtracting Beans

Work Problems 1-14 using the beans to do the subtractions. **In the first column**, give the final result of the addition or subtraction after all white/red pairs are removed or white/red pairs are added. If you ended up with two colors of beans, cancel pairs until you only have one color left. Record the color of the beans you ended up with. **In the second column**, state whether you actually added the original numbers to get the answer or used subtraction. **Leave the third column blank for now.**

	Total and Color	Add/Subtract?	
1. 5 white beans - 2 white beans			
2. 7 red beans - 4 red beans			
3. 7 red beans - 2 white beans			
4. 7 red beans - 8 white beans			
5. 8 red beans - 8 white beans			
6. 8 white beans - 8 red beans			
7. 7 red beans - 9 red beans			
8. 2 white beans - 9 red beans			
9. 2 white beans - 4 red beans			
10. 2 white beans - 4 white beans			

11. 7 red beans - 3 white beans			
12. 7 red beans - 7 white beans			
13. 3 white beans - 8 white beans			
14. 15 red beans - 19 white beans			

15. In problems where the beans were the same color, when did you add the two original numbers to get the answer?

____

16. In problems where the beans were the same color, when did you use subtraction to get the answer?

____

17. In problems where the beans were a different color, when did you add the two original numbers to get the answer?

____

18. In problems where the beans were a different color, when did you use subtraction to get the answer?

____

You have practiced adding beans in Game 1 and subtracting beans in Game 2, so you should now be able to do either.

Remember that to **add** beans, if the beans are the same color, just add the beans and record the color. If the beans are different colors, cancel red/white pairs until there is only one color left and record your answer.

To **subtract** beans, if there are enough of the right color to remove, then just take away the indicated number of beans and record the answer. If there are not enough of the right color to take away, then add zero in the form of red/white pairs until there are enough beans of the needed color to remove. If you brought in too many pairs, you may have two colors left. It this happens, just remove red/white pairs until you have one color of bean and record the answer.

GAME 3--Addition and Subtraction

For each of the following, record in column 2 whether the **question** is stated as addition or subtraction; then use your beans to do the arithmetic. Record the number and color of beans in column 3. Leave the last column blank for now.

	Addition or Subtraction Question	Answer and Color	
1. 5 white beans + 2 red beans	**Addition**	2 WB	
2. 7 red beans + 4 red beans			
3. 7 red beans − 12 white beans			
4. 7 red beans + 8 white beans			
5. 8 red beans + 18 white beans			
6. 8 white beans − 7 red beans			
7. 6 red beans + 9 red beans			

8. 11 red beans − 7 white beans			
9. 2 white beans + 11 red beans			
10. 13 white beans - 6 red beans			
11. 4 white beans - 9 white beans			
12. 7 red beans + 3 white beans			
13. 7 red beans + 7 white beans			
14. 8 white beans - 13 white beans			
15. 4 red beans - 13 red beans			

16. When you **add** beans of the **same color**, do you add, subtract, or sometimes add or sometimes subtract the original numbers to get the answer?

____

17. When you **add** beans of **different colors**, do you add, subtract, or sometimes add and sometimes subtract the original numbers to get the answer?

____

18. When you **subtract** beans of **different colors**, do you add, subtract, or sometimes add or sometimes subtract the original numbers to get the answer?

____

19. When you **subtract** beans of the **same color**, do you add, subtract, or sometimes add or sometimes subtract the original numbers to get the answer?

____

What do the Bean Games have to do with arithmetic? To find out, we are now going to make a new rule of the game. In banking, being "in the red" means that we are overdrawn, or that we owe more money than we have. Since "in the red" refers to a negative balance in banking, we will now arbitrarily refer to red beans as negative and white beans as positive.

Rule 4: White beans represent positive numbers and red beans represent negative numbers.

Example 10:

4 white beans represent the number $+4$ (or just 4).

4 red beans represent the number -4.

Example 11: 2 red beans minus 3 white beans can be modeled as shown.

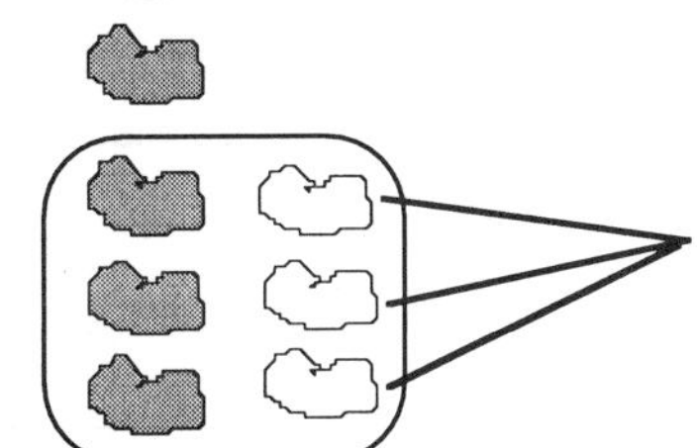

Add zero in the form of three red/white bean pairs as shown in the box.

Remove the 3 white beans. 5 red beans are remaining.

2 RB - 3 WB = 5 RB   Check by adding: 5 RB + 3 WB = 2 RB

$(-2) - (+3) = -5$ Check by adding: $(-5) + (3) = -2$

Other ways to write the subtraction in Example 11 are
-2 - +3 = -5 or -2 - (+3) = -5 or a shorter version is: - 2 - 3 = -5

To finish Games 1, 2, and 3, fill in the third column of the charts in the following way. In the last column in Games 1, 2, and 3 write the arithmetic problem and answer.

Example 12: In Game 1, the first problem was "7 red beans + 9 red beans." The answer was 16 red beans. In the last column, write -7 + (-9) = -16.

3 WB - 5 RB means positive 3 minus negative 5. To separate the minus from the negative sign, we use parentheses and represent "3 - negative 5" as "3 - (-5)."

The tables should look something like this:

	Total and Color	Add/Subtract	
7 red beans + 9 red beans	16 red beans	add	**(-7) + (-9) = -16**
4 white beans - 3 red beans	7 white beans	add	**4 - (-3) = 7**

Exercise 1--Arithmetic and Beans

Purpose of the Exercise
- To transfer the bean games to arithmetic problems

Materials Needed
- Red and white beans--about 30 of each for each group

Write the following arithmetic problems in terms of red and white beans and then find the answer both in terms of beans as well as in terms of numbers. The first problem is done as an example.

	Numeric Answer	Translation to Beans
1. 6 - (-8) =	14	Six white beans minus 8 red beans. We bring in 8 "red-white" pairs and then remove the 8 red beans.
2. - 9 + 27 =	________	________
3. -17 + 23 =	________	________
4. 16 - (-3) =	________	________
5. 14 + (-3) =	________	________
6. 14 - 3 =	________	________
7. - 29 + 27 =	________	________
8. 27 - 29 =	________	________

9. - 19 - (- 27) = _____ ____

____

10. - 19 + 27 = _______ ____

____

11. 354 - (-700) = _____ ____

____

12. (82) - 9 = ________ ____

____

13. 82 - (9) = ________ ____

____

14. - 297 + 89 = _______ ____

____

15. 89 - 297 = ________ ____

____

Now let's summarize what we have learned with the beans.

When we add 2 positive numbers, the addition is the same as we know from money.

4 + 3 = 7

To add 2 negative numbers, add the numbers. The answer is negative.

(−4) + (−3) = (−7)

To add a negative number and a positive number, subtract the smaller number from the larger and keep the sign of the larger number. (Cancel the red/white pairs. The color with the most beans is the one remaining.)

(−4) + 3 = −1

4 + (−3) = 1

In subtraction, change the sign of the number being subtracted and follow the rules of addition.

$5 - (-4)$ Restate as $5 + 4 = 9$ (Bring in 4 red/white pairs; then take away 4 reds.)

$-5 - (-4) = -5 + 4 = -1$

$7 - 8 = 7 + (-8) = -1$

Section 2-2 *Multiplication and Division of Signed Numbers--The Bean Model*

Multiplication, as we learned from Chapter 1, is repetitive addition and division is divvying up. In the multiplication and division processes we studied, all of the numbers involved were positive. Now we have to see how negative signs affect the product or quotient when some of the numbers multiplied or divided are negative. As with addition and subtraction, we will use beans to help us understand the rules of signs.

Example 1: Compute 3 times 4 red beans. Or, stated mathematically, compute $3(-4)$.

Three groups of 4 red beans is 12 red beans.
Written as arithmetic, $3(-4) = -12$.

Example 2: Compute $(-7)\div 3$ by dividing 7 red beans among 3 people.
As in Chapter 1, division means to "divvy" up.

Each person gets **2 red beans** with **1 red bean** left over.
$(-7)\div 3 = -2$ with a remainder of -1. Or stated differently, use the calculator to approximate the division, $(-7)\div 3$ -2.3333.

GAME 4--Multiplying and Dividing Red and White Beans

Do each of the following multiplication or division of beans problems. In the last column, write the problem and answer in mathematical symbolism. The first two are done as examples.

	Answer in Beans	Mathematical Symbolism/Answer
1. 18 red beans divided by 9	2 red beans	$(-18)\div 9 = -2$
2. 3 times 7 red beans	21 red beans	$3(-7) = -21$
3. 18 white beans divided by 9		
4. 4 times 8 red beans		
5. 27 red beans divided by 3		
6. 6 times 4 white beans		
7. 22 times 8 red beans		
8. 255 times 18 white bean		
9. 70 times 76 red beans		
10. 821 times 20 red beans		
11. 19 times 88 red beans		
12. 25 red beans divided by 5		
13. 14 red beans divided by 7		

Generalize your findings. In the following questions, circle the words that correctly completes the sentence.

14. A positive number times a negative number is positive negative
15. A positive number times a positive number is positive negative
16. A positive number divided by a positive number is positive negative
17. A negative number divided by a positive number is positive negative

We have not yet covered how to divide when the divisor is negative or how to multiply a negative number times a negative number. We will save those topics for a later section.

GAME 5--Addition, Subtraction, Multiplication, and Division Together

In each of the following, do the computations inside of the parentheses before other computations. Write the **first step** of each problem in "bean" language and then represent the problem in mathematical notation to compute the answer. The first one is done as an example.

	Answer and Color	Mathematical Symbolism/Answer
1. 4 white beans minus (3 times 7 red beans) 1: 4 white beans minus 21 red beans.	25 white beans	4 - (3)(-7) = 25
2. 3 times 8 red beans minus (2 times 4 red beans) 1:		
3. (3 times 8 red beans) plus (2 times 4 red beans) 1:		
4. 3 times 8 red beans minus (2 times 4 white beans) 1:		
5. (7 times 2 white beans) minus (7 times 2 red beans) 1:		
6. 20 red beans minus (15 red beans divided by 3) 1:		
7. 3 white beans minus (5 times 3) red beans 1:		
8. 3 white beans plus (5 times 3) white beans 1:		
9. 6 red beans divided by (2 plus 3 times 8 white beans) 1:		

10. 17 red beans minus 10 red beans 1:		
11. (14 red beans divided by 2) minus (8 white beans divided by 4) 1:		
12. 18 times 34 red beans 1:		
13. 17 red beans minus (6 times 5 red beans) 1:		

In each of the following, perform the indicated arithmetic operations. If you have difficulty with a problem, convert it to "bean" arithmetic.

Numeric Answer	Translation to Beans
14. 4 - 3(-5) = 19	4 white beans minus three times 5 red beans
	4 white beans minus 15 red beans.
	19 white and 15 red beans minus 15 red beans.
15. 4(-5) + 3(7) = ______	______
16. -8 + 6(-3) = ______	______
17. -8 - 6(-3) = ______	______
18. 7(-5) = ______	______
19. 3(-10) - (-6) = ______	______
20. 75 - (-93) = ______	______
21. -1357 + 7(2) = ______	______
22. $\frac{-16}{2}$ - (-5) = ______	______
23. (14/7) + (-3) = ______	______
24. $\frac{25}{5} - \frac{-32}{8}$ = ______	______

Section 2-3 *Positive and Negative Numbers--The Checkbook Model*

There are several models for understanding positive and negative numbers. In the last section, we used the two colors of beans to represent opposites. In this section, we will work with money and positive and negative balances in bank accounts. Working with money--debt as well as money in the bank--not only furthers our understanding of positive and negative numbers, but it also provides us with real-life circumstances in which signed numbers are used.

In the checkbook model, money in the bank is a positive amount and money owed (or overdrawn) is a negative amount. When a check is written, we subtract money from the account. When a deposit is made, we add money to the account.

Example 1:

1. A $25-deposit is made and a $15-check is written.
2. After the two transactions, the balance in the account has **increased** by $10.
3. Written as an arithmetic statement, 25 - 15 = 10 or 25 + (-15) = 10.

Example 2:

1. A $25-check is written and a $15-deposit is made.
2. After the two transactions, the balance in the account has **decreased** by $10.
3. Written as an arithmetic statement, -25 + 15 = -10.

Example 3:

1. The balance shown in the checkbook is $257 and a check is written for $82
2. After writing the check, the new balance is $175.
3. Written as an arithmetic statement, 257 - 82 =175.

Example 4:

1. The balance in a bank account is $186 and a check is written for $199.
2. After writing the check, the account is **overdrawn** by $13. (We are not including bank overdraft charges.)
3. Written as an arithmetic statement, 186 - 199 = -13 or 186 + (-199) = -13.

Example 5:

1. Deposits are made for $18 and $22.
2. After the two transactions, the balance in the account has **increased** by $40.
3. Written as an arithmetic statement, 18 + 22 = 40.

Example 6:

1. Checks are written for $33 and $14.
2. After the two transactions, the balance in the account has **decreased** by $47.
3. Written as an arithmetic statement, -33 + (-14) = -47 or -33 - 14 = -47.

Example 7:

1. The account is overdrawn by $12 and a $15 check is written.
2. After writing the check, the account is **overdrawn** by $27.
3. Written as an arithmetic statement, -12 - 15 = -27 or -12 + (-15) = -27.

Example 8:

1. The account is overdrawn by $13 and a deposit of $20 is made.
2. After making the deposit, the balance in the account is now $7.
3. Written as an arithmetic statement, -13 + 20 = 7.

Example 9:

1. You owe \$75 to one creditor and \$82 to another creditor.
2. Your total debt is \$157
3. Written as an arithmetic statement, -75 + (-82) = - 157 or -75 - 82 = -157.

Example 10:

1. You have \$1257 in the bank.
2. Bills to be paid add up to \$1467.
3. You are short \$210 needed to pay the bills.
4. Written as an arithmetic statement, 1257 - 1467 = -210.

In the bean model as well as the checkbook model, when adding a positive and a negative number, the answer is sometimes positive and sometimes negative. When adding two positive numbers, the answer is always positive. When adding two negative numbers, the answer is always negative. The following examples demonstrate what happens when we multiply or divide a negative number by a positive number.

Example 11:

1. You owe three different creditors \$250 each.
2. You owe a total of \$750.
3. Written as an arithmetic statement, amount owed is 3(-250) = -750.

Example 12:

1 8 people won a total of \$5000.
2 Each person's share is \$625
3 Written as an arithmetic statement, each person's share is 5000 ÷ 8 = 625.

Example 13:

1 8 people together bet and lost \$5000.
2 Each person owes \$625
3 Written as an arithmetic statement, each person's loss is -5000 ÷ 8 = -625.

To summarize Examples 11, 12 and 13, the product of a positive and a negative number is a negative number. A negative number divided by a positive number is a negative number.

In money, the way we get rid of a debt is to make a payment. Since a debt is considered negative, getting rid of (subtracting) a negative number is the same as adding a positive number. The examples below demonstrate subtracting negative numbers.

Example 14:

1. You owe \$75 to one creditor and \$82 to another creditor. Your total **debt** is \$157.
2. You cancel the \$82 debt by making a payment. Your total debt is now \$75.
3. Written as an arithmetic statement, -157 - (-82) = -75.
4. Notice this is the same end result as -157 + 82 = -75.

Example 15:

1. Your checkbook balance is \$1087. Already accounted for in the balance is a \$250 check that you wrote but it has not yet cleared the bank.
2. You get the check back and tear it up.
3. To correct the bank balance, add the \$250 back in (or cancel the \$250 check).
4. To add the check back in; 1087 + 250 = 1337.
5. To cancel the check; 1087 - (-250) = 1337.

Example 16:

1. You owe $678.85 on a credit card.
2. You cancel part of the debt by paying $200.00.
3. The new balance on the card is $478.85.
4. Written as an arithmetic statement: -678.85 - (-200.00) = - 478.85.

In the above example, the 678.85 is considered a negative value because it is money owed. Subtracting a $200 debt is done by making a payment. In other words,
-678.85 - (-200.00) is the same as -678.85 + 200.00.

Example 17:

1. You owe $315.90 on a credit card.
2. You misread the bill, and canceled the debt by making a payment of $325.90.
3. You now have a credit balance on your card of $10.00.
4. Written as an arithmetic problem; -315.90 - (-325.90) = 10.

In the above example, the 315.90 is considered a negative value because it is money owed. Subtracting a 325.90 debt is done by making a payment. In other words,
-315.90 - (-325.90) is the same as -315.90 + 325.90.

Any subtraction problem can be written as an addition problem.

To subtract a negative number, change "back-to-back" negative signs to a "+" sign and then do the arithmetic.

-13 - (-75) is the same as -13 + 75 = 62
87 - (-16) is the same as 87 + 16 = 103
-45 - (-22) is the same as -45 + 22 = -23

To subtract a positive number, you can change the subtraction to addition of a negative number.

87 - 82 is the same as 87 + (-82) = 5
-26 - 84 is the same as -26 + (-84) = -110

Exercise 1--Checkbook Math

Purpose of the Exercise

- To practice restating subtraction as addition
- To better understand the arithmetic of signed numbers using the checkbook model
- To practice the arithmetic of checkbooks

Materials Needed

- Calculator

In each of the following, rewrite the subtraction as addition and then compute the answer.

	Written as an Addition Problem	Answer		Written as an Addition Problem	Answer
1. - 25 - 89 =	____________	= ________	2. - 25 - (-89) =	________	= ________
3. 82 - 31 =	____________	= ________	4. - 82 - 31 =	________	= ________
5. - 74 - 22 =	____________	= ________	6. - 74 - (-22) =	________	= ________
7. 17 - 23 =	____________	= ________	8. 17 - (-23) =	________	= ________

For each of the following, compute the new balance after the transaction. Show the arithmetic done to compute the balance. A negative balance indicates the account is overdrawn.

Balance in Account	Transaction	Arithmetic Done	New Balance
9. $198	A check for $92 is written.		
10. $235.76	A check for $84.56 is written.		
11. $14.98	A check for $28.76 is written.		
12. -$28.73 (overdrawn)	A check for $10.65 is written.		
13. -$2.98 (overdrawn)	A deposit of $14.00 is made.		
14. $179.36	A deposit of $29.00 is made.		
15. $1876.59	A deposit of $275.86 is made.		
16. $4,786.98	A check for $5000 is written		
17. -$297 (overdrawn)	A deposit of $200 is made.		
18. $278.90	Cancel a check for $17.32 that has already been subtracted.		
19. $58.97	A deposit of $67.90 is made and a check for $102.78 is written.		
20. $1095.63	A deposit of $350 is made and a check for $1296.75 is written.		
21. -$295.67	A check for $400 is written and a deposit of $100 is made.		
22. $78.56	Cancel a check for $83.50 that has already been subtracted.		
23. -$15.73	Cancel a check for $22.00 that has already been subtracted.		
24. $789.22	Cancel a check for $56 that has already been subtracted.		
25. -$2.45	A check for $5.28 is written.		
26. $2.45	A check for $5.28 is written.		
27. -$56.78	A deposit of $125 is made.		

Vocabulary

The **natural,** or **counting,** numbers are 1, 2, 3, 4, ... They are the numbers we count with.

The **whole numbers** are 0, 1, 2, 3, 4, 5, ... The whole numbers include the counting numbers along with 0.

The **integers** are the whole numbers and their opposites.

A **rational** number is a positive or negative number that has no digits after the decimal, or has a repeating pattern of digits after the decimal point, or is a terminating decimal (meaning the digits stop eventually). Fractions, which we will study in Chapter 4, are another form of rational numbers.

The **irrational numbers** are the numbers that are not rational. Irrational numbers can be expressed in decimal form in which the digits after the decimal point continue forever without a repeating pattern. We haven't covered any irrational numbers yet, but they are numbers like and square roots that do not come out even.

All of the numbers above fit in a category called **real numbers**.

All lengths can be expressed as real numbers and all positive real numbers can represent lengths. This allows us to place real numbers along a number line. A ruler is an example of lining up positive numbers. On the ruler below, only the whole numbers are labeled. One way to place numbers other than whole numbers on the ruler is to convert the number to decimal form or to a decimal approximation and then put the number in the correct position on the ruler. A calculator is a handy tool for converting a fraction to decimal form or for finding a decimal approximation of an irrational number.

Example 1: Place $\frac{29}{18}$ on a ruler.

First, convert the fraction to decimal form using your calculator.

Key strokes: 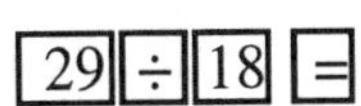

$\frac{29}{18}$ 1.61

so $\frac{29}{18}$ is placed on the ruler between 1 and 2 but a little closer to 2 as shown below.

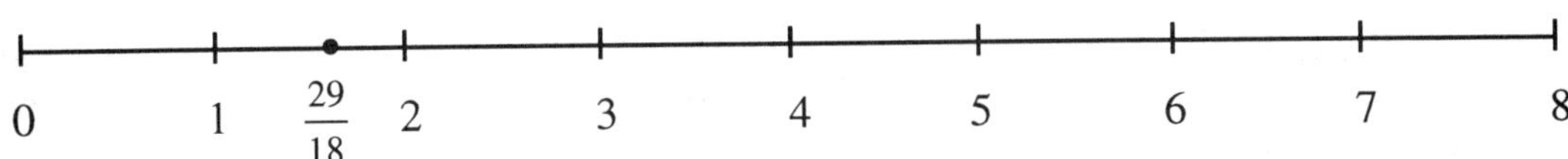

Example 2: Place $\sqrt{63}$ on a ruler.

First, convert $\sqrt{63}$ to decimal form.

Key strokes on a scientific calculator: 63 √ =

Key strokes on a graphics or "heads-up" calculator: √ 65 =

$\sqrt{63}$ 7.94 so is placed on the ruler between 7 and 8, but almost at 8.

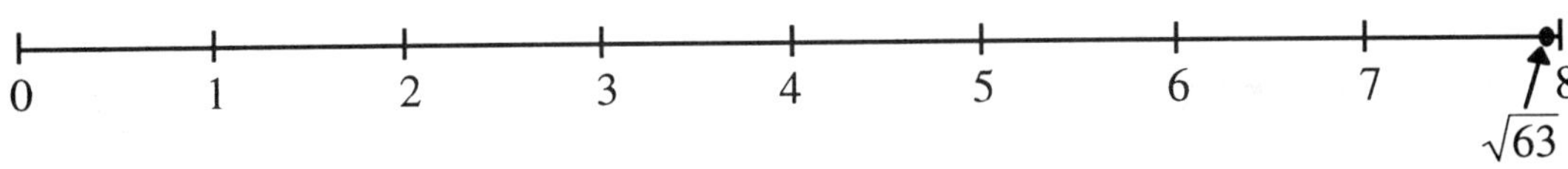

When we extend our ruler to the left of the "0" mark, we refer to it as a number line. We plot the negative real numbers left of 0 as shown in the examples below.

Example 3:

$-\frac{29}{18}$ -1.61 so $-\frac{29}{18}$ is placed on the ruler between -1 and -2 but a little closer to -2 as shown below. $-\sqrt{63}$ -7.94 so is between -7 and -8, but almost at -8.

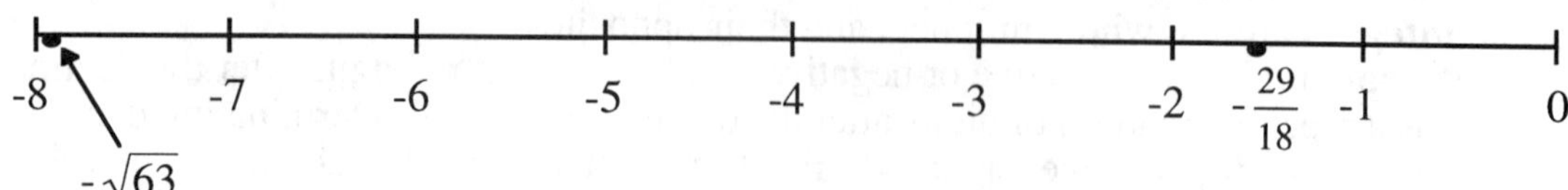

Exercise 1--Placing Real Numbers on a Ruler

Purpose of the Exercise

- To see how positive rational numbers line up on a ruler

Materials Needed

- Calculator

For each of the following numbers, write its decimal representation, rounded to 2 decimal places if needed. Place and label a dot appropriately on the number line below to represent each of the numbers.

1. $\sqrt{7} \approx$ ______________ 2. 29/11 = ______________ 3. 2 ≈ ______________

4. 1/5 = ______________ 5. 19/12 = ______________ 6. $\sqrt{62} \approx$ ______________

7. $\frac{307}{43} \approx$ ______________ 8. 14/3 ______________ 9. 65/10 = ______________

10. $\sqrt{17} \approx$ ______________ 11. $\sqrt{14.6} \approx$ ______________ 12. 325.8/171.4 ≈ ________

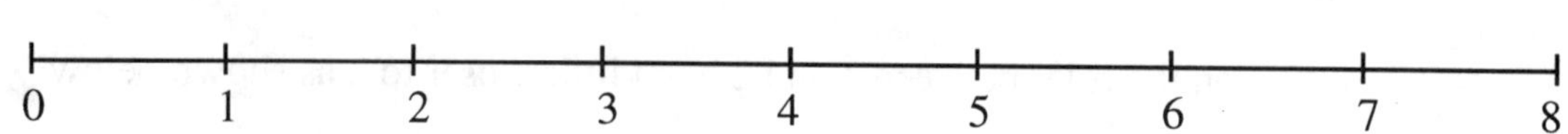

For each of the following numbers, write its decimal representation, rounded to 2 decimal places if needed. Place and label a dot appropriately on the number line below to represent each of the numbers.

13. $-\sqrt{7} \approx$ ______________ 14. - 29/11 = ______________ 15. - 2 ≈ ______________

16. - 1/5 = ______________ 17. - 19/12 = ______________ 18. $-\sqrt{62} \approx$ ______________

19. $-\frac{307}{43} \approx$ ______________ 20. - 14/3 ≈ ______________ 21. - 65/10 = ______________

22. $-\sqrt{17} \approx$ ______________ 23. $-\sqrt{14.6} \approx$ ______________ 24. - 325.8/171.4 ≈ ________

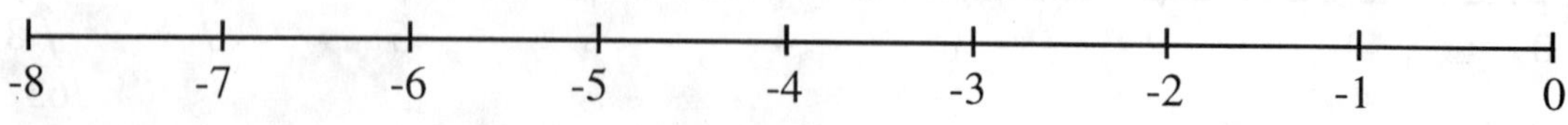

A number line, unlike a ruler, theoretically goes on forever in both directions. Since we cannot draw such a line, we draw whatever part of the line that we need to work with for the current situation. We put arrows at both ends of the number line to indicate that we are only showing part of the line. If we need to represent large numbers, we may also use different **scales** to represent the unit lengths.

The examples below show 2.4 and -2.4 placed on number lines with different scales.

Example 4: In the number line below, the **scale** is 1. 2.4 lies between 2 and 3, but is slightly closer to 2.

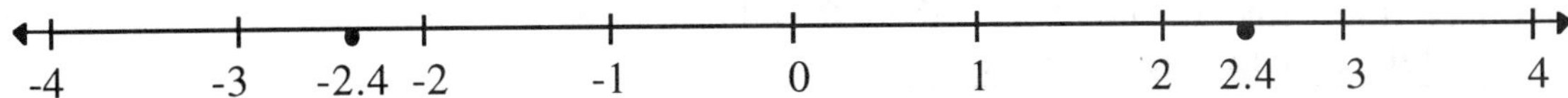

In the above number line below, the **scale** is 2, and 2.4 lies between 2 and 4. Since 3 is halfway between 2 and 4 and 2.4 is closer to 2 than 3, this puts 2.4 much closer to 2 than 4.

-8 -6 -4 -2.4 -2 0 2 2.4 4 6 8

In the number line below, the **scale** is 5, and 2.4 lies between 0 and 5 but is closer to 0.

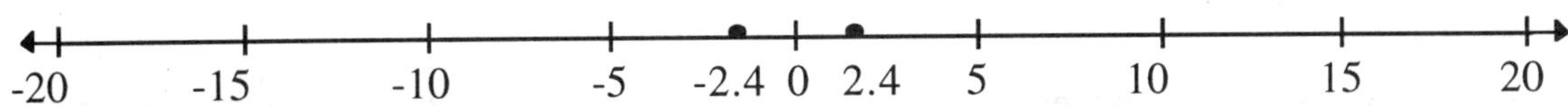

Exercise 2--Plotting Numbers on a Number Line

Purpose of the Exercise

- To see how positive rational numbers line up on a number line
- To practice plotting numbers on number lines with various scales

Materials Needed

- Calculator

For each of the following numbers, write its decimal representation, rounded to 2 decimal places if needed. Place and label a dot appropriately on the number line below to represent each of the numbers.

1. $\sqrt{8} \approx$ __________ 2. 17/5 = __________ 3. 2 $\approx$ __________

4. 1/2 = __________ 5. $(1.2)^2$ = __________ 6. $\sqrt{45} \approx$ __________

7. $-\sqrt{8} \approx$ __________ 8. - 17/5 = __________ 9. - 2 $\approx$ __________

10. - 1/2 = __________ 11. $-(1.2)^2$ = __________ 12. $-\sqrt{45} \approx$ __________

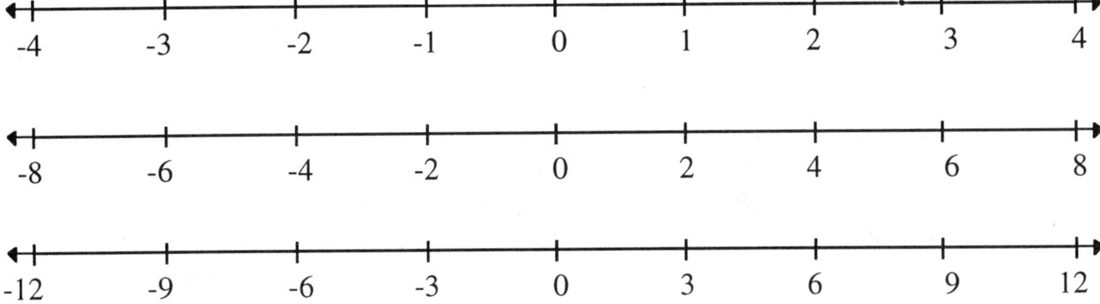

The number line gives us a new way to compare the relative sizes of numbers. With the positive numbers, it is not difficult to determine which of the numbers is greater. Represent both numbers as decimals and compare digits to pick out the larger and smaller numbers. On a number line the smaller positive number is closer to zero and the larger positive number is further away from zero. **The smaller number is to the left of the larger number.**

> **Definition:**
> This generalization applies to comparisons of two negative numbers and a positive and a negative number, as well as to two positive numbers. When two numbers are placed on a number line, the number on the **left** is the **smaller number** and the number on the **right** is the **larger number.**

Exercise 3--Comparisons of Numbers on a Number Line

Purpose of the Exercise

- To make size comparisons between two negative numbers
- To make comparisons between positive and negative numbers
- To gain experience in comparing rational and irrational numbers

Materials Needed

- Calculator

In each of the following, convert the two numbers to their decimal form if needed and plot the two numbers on the number line (label the scale you use). Record which number is larger.

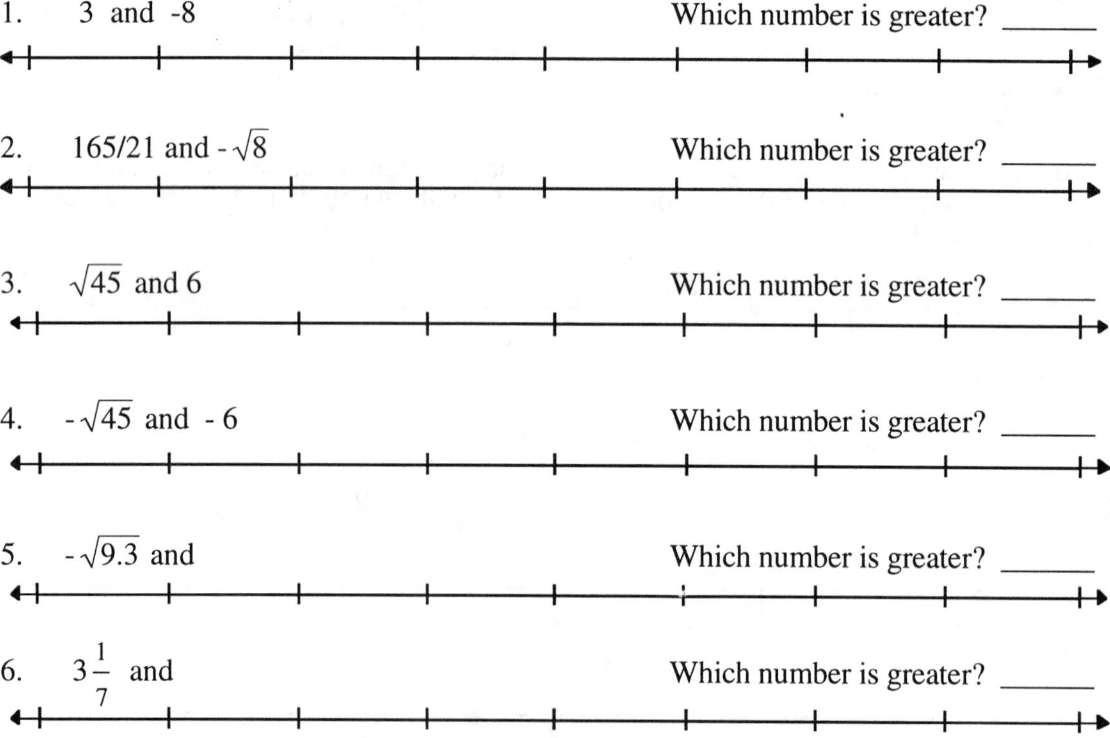

Compute 7x and 9x for the different values of x given. Pay close attention to signs. Compare the products and determine which is the larger number. If needed, draw a number line to make the determination. In each case, determine which is greater, 7x or 9x. The first one is done as an example.

7. x = -12.5 7x = 7(-12.5) = -87.5 9x = -112.5 Which is greater, 9x or 7x?

7x is bigger since -87.5 is to the right of -112.5 on the number line.

8. x = 4 7x = ________ 9x = ________ Which is greater, 9x or 7x?

9. x = 7.3 7x = ________ 9x = ________ Which is greater, 9x or 7x?

10. x = 19 7x = ________ 9x = ________ Which is greater, 9x or 7x?

11. x = 22 7x = ________ 9x = ________ Which is greater, 9x or 7x?

12. x = 7x = ________ 9x = ________ Which is greater, 9x or 7x?

13. x = 8.21 7x = ________ 9x = ________ Which is greater, 9x or 7x?

14. x = -4 7x = ________ 9x = ________ Which is greater, 9x or 7x?

15. x = -7.3 7x = ________ 9x = ________ Which is greater, 9x or 7x?

16. x = - 7x = ________ 9x = ________ Which is greater, 9x or 7x?

17. x = -22 7x = ________ 9x = ________ Which is greater, 9x or 7x?

18. x = -8.1 7x = ________ 9x = ________ Which is greater, 9x or 7x?

19. x = -478.9 7x = ________ 9x = ________ Which is greater, 9x or 7x?

20. Under what circumstances can you say that 7x is less than 9x? ________________

____

____

21. Under what circumstances can you say that 7x is greater than 9x? ________________

____

____

Another Look at Addition and Subtraction of Positive and Negative Numbers

We have approached the rules of adding positive and negative numbers by playing games with beans and by doing checkbook exercises. Still another approach involves the number line.

In the following examples, a positive number is added to another number.

Example 5: Add -6 + 10 using the number line.

1. Start at the -6 position.
2. To add 10, move 10 units to the **right** on the number line.
3. The sum is the end number.

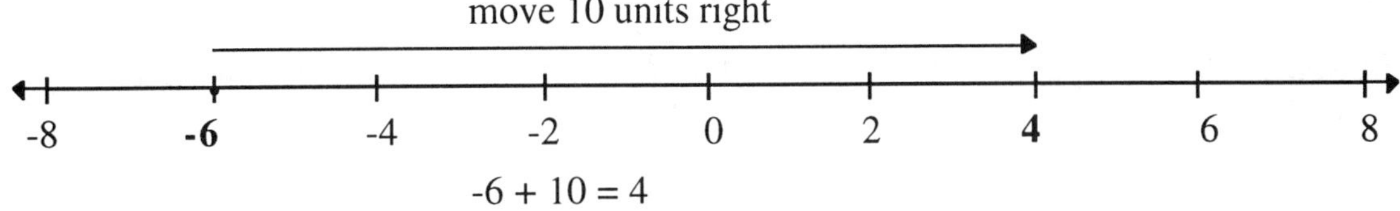

Example 6: Add 20 + 15 on the number line.

1. Start at the position 20.
2. Move 15 units to the **right** on the number line.
3. The sum is the end number.

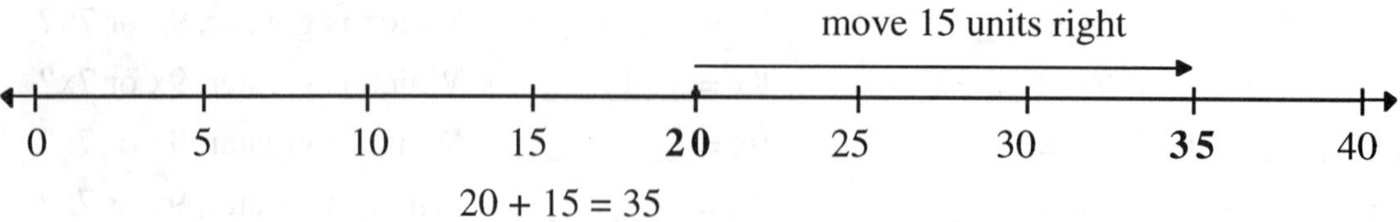

20 + 15 = 35

In the following examples, a positive number is subtracted from another number.

Example 7: Use the number line to subtract -6 - 10.

1. Start at the -6 position.
2. Move 10 units to the **left** on the number line.
3. The difference is the end number.

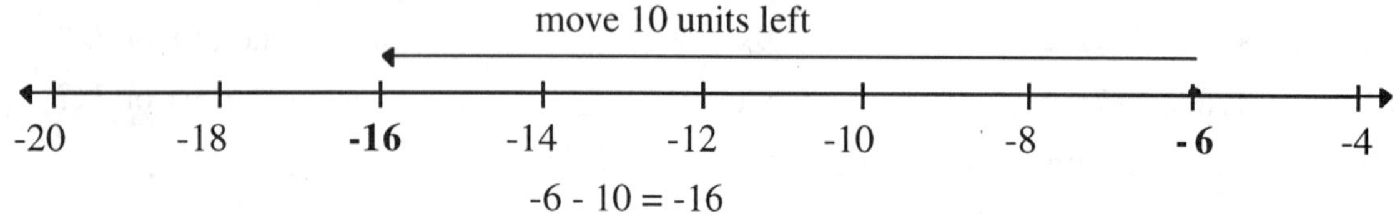

-6 - 10 = -16

Example 8: Subtract 6 - 10.

1. Start at the 6 position.
2. Move 10 units to the **left** on the number line.
3. The difference is the end number.

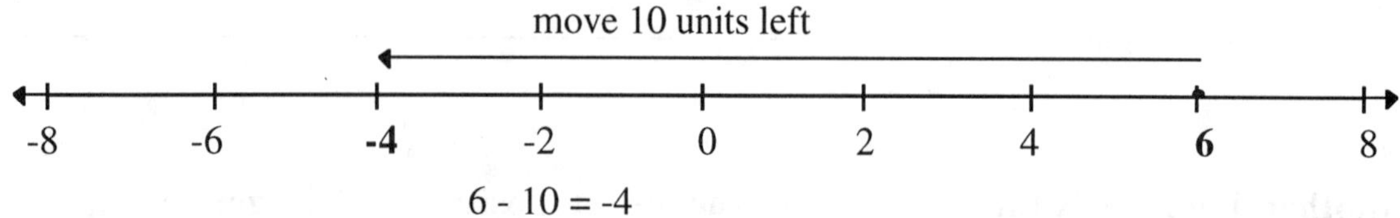

6 - 10 = -4

Example 9: Subtract 8 - 2.

1. Start at the 8 position.
2. Move 2 units to the **left** on the number line.
3. The difference is the end number.

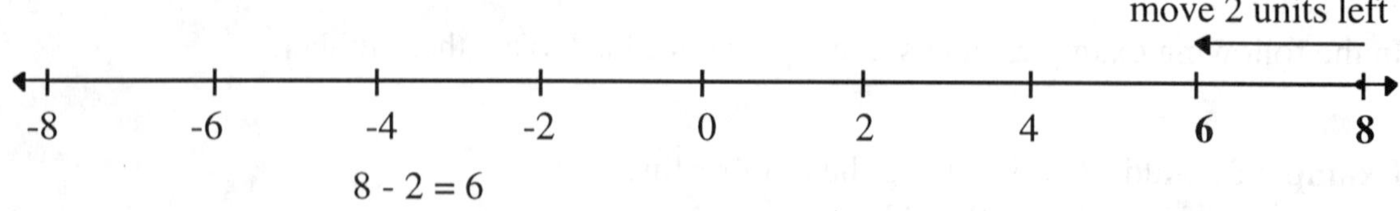

8 - 2 = 6

The concept of the opposite of a number and the "double negative" can also be explained with the number line. The opposite of any number is the number which is the same distance from 0 but on the opposite side of 0 on the number line.

Example 10: The opposite of 7 is -7 because both 7 and -7 are 7 units from 0 on the number line but -7 is 7 units left of 0 and 7 is 7 units right of 0. The opposite of 86 is -86. The opposite of -9 is 9.

To subtract a negative number from another number, we do the **opposite** of subtracting the positive number.

Example 11: Subtract 8 - (-2).

1. Start at the 8 position.
2. Move 2 units to the **right** on the number line. (This is the opposite of subtracting 2, where we would move 2 units left on the number line.) Compare this with adding 2 to 8 to see that 8 - (-2) = 8 + 2.
3. The answer is the end number.

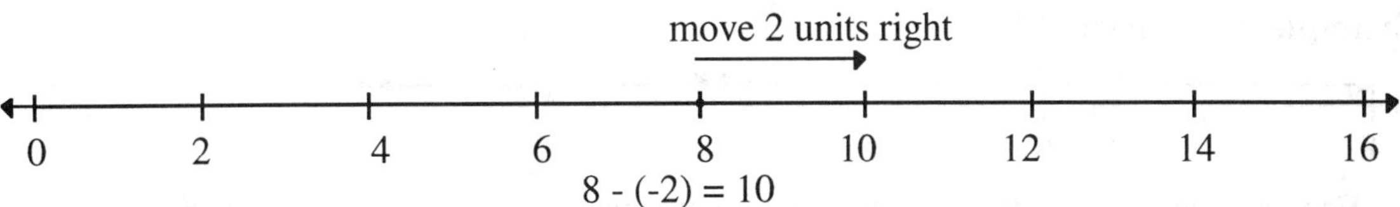

8 - (-2) = 10

Example 12: Subtract -5 - (-3).

1. Start at the -5 position.
2. Move 3 units to the right on the number line--this is the opposite of subtracting a positive number.
3. The answer is the end number.

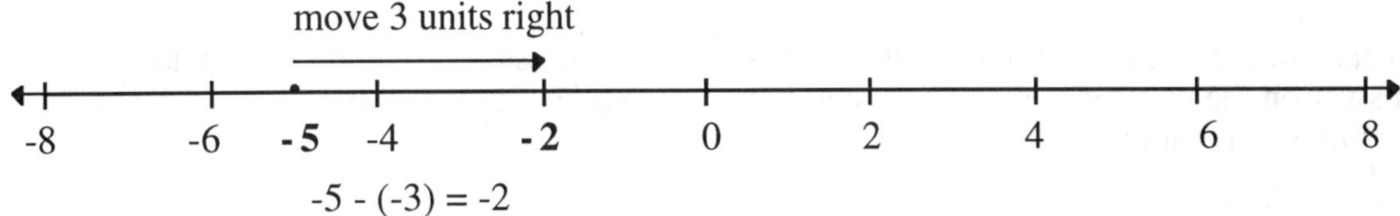

-5 - (-3) = -2

A positive number times a negative number can also be explained with the number line.

Example 13: Multiply 3(-4).

3(-4) can also be interpreted as (-4) + (-4) + (-4).

1. Start at 0.
2. Move left 4 units, then another 4, and then another 4.
3. The product is the end number.

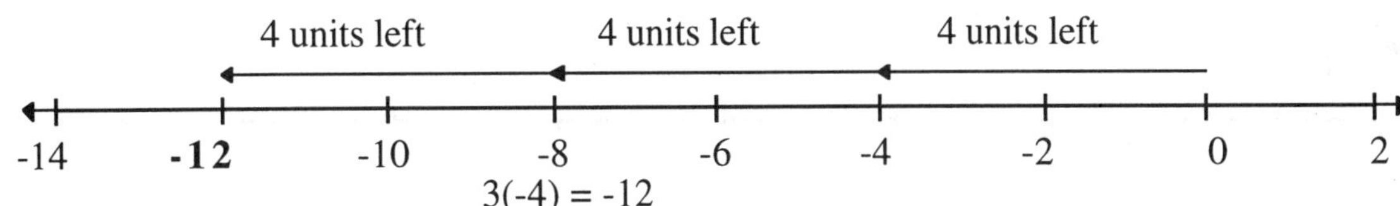

3(-4) = -12

Example 14: Multiply -3(-4).

-3(-4) is the opposite of 3(-4). As observed above, 3(-4) = -12.
The opposite of 3(-4) is the opposite of -12 which is 12.
-3(-4) = 12

Division can also be represented on a number line.

Example 15: Divide $8 \div 4$ on the number line.
Mark 8 on the number line and divide the length from 0 to 8 into 4 equal segments.

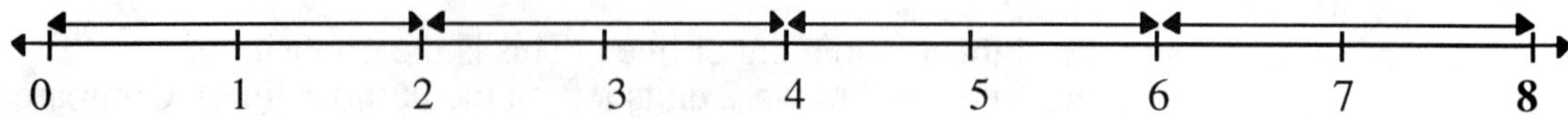

Each segment is 2 units long, so $8 \div 4 = 2$.

Example 16: Divide $(-8) \div 4$.

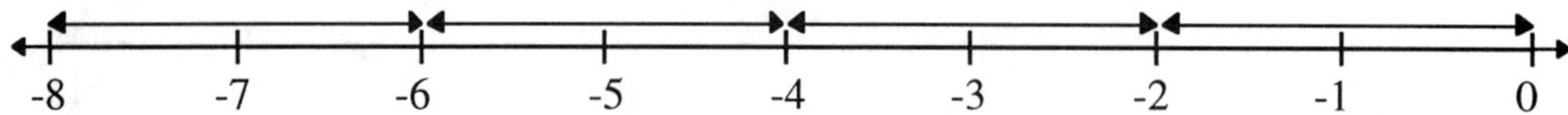

Each segment is 2 units long, but they are on negative parts of the number line, so we treat them as negative. $(-8) \div 4 = -2$.

Exercise 4--Arithmetic on the Number Line

Purpose of the Exercise

- To practice arithmetic with signed numbers, using the number line as an aid

In each of the following do the arithmetic as indicated, illustrating the problem and the answer on the number line. Choose and label your scale appropriately to represent the arithmetic operation.

1. $-4 - 7$

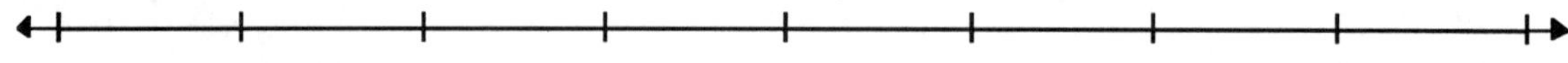

2. $-4 - (-7)$

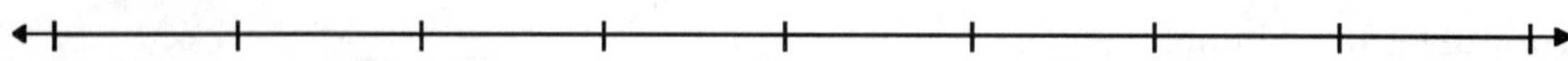

3. $5(-2)$

4. $6 - 8$

5. $8 - 6$

6. $8 - (-6)$

7. $(-6) \div 3$

Section 2-5 *Absolute Value and Rules for Signed Numbers*

The Absolute Value of a Number

Vocabulary

The **absolute value** of a number is the value of the number without regard to sign. Since we do not include the sign, the absolute value of a number is always positive or zero.

Symbolically, the absolute value of x is written **|x|.**

In the bean model, the absolute value of a number is the number of beans of a particular color, without considering the color of the beans. In the checkbook model, the absolute value of a number is the amount of money involved without considering whether the money is a check or deposit. On the number line, the absolute value of a number is the distance of that number from 0.

Example 1: The absolute values of the numbers, 3, -8, -22.3, , -2 , $\sqrt{879}$, $-\sqrt{879}$, and 0 are as shown.

$|3| = 3$ $\quad$ $|-8| = 8$ $\quad$ $|-22.3| = 22.3$ $\quad$ $|\ | =$

$|-2| = 2$ $\quad$ $|\sqrt{879}| = \sqrt{879}$ $\quad$ $|-\sqrt{879}| = \sqrt{879}$ $\quad$ $|0| = 0$

The opposite of the absolute value of a number is always negative or zero.

Example 2: The opposite of the absolute values of the numbers, 3, -8, -22.3, , -2 , $\sqrt{879}$, $-\sqrt{879}$, and 0 are as shown.

$-|3| = -3$ $\quad$ $-|-8| = -8$ $\quad$ $-|-22.3| = -22.3$ $\quad$ $-|\ | = -$

$-|-2| = -2$ $\quad$ $-|\sqrt{879}| = -\sqrt{879}$ $\quad$ $-|-\sqrt{879}| = -\sqrt{879}$ $\quad$ $|0| = 0$

When computations appear within absolute value signs, do the computation first; then take the absolute value of the number. When absolute values appear along with other computations, compute the absolute value(s) first and then do the other computations.

Example 3: Computations involving absolute values.

$$|7 - 9| = |-2| = 2$$

$$|(-4)(7)| = |-28| = 28$$

$$7 - |-9| = 7 - 9 = -2$$

$$-4|10 + 3| = -4\,|13| = -52$$

$$-5 - |(-3) + (-8)| = -5 - |-11| = -5 - 11 = -16$$

$$6 + |-9| = 6 + 9 = 15$$

$$6 - |-9| = 6 - 9 = -3$$

$$-|8 - 19 + 3| = -|-8| = -8$$

$$\left|\frac{12}{-25}\right| = \left|\frac{12}{25}\right| = \frac{12}{25}$$

$$-\left|-\frac{12}{25}\right| = -\frac{12}{25}$$

Exercise 1--Computations Involving Absolute Values

Purpose of the Exercise

- To practice arithmetic involving absolute values
- To discover rules of absolute values

Do the computations indicated in each of the following. Show your steps where appropriate.

1. $|-17| =$ ______________________
2. $|2789.34| =$ ______________________
3. $|9| - 8 =$ ______________________
4. $|9 - 8| =$ ______________________
5. $|4 - 11| =$ ______________________
6. $|4| - |11| =$ ______________________
7. $-5|7| =$ ______________________
8. $-5|-7| =$ ______________________
9. $13 - |-21| =$ ______________________
10. $13 - (-21) =$ ______________________
11. $\left|\frac{-8}{-5}\right| =$ ______________________
12. $-\left|\frac{8}{-5}\right| =$ ______________________

In the following exercises, you are to observe when patterns appear to be emerging.

13. $|(4)(5)| =$ __________ $|4|\,|5| =$ __________ True False $|(4)(5)| = |4|\,|5|$

14. $|(-6)(3)| =$ __________ $|-6|\,|3| =$ __________ True False $|(-6)(3)| = |-6|\,|3|$

15. $|(-9)(-2)| =$ __________ $|-9|\,|-2| =$ __________ True False $|(-9)(-2)| = |-9|\,|-2|$

16. From your observations in Problems 13 - 15, discuss the meaning of the following formula.
 $|xy| = |x|\,|y|$

Do each of the following computations to observe a pattern.

17. $|4 + 5| =$ __________ $|4| + |5| =$ __________ True False $|4 + 5| = |4| + |5|$

18. $|4 - 5| =$ __________ $|4| - |5| =$ __________ True False $|4 - 5| = |4| - |5|$

19. $|8 + (-3)| =$ __________ $|8| + |-3| =$ __________ True False $|8 + (-3)| = |8| + |-3|$

20. From your observations in Problems 17 - 19, discuss why the following formulas are false.
 $|x + y| = |x| + |y|$ $|x - y| = |x| - |y|$

21. From your observations in Problems 17 - 19, discuss why the following formula is true.
 $|x + y| \leq |x| + |y|$

Rules of Absolute Values

$|xy| = |x|\,|y|$	The product rule for absolute values.
$\left|\frac{x}{y}\right| = \frac{|x|}{|y|}$	The quotient rule for absolute values.
$|x + y| \leq |x| + |y|$	The summation rule for absolute values.

In previous sections we covered three different models for arithmetic involving positive and negative numbers. In this section, we will summarize the rules for signed numbers that we have discovered.

Rules for Addition

To add **two positive numbers**, just add the numbers. The sum is positive.

$7 + 9 = 16$

To add **two negative numbers**, add the absolute values of the two numbers. The sum is negative.

Add: $-7 + (-9)$

$$|-7| + |-9| = 7 + 9$$
$$= 16$$
$$-7 + (-9) = -16$$

When adding a **positive number and a negative number**, subtract the number with the smaller absolute value from the number with the larger absolute value and keep the sign of the number with the larger absolute value.

Add: $-9 + 15$

$$|15| - |-9| = 15 - 9$$
$$= 6$$ 15 has the larger absolute value so the sum is positive.
$$-9 + 15 = 6$$

Add: $9 + (-15)$

$$|-15| - |9| = 15 - 9$$
$$= 6$$ -15 has the larger absolute value so the sum is negative.
$$9 + (-15) = -6$$

Rules for Subtraction

Rewrite subtraction as addition and follow the rules for addition.

$$-9 - 20 = -9 + (-20) = -29$$
$$17 - (-8) = 17 + 8 = 25$$

Rules for Multiplication

Multiply the absolute values of the number and then determine whether the product is positive or negative.

A **positive number times a positive number** is a **positive** number.

$$(5)(9) = 45$$

The product of a **positive number and negative number** is negative.

$$(5)(-9) = -45$$
$$(-9)(5) = -45$$

A **negative number times a negative number** is a **positive** number.

$$-9(-5) = 45$$

Rules for Division

Divide the absolute value of the dividend by the absolute value of the quotient and then determine whether the answer is positive or negative.

A **positive number divided by a positive number** is **positive**.

$$\frac{15}{3} = 5$$

A **negative number divided by a positive number** is **negative.**

$$\frac{-15}{3} = -5$$

A **positive number divided by a negative number is negative.**

$$\frac{21}{-7} = -3$$

A **negative number divided by a negative number** is **positive.**

$$\frac{-21}{-7} = 3$$

Example 4: Use your calculator to compute (-357.8) ÷ (47.8) correct to 3 decimal places.

A negative number divided by a positive number is negative, so we will input positive numbers and then put in the negative sign at the end.

To use a calculator to find a quotient, input the dividend first, then the "÷", then the divisor, and finally the "=."

[357.8] [÷] [47.8] [=] The calculator displays 7.485355649.

Round to 3 decimal places and put a negative sign in front of the rounded quotient.

(-357.8)÷(47.8) ≈ - 7.485

Another way to approach the same problem is to input the negative dividend along with the string operations.

On scientific calculators that display one number at a time:

[357.8] [+/-] [÷] [47.8] [=] The calculator displays - 7.485355649.

On scientific or graphics calculators that display the entire problem on the screen:

[(-)] [357.8] [÷] [47.8] [=] The calculator displays - 7.485355649.

However you use the calculator, you must be careful with signs and input the dividend first and divisor second. Even if you use string operations to key in the negative signs "in line," it is a good idea to determine ahead of time whether the answer is positive or negative and estimate what size answer is reasonable.

Since division can be particularly confusing, we will work with signs regarding division in the exercise set.

Exercise 2--Division and Signs

Purpose of the Exercise
- To drill division and division terminology
- To drill the rules of signs pertaining to division

Materials Needed
- Calculator

In each of the following, record the divisor and the dividend and determine if the quotient is positive or negative. Use your calculator to compute the quotient and round off to 3 decimal places. The first one is done as an example.

1. 149÷75   divisor = __74__   dividend = __149__   (pos) neg   quotient __1.987__
2. 68/97   divisor = ______   dividend = ______   pos neg   quotient ________
3. $\frac{-987}{765}$ divisor = ______   dividend = ______   pos neg   quotient ________
4. $\frac{80}{17}$ divisor = ______   dividend = ______   pos neg   quotient ________
5. $2710\overline{)387.5}$ divisor = ______   dividend = ______   pos neg   quotient ________

6. $83.4\overline{)4789}$ divisor = ______ dividend = _______ pos neg quotient _________

7. (-187)÷467 divisor = ______ dividend = _______ pos neg quotient _________

8. 13.09÷(-35) divisor = ______ dividend = _______ pos neg quotient _________

9. $\frac{178}{-38}$ divisor = ______ dividend = _______ pos neg quotient _________

10. $17\overline{)467}$ divisor = ______ dividend = _______ pos neg quotient _________

11. $\frac{-29}{-35}$ divisor = ______ dividend = _______ pos neg quotient _________

12. 376÷(-25) divisor = ______ dividend = _______ pos neg quotient _________

13. (-800)÷(-80) divisor = ______ dividend = _______ pos neg quotient _________

14. $-25\overline{)-400}$ divisor = ______ dividend = _______ pos neg quotient _________

15. $25\overline{)-400}$ divisor = ______ dividend = _______ pos neg quotient _________

16. 88/(-11) divisor = ______ dividend = _______ pos neg quotient _________

17. (-27)÷(3) divisor = ______ dividend = _______ pos neg quotient _________

Exercise 3--Parentheses and the Rules of Signs

Purpose of the Exercise

- To practice rules of signs
- To practice doing arithmetic with a calculator

Materials Needed

- Calculator

In each of the following, indicate which arithmetic operation is to be done: addition (+), subtraction (-), multiplication (x), division (x). Then do the arithmetic. Use a calculator as needed and round off where appropriate to 3 decimal places. The first one is done as an example.

Arithmetic Expression	Operation	Computation and Answer
1. (-5) - (-19)	-/+	(-5) + 19 = **14**
2. (-5) (-19)		
3. (-5) + (-19)		
4. 27÷ 3		
5. 27 + 3		
6. $\frac{17}{5}$		

7. 85 - 30		
8. (85)(-30)		
9. (85) - (30)		
10. 40 + (-1.4)		
11. 40 ÷ (-1.4)		
12. -127 ÷ (-2)		
13. 321.56(-0.345)		
14. 17.2 - (-27.96)		
15. -28 - (-15.8)		
16. (-28)(765)		
17. -478÷(-18)		
18. (-478) + (-18)		
19. -1.4(82)		

In each of the following, determine whether the arithmetic operation is addition, subtraction, multiplication, or division. Since subtraction can be treated as addition, either addition or subtraction could be correct in some problems. Decide whether the numerical answer will be positive or negative and then calculate the answer using a calculator if needed. The first one is done as an example.

Problem	Operation	Positive or Negative?	Answer
20. (-537.76)(12.1)	x	neg.	-6506.896
21. -378 + 278.97			
22. -536.9 + (278.97)			
23. -302.4(-100)			
24. $\frac{-90}{-22}$			
25. 14(-15.9)			
26. 14 - 15.9			
27. 14 - (-15.9)			
28. $\frac{-90}{22}$			

29. 907.890 - 567			
30. 86 ÷ (-10)			
31. -87.5 - (- 280)			
32. 10 - 22			
33. 10(-22)			
34. (-96) - (-28)			
35. -9876.7 ÷ 100			
36. -299.3(14)			
37. (-18.278)(-22)			

In each of the following problems, determine whether the answer is positive or negative, and find the answer.

	Positive or Negative?	Answer
38. 7(- 8) =	________	________
39. - 7(- 8) =	________	________
40. - (7)(- 8) =	________	________
41. - (- 7)(- 8) =	________	________
42. (- 7)(8) =	________	________
43. -8(-7) =	________	________
44. - (- 7)(8) =	________	________
45. 7 - 8 =	________	________
46. - 7 - 8 =	________	________
47. - 7 - (- 8) =	________	________
48. - 7 ÷ (- 8) =	________	________
49. - 7 + (- 8) =	________	________
50. 8 - (-7) =	________	________
51. 8 - 7 =	________	________
52. 22/11 =	________	________
53. (-22)/11 =	________	________
54. - (-22)/11 =	________	________
55. -22/(-11) =	________	________
56. 29/(-12) ≈	________	________

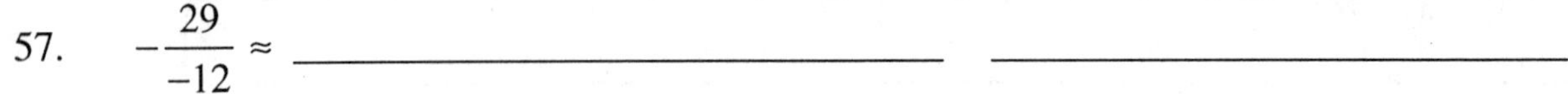

57. $-\frac{29}{-12} \approx$ ______________________ ______________________

58. -29(-12) = ______________________ ______________________

59. -29/(-12) ≈ ______________________ ______________________

In each of the following state whether the answer will be positive, negative or whether there is not enough information to tell.

60. A positive number times a negative number ______________________
61. A positive number times a positive number ______________________
62. A negative number times a negative number ______________________
63. A negative number times a positive number ______________________
64. A negative number plus a negative number ______________________
65. A negative number minus a negative number ______________________
66. The opposite of (a positive number times a negative number) ______________________
67. The opposite of (a positive number times a positive number) ______________________
68. The opposite of (a negative number times a negative number) ______________________
69. The opposite of (a negative number times a positive number) ______________________
70. The opposite of a negative number ______________________
71. The quotient of a positive number divided by a negative number ______________________
72. The quotient of two negative numbers ______________________
73. Give an example of an arithmetic problem where a positive number plus a negative number is positive. ______________________
74. Give an example of an arithmetic problem where a positive number plus a negative number is negative. ______________________

We have used the minus sign in three different ways. Let's summarize those three different meanings.

1. The minus sign is used to indicate that a number is **negative**, as in "-5," which is read as "negative 5."

2. The minus sign is used to indicate the **opposite** of a particular number, as in "-(-5)." In this case, the first minus sign means opposite, the second one means negative. The opposite of -5 is 5 so -(-5) = 5.
   A minus sign in front of a variable means the opposite of the value of the variable.
   If T = -6, then -T is the **opposite of -6** and -T = 6
   If T = -π, then -T is the **opposite of -π** and -T =
   If T = 467.3, then -T is the **opposite of 467.3** and -T = -467.3

3. The minus sign is used to indicate **subtraction**.
   7 - 5 = 2, or subtract 5 from 7.

In the past, we defined a variable as a letter that represents a number. We can also view a variable as a letter that is a name of a storage location that holds a number. This is what a variable is on a computer. Looking at a variable from this perspective can help us understand and remember where to place parentheses when substituting numbers into algebraic expressions.

To assign a value to a variable, we use a statement like: X = 3

This statement puts the number 3 into the storage location called X.

When using variables and arithmetic, it is important to understand the notation. Let's consider the following statements and what they mean in the context of a variable as a storage location.

T - X	represents subtraction. Subtract the contents of X from the contents of T. (This is the same as saying subtract the value of X from the value of T.)
T + M	represents addition. Add the contents of T and M (or add the values of T and M).
(R)(S)	represents multiplication. Multiply the contents of R by the contents of S.
RS	represents multiplication. Multiply the contents of R by the contents of S.
H/K	represents a division. Divide the contents of H by the contents of K.
2y	represents multiplication. Multiply the contents of y by 2.
3x - 4y	involves multiplication and subtraction. Multiply the contents of x by 3; multiply the contents of y by 4; then subtract 4y from 3x.
5 - (-x)	represents subtraction. Subtract the <u>opposite</u> of the contents of x from 5.

Visually, if we picture the storage location as a barrel, it can help us in substituting correctly into a formula using variables.

Example 5: Evaluate 7t - x when x = -4 and t = 13.

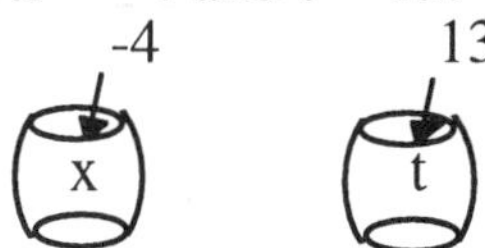

Put the -4 into the x-location and 13 into the t-location. Now when evaluating a formula, put the barrel, x, for the x's and the barrel called t for the t's.

7t - x becomes 7(t) - (x) = 7(13) - (-4) = 7(13) + 4 = 91 + 4 = 95

Take off the top and the bottom of the barrel to reveal the contents. This process helps visually to get the parentheses in the correct place.

Example 6: Evaluate -tm + 3t when t = -5 and m = 6.

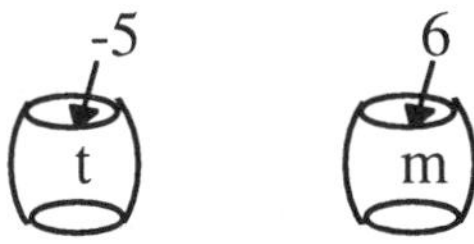

Put -5 into the t-barrel and 6 into the m-barrel.

-tm + 3t becomes - (t)(m) + 3(t) = -(-5)(6) + 3(-5)

= (5) (6) + (-15) = 30 - 15 = 15

Take off the top and bottom of the barrel to reveal the contents and then do the arithmetic.

Exercise 4--Evaluating Mathematical Expressions with Variables.

Purpose of the Exercise

- To practice substitution of numbers for variables
- To recognize the symbols indicating the operations of arithmetic

Materials Needed

- Calculator

In each of the following, state whether the operation indicated is addition, subtraction, multiplication or division.

1. (257.8)(-87.5) ________ 2. 395a ____________ 3. a - 4 ____________

4. x/38.5 ______________ 5. (K) - (-R) _________ 6. K)(-R) _________

7. d __________________ 8. x-y ______________ 9. 87.5M __________

10. xy _______________ 11. 8000y ___________ 12. T÷r ____________

Compute each of the following if M = -4 and K = 3. The first two are done as examples.

13. (12)(M) = (12)(-4)=- 48 14. M/2 = (-4)/2 = -2 15. 5K = __________

16. 5M = _____________ 17. (K) - (M) = ______ 18. (K)(M) = _______

19. K - M = ___________ 20. K + M = ________ 21. K - (-M) = ______

22. -7M = ____________ 23. -7K = __________ 24. $\frac{-K}{M}$ = ___________

25. 4 - M = ___________ 26. (4) - (M) = _______ 27. (4)(-M) = ________

28. (K) - (-8) = ________ 29. (K)(-8) = _________ 30. (K) + (-8) = _______

31. M÷K =____________ 32. 20÷M = __________ 33. 20 - M = _________

Compute each of the following using a calculator as needed. In the problems involving , use the decimal approximation for given by your calculator. Round the final answer correct to 2 decimal places. Let x = 85.3, y = -7.2, r = 16.8, and d = 33.6.

34. xy = ______________________ 35. 2πr = ______________________

36. πd = ______________________ 37. x/y = ______________________

38. 3.5y = _____________________ 39. 2yr = ______________________

40. r - d = _____________________ 41. r - (-d) = ___________________

42. x - y = _____________________ 43. -xy = ______________________

44. x - (-y) = ___________________ 45. r(-y) = _____________________

Section 2-6 *The Commutative and Associative Laws*

To "commute" means to "move or travel from one place to another." A commuter train transports people. In mathematics, the commutative laws of arithmetic have to do with switching positions of numbers within a problem.

Example 1: 3 x 4 literally means 3 four's.

O O O O
O O O O 3 x 4 = 12
O O O O

4 x 3 means 4 three's.

O O O
O O O 4 x 3 = 12
O O O
O O O

It does not matter which number we write first, 4 x 3 = 3 x 4. We can switch the positions of the numbers and get the same answer.

A similar thing happens with addition. We represented multiplication with objects. We will represent addition with the number line.

Example 2:

3 + 5 = 8

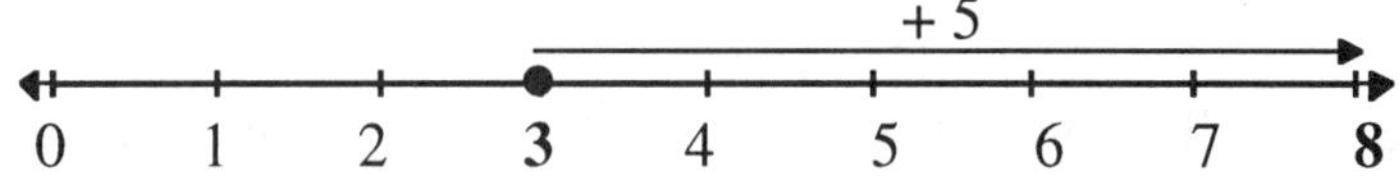

5 + 3 = 8

+ 3
0 1 2 3 4 **5** 6 7 **8**

3 + 5 = 5 + 3. Reversing the order of the terms gives the same sum.

Example 3:

(-3) + 4 = 1

+ 4
-3 -2 -1 0 **1** 2 3 4 5

4 + (-3) = 1

+ (-3)
-3 -2 -1 0 **1** 2 3 **4** 5

(-3) + 4 = 4 + (-3). Reversing the order of the terms gives the same sum.

Exercise 1--The Commutative Laws

Purpose of the Exercise

- To understand the commutative laws and determine whether they can be expanded to subtraction and division

In each of the following, illustrate the addition or subtraction on a number line, labeling an appropriate scale. Determine whether each pair has the same sum or difference.

1. (-5) + 4 = __________ 

   4 + (-5) = __________

   True False (-5) + 4 = 4 + (-5)
   True False The **sum** of -5 and 4 is the same computed in either order.

2. 5 - 4 = ____________

   4 - 5 = ___________

   True False 5 - 4 = 4 -5
   True False The **difference** of 5 and 4 is the same computed in either order.

3. 3 + 12 = ____________

   12 + 3 = ___________

   True False 3 + 12 = 12 + 3
   True False The **sum** of 12 and 3 is the same computed in either order.

4. 3 - 12 = ____________

   12 - 3 = ___________

   True False 3 - 12 = 12 - 3
   True False The **difference** of 12 and 3 is the same computed in either order.

5. From your observations, does addition appear to be commutative? (In other words, can you add two terms in either order and get the same sum?) ____________

6. From your observations, does subtraction appear to be commutative? (In other words, can you subtract terms in either order and get the same difference?) ______

Compute the following products and quotients and answer the true/false questions.

7. (5)(12) = ____________ (12)(5) = ____________________

   True False (5)(12) = (5)(12)
   True False The **product** of 5 and 12 is the same computed in either order.

8. (6)(13) = ____________ (13)(6) = ____________________

   True False (6)(13) = (13)(6)
   True False The **product** of 6 and 13 is the same computed in either order.

9. 12 ÷ 4 = ____________ 4 ÷ 12 = ____________________

   True False 12 ÷ 4 = 4 ÷ 12
   True False Reversing the divisor and dividend gives the same **quotient**.

10. $8 \div 5 =$ ____________ $5 \div 8 =$ ____________________

True False $8 \div 5 = 5 \div 8$
True False Reversing the divisor and dividend gives the same **quotient**.

11. From your observations, does multiplication appear to be commutative? (In other words, does reversing the factors give you the same product?) ________________

12. From your observations, does division appear to be commutative? (In other words, does reversing the divisor and dividend give you the same quotient?) ________

As you observed in the problems above, it appears that we can add two numbers in either order and get the same sum and we can multiply two numbers in either order and get the same product. The same is not true for division and subtraction. The commutative laws for addition and multiplication are stated symbolically below. "ab" means to multiply a times b.

Commutative Law for Addition
$a + b = b + a$ for all possible values for a and b.
Commutative Law for Multiplication
$ab = ba$ for all possible values for a and b.

Although subtraction is not commutative, we can restate subtraction as addition and then apply the commutative law for addition.

Example 4: 17 - 25 is the same as 17 + (-25), so although 17 - 25 is not the same as 25 - 17, we can write the difference as a sum and then change the order and add.
$17 + (-25) = (-25) + 17$

When we add three numbers, we can add any two of the numbers and then add the sum of these two numbers to the third number. A similar process is true for multiplication. As it turns out, it does not matter which two numbers we add or multiply first. In the examples below, parentheses are used to show which multiplication or addition to do first.

Example 5: Multiply 7 x 5 x 9.
First grouping: (7 x 5) x 9 = 35 x 9 = 315
Second grouping: 7 x (5 x 9) = 7 x 45 = 315

Example 6: Add (-5) + 8 + 9.
First grouping: (- 5 + 8) + 9 = 3 + 9 = 12
Second grouping: (-5) + (8 + 9) = -5 + 17 = 12

In Examples 5 and 6, both groupings give the same answer. This is always true for multiplication of three numbers or addition of three numbers. The properties of arithmetic demonstrated in the examples are the associative laws for addition and multiplication and are stated symbolically below.

Associative Law for Addition
$(a + b) + c = a + (b + c)$ for all possible values for a, b, and c.
Associative Law for Multiplication
$(a \cdot b) \cdot c = a \cdot (b \cdot c)$ for all possible values for a, b, and c.

When we apply both the associative law and the commutative law for multiplication and expand the laws to the product of more than three numbers, we can change the order and regroup the numbers as we wish. The same is true for sums of more than three numbers. When we use the associative and commutative laws over and over in the same problem, we can reorder and group the terms any way we choose.

Notation

When working with positive and negative numbers in multiplication, we use parentheses for clarity to show multiplication. For further clarity, we use brackets,[ ] or { }, to indicate groupings. Mathematically, brackets mean the same thing as parentheses. Using both symbols helps us to distinguish groupings.

The following examples show that any way we change the order and groupings within multiplication, the final product is the same.

Example 7: Multiply 9 and 7 and 3 using different groupings and different orders.

1. $(9 \cdot 7)3 = (63)(3) = 189$
2. $9(7 \cdot 3) = (9)(21) = 189$
3. $(9 \cdot 3)(7) = (27)(7) = 189$
4. $3(7 \cdot 9) = (3)(63) = 189$

Example 8: Multiply -5 and 7 and -13 using different groupings and different orders.

1. $[(-5)(7)](-13) = (-35)(-13) = 455$
2. $(-5)[(7)(-13)] = (-5)[-91] = 455$
3. $[(-13)(-5)](7) = (65)(7) = 455$
4. $(7)[(-5)(-13)] = (7)(65) = 455$

Example 9: Add -2 and 5 and -7 and -9 using different groupings and different orders.

1. $[(-2) + 5]+[(-7)+(-9)] = (3)+(-16) = -13$
2. $\{[(-2) +5]+[-7]\}+(-9) = \{3 + (-7)\} +(-9) = -4 + (-9) = -13$
3. $[(5)+(-7)]+[(-9)+(-2)] = [-2]+[-11] = -13$
4. $[(-9)+(5)]+[(-2)+(-7)] = [-4] + [-9] = -13$

Exercise 2--Associative and Commutative Laws for Multiplication

Purpose of the Exercise

- To practice order and grouping of factors
- To discover rules of signs for multiple factors

Materials Needed

- Calculator

For each of the following products, show three different ways to group and order the numbers to find the product. Show the arithmetic for each grouping or ordering. If you get different answers for different groupings or orders, check your work. In the last two columns, list how many factors in the product are negative and whether the answer is positive or negative.

The Numbers and Setup	Show the Computations = Answer	No. of Neg. Factors	Pos/Neg Answer
1. (-3)(-5)(-9)(-21)(8) 1. [(-3)(-5)][(-9)(-21)](8) 2. 3.	[15][189](8) = {[15][189]}(8) = {2835}(8) = 22,680	4	+
2. (-1.9)(-3.5)(-12) 1. 2. 3.			
3. (-8)(-1.6)(-9.3) 1. 2. 3.			
4. (22)(-13.4)(17)(3)(2) 1. 2. 3.			

For each of the following, determine whether the answer will be positive or negative and compute the product, arranging the order and groups any way you choose.

	Positive? Negative?	Answer
5. (8)(-7)(-1) =	______	______
6. (-8)(-7)(-1) =	______	______
7. (-3)(-2)(-5)(-6) =	______	______
8. (-3)(2)(-5)(-6) =	______	______
9. (-3)(-2)(5)(-6) =	______	______
10. (-3)(-2)(-5)(-6) =	______	______
11. (-3)(2)(5)(-6) =	______	______
12. (3)(-2)(-5)(6) =	______	______
13. (3)(-2)(5)(-6) =	______	______
14. (3)(2)(5)(-6) =	______	______

In each of the following determine whether the product would be positive or negative.

15. The product of 5 negative numbers	Pos	Neg
16. The product of 4 negative numbers and 3 positive numbers	Pos	Neg
17. The product of 3 negative numbers and 8 positive numbers	Pos	Neg
18. The product of an odd number of negative numbers	Pos	Neg
19. The product of an even number of negative numbers	Pos	Neg

For each of the following rewrite the subtractions as addition and show three different ways to group and order the terms to find the sum. Show the steps to getting the answer.

The Problems Rewritten and Showing the Groupings and Order.	Steps of Work and Answer
20. -3 - 5 - 21 + 8 Rewritten as addition: 1. 2. 3.	
21. -1.9 + 3.5 - 12 Rewritten as addition: 1. 2. 3.	

22. 8 + 1.6 - 9.3 Rewritten as addition: 1. 2. 3.	
23. 11.3 + 7.1 + 9.1 + 1 1. 2. 3.	
24. -5 - 7 -11 - 2 Rewritten as addition: 1. 2. 3.	

In each of the following determine whether the sum would be positive or negative or whether there is not enough information to tell.

25. The sum of 3 negative numbers and 1 positive number	Pos	Neg	Can't tell
26. The sum of 5 negative numbers	Pos	Neg	Can't tell
27. The sum of 1 negative number and 2 positive numbers	Pos	Neg	Can't tell
28. The sum of 14 positive numbers	Pos	Neg	Can't tell

The Commutative and Associative Laws and Combining Like Terms

In Chapter 1, we learned that to combine like terms we first identify the like terms and then add or subtract (as indicated) the coefficients of the like terms. The procedure is still the same, but now that we know how to add and subtract positive and negative numbers in all possible combinations, we will practice combining like terms again.

Example 10: 2x - 7 - 9x + 15x - 3 can be restated as addition as 2x + (-7) + (-9x) + 15x + (-3). Now, we can reorder and regroup the terms however we want. In this case, we will put the like terms together within one set of brackets.

First: Put the x-terms in one set of brackets and the constants in another.
2x + (-7) + (-9x) + 15x + (-3) = [2x + (-9x) + 15x] + [(-7) + (-3)]
Second: Rearrange the x-terms, putting the positive terms together. Add the constants.
= [2x + 15x + (-9x)] + [-10]
Third: Add 2x + 15x by adding the coefficients, 2 and 15.
= [17x + (-9x)] + [-10]
Fourth: Add 17x and -9x by adding the coefficients, 17 + (-9) = 8. Rewrite +[-10] as -10.
= 8x - 10

Exercise 3--Combining Like Terms

Purpose of the Exercise

- To apply the commutative and associative laws for combining like terms when some terms are positive and some are negative

For each of the numeric expressions below, first perform the computations in the order written. Then rewrite and do the computations, grouping the negative terms together and the positive terms together.

For each of the expressions that involve variables, rewrite the subtracted terms as negative terms added; then group the like terms together before simplifying the expression by combining like terms. Show the computations in the last column. The first two are done as examples.

1. -5 - 11 + 19 - 3 + 22 Regrouped: [(-5) + (-11) + (-3)] + [19 + 22]	= -16 + 19 - 3 + 22 = 3 - 3 + 22 = 0 + 22 = 22 = [-16 + (-3)] + [41] = [-19] + [41] = 22
2. -14x + 14 - 3 - 5x + x - 4 Rewritten: (-14x) + 14 + (-3) + (-5x) + x + (-4) Regrouped: [(-14x) + (-5x) + x] + [14 + (-3) + (-4)]	= [(-19x) + x] + {14 + [(-3) + (-4)] = [-18x] + {14 + [-7]} = -18x + 7
3. 22 - 19 - 18 + 11 - 21 Regrouped:	

4. -9 - 9 + 27 + 3 - 18 Regrouped:	
5. -5x - 12y + 22x + y Rewritten: Regrouped:	
6. 17a - 17 + 11a Rewritten: Regrouped:	
7. 98 - 92 - 15 - 11 - 20 Rewritten: Regrouped:	
8. 25b - 13b - 13b + b + 1 Rewritten: Regrouped:	
9. -9t + 9 + 11 - 13t + 13 Rewritten: Regrouped:	
10. -9 + 9 + 11 - 13 - 13 Rewritten: Regrouped:	

Section 2-7 *Exponents and the Order of Operations*

In Chapter 1, we worked with the powers of 10 when we wrote numbers in scientific notation. In this section, we will expand our definition of powers or exponents of base numbers other than 10. We will also learn the "order of operations," or the priorities indicating which operations are done first, which are done second, and so on.

Symbolism and Vocabulary

In 3^4, the number 3 is called the **base** and the number 4 is called the **power**, or **exponent**, of 3. When the base is negative, we use parentheses to include the sign as part of the base. For example, in the number $(-8)^5$, the base is -8 and the power or exponent is 5.

Exercise 1--Exponents

Purpose of the Exercise

- To understand the symbolism and terminology of exponents

In each of the following problems, state what is the base and the power of the number.

1. 7^2 base = _____ power = _____ 2. 85^7 base = _____ power = _____

3. 29^5 base = _____ power = _____ 4. 17^2 base = _____ power = _____

5. $(-7)^2$ base = _____ power = _____ 6. $(-2)^4$ base = _____ power = _____

7. $(1.6)^3$ base = _____ power = _____ 8. $(-0.7)^4$ base = _____ power = _____

9. $(-0.12)^2$ base = _____ power = _____ 10. 756^5 base = _____ power = _____

11. 0^{15} base = _____ power = _____ 12. 1^{215} base = _____ power = _____

In each of the following, write the number in exponential form. The first two are done as examples.

13. The base is 879 and the exponent is 8. The number written in exponential form is $\underline{879^8}$.
14. The base is -91 and the power is 4. The number written in exponential form is $\underline{(-91)^4}$.
15. The base is 18 and the power is 3. The number written in exponential form is ______
16. The base is -8 and the exponent is 3. The number written in exponential form is ____
17. The base is 457 and the power is 9. The number written in exponential form is _____
18. The base is 1 and the power is 6. The number written in exponential form is _______
19. The base is -6 and the exponent is 7. The number written in exponential form is _____
20. The base is -1 and the power is 4. The number written in exponential form is ______
21. The base is 5,789.3 and the power is 4. The number in exponential form is _______
22. The base is 17.4 and the power is 5. The number in exponential form is _________
23. The base is -0.27 and the power is 2. The number in exponential form is _________

24. The base is 0.5 and the exponent is 3. The number in exponential form is ________

25. The base is -11 and the power is 6. The number in exponential form is ________

26. The base is -3.45 and the exponent is 7. The number in exponential form is ________

27. The base is 0 and the power is 85. The number in exponential form is ________

4^3 is a shortcut way of writing $4 \cdot 4 \cdot 4$. In general, the power or the exponent is a count of how many times the base is multiplied by itself.

Example 1: Positive and negative bases.

$4^3 = 4 \cdot 4 \cdot 4 = 16 \cdot 4 = 64$
$(-4)^3 = (-4)(-4)(-4) = (16)(-4) = -64$

$4^2 = 4 \cdot 4 = 16$
$(-4)^2 = (-4)(-4) = 16$

Example 2: Negative bases.

$(-5)^2 = (-5)(-5) = 25$
$(-5)^3 = (-5)(-5)(-5) = -125$
$(-5)^4 = (-5)(-5)(-5)(-5) = 625$
$(-5)^5 = (-5)(-5)(-5)(-5)(-5) = -3125$

Example 3: Negative bases.

$(-3)^2 = (-3)(-3) = 9$
$(-3)^3 = (-3)(-3)(-3) = -27$
$(-3)^4 = (-3)(-3)(-3)(-3) = 81$
$(-3)^5 = (-3)(-3)(-3)(-3)(-3) = -243$

Example 4: Negative bases.

$(-2)^2 = (-2)(-2) = 4$
$(-2)^3 = (-2)(-2)(-2) = -8$
$(-2)^4 = (-2)(-2)(-2)(-2) = 16$
$(-2)^5 = (-2)(-2)(-2)(-2)(-2) = -32$

Rules of Signs and Exponents

- When the base is positive, the base raised to any power is positive.
- When the **base** of an exponent is **negative**
  and the **power is an even number**, the **result is positive**.
  and the **power is an odd number**, the **result is negative**.

Exercise 2--Exponents

Purpose of the Exercise

- To practice simplifying exponential expressions

In each of the following, write the exponential expression as a product of the bases. Without a calculator, compute the product.

Problem	Multiplication Form	Answer	Problem	Multiplication Form	Answer
1. 5^2			2. $(-5)^2$		
3. 10^3			4. 12^2		
5. $(-4)^1$			6. $(-3)^3$		
7. 2^4			8. 2^5		
9. $(-2)^4$			10. $(-2)^5$		
11. 1^2			12. 1^5		
13. 0^3			14. 0^{12}		
15. $(-1)^2$			16. $(-1)^3$		
17. $(-1)^5$			18. $(-1)^6$		
19. 10^3			20. 10^4		
21. 10^5			22. 10^6		
23. $(-4)^2$			24. $(-10)^3$		

In each of the following determine whether the result is positive or negative.

25. $(-23.4)^5$	positive	negative	26. $(-287)^{10}$	positive	negative
27. 0.578^{25}	positive	negative	28. $(-0.78)^{12}$	positive	negative
29. $(-719)^{15}$	positive	negative	30. $(-27)^9$	positive	negative
31. $(-14)^5$	positive	negative	32. $(-14)^8$	positive	negative
33. 1^{45}	positive	negative	34. $(-1)^{45}$	positive	negative
35. $(8)^3$	positive	negative	36. $(-73)^6$	positive	negative
37. $(-10)^7$	positive	negative	38. $(-10)^{10}$	positive	negative
39. $(10)^3$	positive	negative	40. $(-1.78)^2$	positive	negative
41. $(-1.78)^3$	positive	negative	42. $(-1.78)^{10}$	positive	negative
43. $(1.78)^3$	positive	negative	44. $(-100)^2$	positive	negative

Symbolism

In each of the examples above, when the base was negative, the base was written inside of parentheses. There is a difference between $(-5)^2$ and -5^2.

$(-5)^2 = (-5)(-5) = 25$ The base is -5 and the power is 2.

$-5^2 = -(5)(5) = -25$ The base is 5 and the power is 2.
The minus sign means **the opposite** 5^2.

Example 5: Symbolism of parentheses.

$(-4)^3 = (-4)(-4)(-4) = -64$ The product of 3 negative numbers is negative.
$-4^3 = -(4)(4)(4) = -64$ The opposite of 4^3 is negative.

$(-3)^4 = (-3)(-3)(-3)(-3) = 81$ The product of 4 negative numbers is positive.
$-3^4 = -(3)(3)(3)(3) = -81$ The opposite of 3^4 is negative.

Example 6: Using a scientific calculator that displays one number at a time to compute a base raised to a power.

Problem	Key Strokes	Answer
$(-5)^2$	[5] [+/-] [x^2] or [5] [+/-] [x^y] [2] [=]	25
-5^2	[5] [x^2] [+/-]	-25
$(-4)^3$	[4] [+/-] [x^y] [3] [=]	-64
10.2^3	[10.2] [x^y] [3] [=]	1061.208
$(-10.2)^3$	[10.2] [+/-] [x^y] [3] [=]	-1061.208
-10.2^3	[10.2] [x^y] [3] [=] [+/-]	-1061.208
-2^4	[2] [x^y] [4] [=] [+/-]	-16
$(-2)^4$	[2] [+/-] [x^y] [4] [=]	16

If you have a graphics or scientific calculator that displays the whole expression at the same time, the key strokes are closer to how we write the expression.

Problem	Key Strokes	Answer
$(-5)^2$	[(] [(-)] [5] [)] [x^2] [=]	25
-5^2	[(-)] [5] [x^2]	-25
$(-4)^3$	[(] [(-)] [4] [)] [^] [3] [=]	-64
10.2^3	[10.2] [^] [3] [=]	1061.208
$(-10.2)^3$	[(] [(-)] [10.2] [)] [^] [3] [=]	-1061.208
-10.2^3	[(-)] [10.2] [^] [3] [=]	-1061.208
-2^4	[(-)] [2] [^] [4] [=]	-16
$(-2)^4$	[(] [(-)] [2] [)] [=]	16

If you do not get the same results as above using your calculator, refer to the manual or get help from the instructor to learn how your calculator works. **Calculators do not all work the same way.**

Vocabulary

Two numbers or variables multiplied are called **factors**.
Two numbers or variables added are called **terms**.
Algebraic expressions are any combination of variables with arithmetic operations.

Example 7: Factors and terms.

In the product (-7.3)(-9.29) there are **two factors**. -7.3 is a factor and -9.29 is a factor.
In the sum (-7.3) + (-9.29) there are **two terms**. -7.3 is a term and -9.29 is a term.
In the expression 4x - 4y there are **two terms**. One term is 4x; the other term is 4y.
In the expression 5y, there are **two factors**. 5 is a factor and y is a factor.
In the expression 4xy, there are **three factors**, 4, x, and y.

Exercise 3--Terms, Factors, and Exponents

Purpose of the Exercise

- To better understand the symbols of exponential expressions
- To recognize terms and factors

In each of the following, determine whether the negative sign is part of the base and whether the answer is positive or negative. **Do not use a calculator.**

Problem	Is the negative sign part of the base?	Is the result positive or negative?
1. $(-14.2)^3$	yes no	positive negative
2. -22.4^2	yes no	positive negative
3. $-(78)^3$	yes no	positive negative
4. -89^5	yes no	positive negative
5. $(-17)^2$	yes no	positive negative
6. -17^2	yes no	positive negative
7. $(-17)^3$	yes no	positive negative
8. -17^3	yes no	positive negative
9. $-(45)^4$	yes no	positive negative
10. $(-45)^4$	yes no	positive negative
11. $(-467)^5$	yes no	positive negative
12. -467^5	yes no	positive negative
13. -467^2	yes no	positive negative

In each of the following, determine whether there are two terms or two factors.

14. (-987.2)(28)	terms/factors?	15. 3x - 5y	terms/factors?
16. xy	terms/factors?	17. (-987.2) - (28)	terms/factors?
18. 75.9 + x	terms/factors?	19. 87.9(-14)	terms/factors?
20. 84.2 - T	terms/factors?	21. 84.2x	terms/factors?
22. mt	terms/factors?	23. m - t	terms/factors?
24. (-45.3)t	terms/factors?	25. (-45.3) + t	terms/factors?

When more than one operation appears in the same expression, a set order of which to do first, second, third, and so on, has been adopted. This set hierarchy is called **order of operations**.

Order of Operations

The basic hierarchy is

All **powers** are computed first.
Multiplication and **division** are done next (from left to right in an expression).
Addition and **subtraction** are done last.

Note: Multiplication is done before division if multiplication comes first in the expression (or term.) Division is done before multiplication if division comes first in the expression (or term.)

With addition and subtraction, although we may work from left to right through a problem, the order can be rearranged using the commutative and associative laws for addition. (Remember to restate subtraction as addition before applying the associative and commutative laws for addition.)

Since addition and subtraction are done last in the hierarchy, each **term** can be computed as an isolated group.

To change the basic hierarchy:

Parentheses can be used to change the basic hierarchy. When there are arithmetic operations inside parentheses, those operations should be done first before performing operations not in parentheses.

When using fraction notation with the **numerator** (dividend) on top and the **denominator** (divisor) on the bottom, the hierarchy is altered. We do the operations in the numerator and denominator first and then the division last. There is an implied set of parentheses around the entire numerator and around the denominator. (We will talk more about the numerator and denominator when we cover fractions in Chapter 4.)

It is important to recognize when parentheses include arithmetic operations and when they do not. Remember, it is sometimes confusing to use a dot or times sign to indicate multiplication. When in doubt, we use parentheses to distinguish factors.

Example 8:

3(5 - 8)	First, 8 is subtracted from 5 inside of the parentheses. (The parentheses change the basic hierarchy. Subtraction is done first in this case.) 3 is multiplied by the difference of 8 and 5.
3(-3)	After doing the subtraction inside of the parentheses, we retain the parentheses to indicate that the 3 and the -3 are multiplied but there are no computations to be done within the parentheses.
-9	This is the result.

Example 9: There are implied parentheses around the entire numerator and entire denominator below. We finish the calculations in the numerator and denominator before doing division as the final step. There are two **terms** in the numerator and two terms in the denominator so we will do the computations in each term and then add or subtract the terms as indicated.

$\frac{5(7+13)-4^2}{5(-4)+4^2}$	First, in the numerator, add the 7 and 13 inside of the parentheses and square the 4. In the denominator, multiply 5(-4) and compute 4^2.
$\frac{5(20)-16}{-20+16}$	The parentheses are retained from doing the previous step and indicate that the 5 is multiplied by 20. We still have two terms in both the numerator and denominator
$\frac{100-16}{-4}$	Subtract 16 from 100.
$\frac{84}{-4}=-21$	The quotient is -21. The quotient is negative, because as we already learned from rules of signs, a positive number divided by a negative number is negative.

Exercise 4--Order of Operations

Purpose of the Exercise

- To associate order of operations with symbols of arithmetic

For each of the following, fill in numbers in the blanks below the expression to indicate which operations are done first, second, third, and so on. The first one is done as an example.

1. $5(4.7) - 9.5(-3.2) + 14 \div 3.6 \cdot 7.5$

   $\underline{1}$ $\underline{5}$ $\underline{2}$ $\underline{6}$ $\underline{3}$ $\underline{4}$

   There are no exponents or operations inside of parentheses. 5(4.7) and 9.5(-3.2) are both products. Steps 1, 2, 3, and 4 are the multiplication and division in the order they appear in the expression. Steps 5 and 6 then are the subtraction and addition.

2. $5(4.7) - 9.5(3.2) + 14 \div (3.6 \cdot 7.5)$

   ___ ___ ___ ___ ___ ___

3. $543.6 \cdot (-82.6) + 35.8 \div 17.3 - (14.57 \cdot 56.3 \cdot 27 - 19 \div 2 \cdot 3$

   ___ ___ ___ ___ ___ ___ ___ ___ ___

4. $(39.4 - 27.3) \cdot (24.6 - 17.3) - (6.3 \cdot 3.6)$

   ___ ___ ___ ___ ___

5. In the fraction, $\dfrac{3 \cdot 5 - 2 \cdot 4}{6 \cdot 2 + 2 \cdot 3}$, although there are no parentheses, division gets done last. Explain why this happens.

   ____

String Operations on a Calculator

When computing arithmetic, whether on paper or with a calculator, it is important to understand the **order of operations and the notations** involved. One way to do the arithmetic is a step-by-step approach rewriting the problem each time a computation is done. This is how we would do the problem by hand, but it also works with a calculator.

Example 10: Let's compute Problem 1 in the exercises above on a calculator. In general, rounding individual computations in a problem can cause large rounding errors in the final answer. Without knowing the kind of accuracy required, we have chosen to round the quotient below to four decimal places. This was an arbitrary decision.

5(4.7) - 9.5(-3.2) - 14 ÷ 3.6 • 7.5	Do the multiplication and division first.
23.5 - (- 30.4) - 3.8889 • 7.5	In the third term, there is both multiplication and division so we perform the left operation first. 14÷3.6 does not come out even, so we have rounded the answer.
23.5 + 30.4 - 29.16675	Multiplication in the third term is done next. Subtracting a negative number is the same as adding a positive number. (Subtracting - 30.4 is the same as adding 30.4.)
53.9 - 29.16675	Addition and subtraction are done last.
24.73325	This is the approximate answer. It is not exact because we rounded some numbers.

Scientific calculators and **graphics calculators** are designed to perform the order of operations in the set hierarchy of arithmetic. As a result, we can set up an expression to be performed in a string of operations on the calculator. An important thing to know about a scientific calculator is how to key in a negative number. The key, [-] , is reserved for subtraction. The subtraction key is <u>not</u> used to key in a negative number.

To key in a negative number,
 On a scientific calculator that displays one number at a time,
  first enter the number, then push the key [+/-] .
 On a scientific or graphics calculator that displays the entire problem on the screen,
  first push the [(-)] key for the negative and then enter the number.

Example 11: Evaluate the expression in Example 1 using string operations.

On a scientific calculator that displays one number at a time,

5(4.7) - 9.5(-3.2) - 14 ÷ 3.6 · 7.5

5 [x] 4.7 [-] 9.5 [x] 3.2 [+/-] [-] 14 [÷] 3.6 [x] 7.5 [=]

On a scientific or graphics calculator that displays the entire expression at once, the key strokes look just like the expression, except we have to distinguish between the minus (subtract) sign "−" and the negative sign, "(-)."

5(4.7) − 9.5(-3.2) − 14 ÷ 3.6 • 7.5

5 [x] 4.7 [-] 9.5 [(-)] 3.2 [-] 14 [÷] 3.6 [x] 7.5 [=]

Try the above key strokes. You should get 24.733333 . . . on your calculator. The ". . ." just indicates that different calculators will show a different number of 3's on the screen. At any place that this answer is terminated, the answer is an approximation. In real life, however, the approximation is no doubt accurate enough for the practical situation.

Be aware that some calculators round off the last digit and some do not--some just truncate, that is, drop the following digits without rounding.

The answer in Example 10 was 24.73325 but the answer to the same problem in Example 11 was 24.733333 . . . . These numbers are close but not exactly the same. The discrepancy occurred in the division step in Example 10 where the numbers did not divide "evenly" to a particular decimal place and we rounded the quotient. That round-off error carried through to the other calculations. Since the calculator retains more decimal places than we would write down in step-by-step processes, the answer obtained when we do the string operations is generally more accurate than when we perform the operations one step at a time and record rounded-off answers earlier in the problem.

Example 12: Calculate the following first one step at a time, recording each result. Then check the answer using string operations on the calculator.

$$\frac{-75(22.3)+3.4(-14.5)}{189-(-22.3)}$$

Compute the entire numerator and the entire denominator and then divide the numerator by the denominator. There are two terms in the numerator and two terms in the denominator. There are computations to do within the terms of the numerator, and these should be done first.

$$\frac{-1672.5+(-49.3)}{189+22.3}$$

Now we are ready to add terms in the numerator and in the denominator. In the numerator, -1672.5 + (-49.3) = -1721.8 In the denominator, just add the numbers.

$$\frac{-1721.8}{211.3} \quad - 8.148603881$$

Since the calculator has a limit of how many digits it can write in a number, this is an approximate answer. With the degree of accuracy we get from string operations, however, this is generally more accuracy than we need when using mathematics.

To compute the same expression with string operations, we place a set of brackets around the entire numerator and around the entire denominator so that the computations in both the numerator and denominator are done before performing the division indicated by the fraction bar.

$$\frac{[-75(22.3)+3.4(-14.5)]}{[189-(-22.3)]}$$

On a scientific calculator that displays one number at a time:

[(] 75 [+/-] [x] 22.3 [+] 3.4 [x] 14.5 [+/-] [)] [÷] [(] 189 [-] 22.3 [+/-] [)] [=]

On a scientific or graphics calculator that displays the entire expression at once:

[(] [(-)] 75 [x] 22.3 [+] 3.4 [x] [(-)] 14.5 [)] ÷ [(] 189 [-] [(-)] 22.3 [)] [=]

The answer, again, is -8.148603881.

Exercise 5--Order of Operations and the Calculator

Purpose of the Exercise

- To drill on order of operations
- To drill using the calculator to do string operations

Materials Needed

- Calculator

In Problems 1-4, write in the order of operations. Following the order of operations, use a calculator for the computations, doing each part individually. Then compute the expression using string operations on the calculator. Record the key strokes you use to type in the problem with string operations. Compare the answers obtained both ways. If the answers are not approximately the same, do the computations again.

1. $5(4.7) - 9.5(3.2) + 14 \div (3.6)(7.5) =$ ______________________________

   __ __ __ __ __ __ Write the order of operations.

   Write the key strokes for string operations: ______________________________

2. $543.6\ (-82.6) + 35.8 \div 17.3 - (14.57)(\ 56.3 \div 27) - 19 \div 2(3)\ =$ ______________

   ___ ___ ___ ___ ___ ___ __ __ __

   Write the key strokes for string operations: ______________________________

3. $(39.4 - 27.3)(24.6 - 17.3) - (6.3)(3.6)\ =$ ______________________________

   ___ ___ __ ___ ___

   Write the key strokes for string operations: ______________________________

4. $\dfrac{(3)(5)-(2)(4)}{(6)(2)+(2)(3)} =$ ______________________________

   Write the key strokes for string operations: ______________________________

Compute each of the following using a calculator. In the expressions that involve division, round the final answer to 2 decimal places.

5. $3(-5) - (-5 \div 2) =$ ______________

6. $(28.3)(-5) - (22 - 87) =$ ______________

7. $(27.5 - 358.9) - 14 \div 2 \bullet 5 =$ ______

8. $(27.5 - 358.9) - 14 \div (2 \bullet 5) =$ ________

9. $4(88 + 25) - 3.4(253 + 3.6) =$ _______

10. $100(38.7 - 90) + 15/4 =$ ______________

11. $0.001(5000 - 4000) + 1 =$ _________

12. $30 \div 0.1 \bullet 4 - 12/(-3) =$ ____________

13. $\dfrac{2.1(-3.2)-4(6.1))}{100 \div (0.01)} =$ ______________

14. $\dfrac{2.1(-3.2)-4(6.1))}{100(0.01)} =$ ______________

Section 2-8 *The Distributive Law*

In previous sections, we practiced using the commutative and associative laws for addition and multiplication. The commutative laws allowed us to change the order of terms or factors and the associative laws allowed us to regroup the terms or factors. The distributive law allows us to **distribute** multiplication over addition. To understand the concept, we will work with numbers first and then generalize to write the generic rule.

Exercise 1--The Distributive Law, Discovering the Concept

Purpose of the Exercise

- To discover what it means to distribute multiplication over addition

Materials Needed

- 20 small gray squares, 20 large gray squares, 20 small white squares, 20 large white squares, 20 red beans, and 20 white beans per group

In each of the following, compute the first expression by doing the calculations within the parentheses first and then multiply the resulting factors. Compute the second expression by multiplying the factors first; then add or subtract the result as indicated. Determine from your results if the last statement is true or false.

1. 7(9 + 8) = ______________________ = ______________

   7(9) + 7(8) = ______________________ = ______________

   True False 7(9 + 8) = 7(9) + 7(8)

2. 7(9 - 8) = ______________________ = ______________

   7(9) - 7(8) = ______________________ = ______________

   True False 7(9 - 8) = 7(9) - 7(8)

3. 14(9 - 15) = ______________________ = ______________

   14(9) - 14(15) = ______________________ = ______________

   True False 14(9 - 15) = 14(9) - 14(15)

4. 10(6 + 5 + 3) = ______________________ = ______________

   10(6) + 10(5) + 10(3) = ______________ = ______________

   True False 10(6 + 5 + 3) = 10(6) + 10(5) + 10(3)

5. 4(3 - 5 + 2) = ______________________ = ______________

   4(3) + 4(-5) + 4(2) = ______________ = ______________

   True False 4(3 - 5 + 2) = 4(3) + 4(-5) + 4(2)

6. -1(9 + 8) = ______________________ = ______________

   (-1)(9) + (-1)(8) = ______________ = ______________

   True False -1(9 + 8) = (-1)(9) + (-1)(8)

7. -1(9 - 8) = ______________________ = ______________

   (-1)(9) - (-1)(8) = ______________ = ______________

   True False -1(9 - 8) = (-1)(9) - (-1)(8)

8. -2(4 + 5) = ______________________ = __________________

(-2)(4) + (-2)(5) = __________________ = __________________

True False -2(4 + 5) = (-2)(4) + (-2)(5)

9. -2(4 - 5) = ______________________ = __________________

(-2)(4) - (-2)(5) = __________________ = __________________

True False -2(4 - 5) = (-2)(4) - (-2)(5)

10. 2(4 + 3 - 7 + 6) = ________________ = __________________

2(4) + 2(3) + 2(-7) + 2(6) = __________ = __________________

True False 7(9 + 8) = 7(9) + 7(8)

For the following exercises, write x on each small white square, -1x on each small gray square, y on each large white square and -1y on each large gray square. As in Sections 1 and 2, white bean represent positive numbers and a red bean represent negative numbers.

| x | | x | | x | represents 3x | | -1x | | -1x | | -1x | represents -3x

11. Show 3(2x + 3y + 5) by representing 2x + 3y + 5 three times with squares and beans. Rearrange the squares and beans (combining like items) to write 3(2x + 3y + 5) in a different form without parentheses.
3(2x + 3y + 5) = ____

12. Show 4(-3x + 2y - 5) by rewriting the subtraction as addition and representing -3x + 2y - 5 four times with squares and beans. Rearrange the squares and beans (combining like items) to write 4(-3x + 2y - 5) in a different form without parentheses.
4(-3x + 2y - 5) = ____

13. Show 2(-5 + 6y) by representing -5 + 6y two times with squares and beans. Rearrange the squares and beans (combining like items) to write 2(-5 + 6y) in a different form without parentheses.
2(-5 + 6y) = ____

14. Show 5(2x - y + 1) by rewriting the subtraction as addition and representing 2x - y + 1 five times with squares and beans. Rearrange the squares and beans (combining like items) to write 5(2x - y + 1) in a different form without parentheses.
5(2x - y + 1) = ____

In the expressions above, when a number was multiplied by two or more terms inside of parentheses, the computations turned out to be the same whether the terms were added first and then multiplied by the first factor, or the factor was multiplied individually by each term inside the parentheses. We get an equivalent expression when we multiply the number outside of the parentheses by each term inside the parentheses. This is called "distributing multiplication over addition." The **distributive law** is stated symbolically below.

The **distributive law** for multiplication over addition is a(b + c) = ab + ac   for all possible values of a, b, and c.

Since subtraction can be written as addition, multiplication also distributes over subtraction.
a(b - c) = a[b + (-c)] = a(b) + a(-c) = ab + (-ac) = ab - ac.
a(b - c) = ab - ac

Example 1: Compute 7(11 - 3) two different ways.

1. Do the computation inside the parentheses first; then multiply the answer by 7.
   $7(11 - 3) = 7(8) = 56$
2. Apply the distributive law first; then add or subtract.
   $7(11 - 3) = 7(11) - 7(3) = 77 - 21 = 56$

Example 2: Compute -7(11 - 3) two different ways.

1. Do the computation inside the parentheses first; then multiply the answer by -7.
   $-7(11 - 3) = -7(8) = -56$
2. Apply the distributive law first; then subtract.

$$\begin{aligned} -7(11 - 3) &= (-7)(11) - (-7)(3) \\ &= -77 - (-21) \\ &= -77 + 21 \\ &= -56 \end{aligned}$$

If the second way was confusing, we can rewrite the subtraction as addition and apply the distributive law as follows:

$$\begin{aligned} -7(11 - 3) &= -7[11 + (-3)] \\ &= (-7)(11) + (-7)(-3) \\ &= -77 + 21 \\ &= -56 \end{aligned}$$

Example 3: Multiply -6(-x - 3) using the distributive law:

1. Using the distributive law of multiplication over addition:
   Restate the problem as addition: $-6[(-x) + (-3)]$
   $= (-6)(-x) + (-6)(-3)$
   $= 6x + 18$
2. Using the distributive law of multiplication over subtraction:

$$\begin{aligned} -6(-x - 3) &= -6(-x) - (-6)(3) \\ &= 6x - (-18) \\ &= 6x + 18 \end{aligned}$$

Exercise 2--The Distributive Law

Purpose of the Exercise

- To practice using the distributive law for numeric as well as algebraic expressions

Materials Needed

- Calculator

In each of the following **numeric** expressions, first evaluate the expression by doing the operations inside parentheses and then multiplying the result by the number outside of the parentheses. Second, evaluate the expression by applying the distributive law first and then adding or subtracting the resulting terms. If the two methods give different answers, check your work.

In each of the following **mathematical expressions**, use the distributive law to multiply the expression. Combine like terms if applicable. Use a calculator as needed and be careful with the rules of signs. The first two are done as examples.

1. $-4(5 - 11) =$

   (1) $-4(-6) = \underline{24}$

   (2) $(-4)(5) - (-4)(11) = -20 - (-44)$
   $= -20 + 44$
   $= \underline{24}$

2. $-4(5x - 11) =$

   $(-4)(5x) - (-4)(11)$
   $= -20x - (-44)$
   $= \underline{-20x + 44}$

3. 6(5 - 17) =

(1)

(2)

4. 6(5c - 17) =

5. -0.5(-0.9 + 1) =

(1)

(2)

6. -0.5(-0.9 + x) =

7. 789(-287 - 78) =

(1)

(2)

8. 789(-287t - 78) =

9. -1(3 - 11) =

(1)

(2)

10. -1(3w - 11) =

Exercise 3--Removing Parentheses, Combining Like Terms

Purpose of the Exercise

- To practice removing parentheses by using the distributive law
- To practice combining like terms
- To understand numerically what is meant by a correct simplification of an algebraic expression

Simplify each of the following expressions by using the distributive law and combining like terms. Choose the answer that you think is right; then check your choice by substituting the values for the variables into the original expression and the simplified expression. If substitution shows that you picked the wrong answer, then choose another answer and test it. If the answer is not on the list, see if you can get the right answer and check it by numerical substitution. The first one is done as an example.

1. 3 -(4 + x) = (-1 + x) -1 - x 1 - x

Simplification: 3 -(4 + x) = 3 - 4 + x = -1 + x

If x = 4
3 - (4 + x) = 3 - (4 + 4) = 3-8 = -5

Substitute x = 4 into the answer you chose. -1 + x = -1 + (4) = 3

Substituting $x = 4$ into the formula on the left gives -5. Substituting $x = 4$ into the answer we chose on the right gave 3. Since the simplified answer computes differently from the original expression, there must be a mistake. Let's make another choice.

Retry $3 -(4 + x) =$ $-1 + x$ (circled: $-1 - x$) $1 - x$

If $x = 4$
$3 - (4 + x) =$ $\underline{3 - (4 + 4) = 3 - 8 = -5}$

Substitute $x = 4$ into the answer you chose. $\underline{-1 - x = -1 - 4 = -5}$

The two answers are now the same, so it looks as if $3 - (4+x) = -1 -x$ might be the correct simplification. Let's see if the computations give the same answer for two other substitutions. Let us check the simplification again.

Simplification: $3 -(4 + x) = 3 - 4 - x = -1 - x$. Before we forgot to subtract the x inside of the parentheses as well as subtract the 4.

If $x = 18$
$3 - (4 + x) =$ $\underline{3 - (4 + 18) = 3 - 22 = -19}$

Substitute $x = 18$ into the answer you chose. $\underline{-1 - x = -1 - 18 = -19}$

The two answers are now the same, so it looks as if $3 - (4+x) = -1 -x$ might be the correct simplification. Let's check with one more number.

If $x = -5$
$3 - (4 + x) =$ $\underline{3 - (4 + (-5)) = 3 - (-1) = 3 + 1 = 4}$

Substitute $x = -5$ into the answer you chose. $\underline{-1 - x = -1 - (-5) = -1 + 5 = 4}$
The two answers are now the same, so it looks as if $3 - (4+x) = -1 -x$ is the correct simplification.

2. $2x - (7 - 3x) =$ $-2x$ $-x - 7$ $5x - 7$

If $x = 4$
$2x - (7 - 3x) =$ ____________________

Substitute $x = 4$ into the answer you chose. ____________

If the answers are not the same, redo your algebra and check the new answer.

If $x = 18$
$2x - (7 - 3x) =$ ____________________

Substitute $x = 18$ into the answer you chose. ____________

If the answers are not the same, redo your algebra and check the new answer.

If $x = -5$
$2x - (7 - 3x) =$ ____________________

Substitute $x = -5$ into the answer you chose. ____________

If the answers are not the same, redo your algebra and check the new answer.

3. $5 - 5x + 8 =$ $\quad 8 \quad -5x - 3 \quad -5x + 13$

If $x = 4$
$5 - 5x + 8 =$ ______________________________

Substitute $x = 4$ into the answer you chose. ______________

If $x = 18$
$5 - 5x + 8 =$ ______________________________

Substitute $x = 18$ into the answer you chose. ______________

If $x = -5$
$5 - 5x + 8 =$ ______________________________

Substitute $x = -5$ into the answer you chose. ______________

4. $5 - 7m - 8 + (3-m) =$ $\quad -6m \quad -6 - 6m \quad -8m$

If $m = 4$
$5 - 7m - 8 + (3-m) =$ ______________________________

Substitute $m = 4$ into the answer you chose. ______________

If $m = 18$
$5 - 7m - 8 + (3-m) =$ ______________________________

Substitute $m = 18$ into the answer you chose. ______________

If $m = -5$
$5 - 7m - 8 + (3-m) =$ ______________________________

Substitute $m = -5$ into the answer you chose. ______________

5. $5 - 3(t - 4) =$ $\quad 1 - 3t \quad 17 - 3t \quad 9 - 3t$

If $t = 4$
$5 - 3(t - 4) =$ ______________________________

Substitute $t = 4$ into the answer you chose. ______________

If $t = 18$
$5 - 3(t - 4) =$ ______________________________

Substitute $t = 18$ into the answer you chose. ______________

If $t = -5$
$5 - 3(t - 4) =$ ______________________________

Substitute $t = -5$ into the answer you chose. ______________

6. $9a + 6(-3 - 2a) =$ $\quad 7a - 18 \quad 21a - 18 \quad -3a - 18$

If $a = 4$
$9a + 6(-3 - 2a) =$ ______________________________

Substitute $a = 4$ into the answer you chose. ______________

If $a = 18$

$9a + 6(-3 - 2a) =$ ______________________

Substitute $a = 18$ into the answer you chose. ______________

If $a = -5$

$9a + 6(-3 - 2a) =$ ______________________

Substitute $a = -5$ into the answer you chose. ______________

7. $-2y - 4(3 - y) =$ $\quad$ $-y - 12$ $\quad$ $-3y - 12$ $\quad$ $12 + 2y$

If $y = 4$

$-2y - 4(3 - y) =$ ______________________

Substitute $y = 4$ into the answer you chose. ______________

If $y = 18$

$-2y - 4(3 - y) =$ ______________________

Substitute $y = 18$ into the answer you chose. ______________

If $y = -5$

$-2y - 4(3 - y) =$ ______________________

Substitute $y = -5$ into the answer you chose. ______________

8. $-3(2x - 1) - 4(x + 2)$ $\quad$ $-10x + 11$ $\quad$ $-10x - 5$ $\quad$ $-10x + 5$

If $x = 4$

$-3(2x - 1) - 4(x + 2) =$ ______________________

Substitute $x = 4$ into the answer you chose. ______________

If $x = 18$

$-3(2x - 1) - 4(x + 2) =$ ______________________

Substitute $x = 18$ into the answer you chose. ______________

If $x = -5$

$-3(2x - 1) - 4(x + 2) =$ ______________________

Substitute $x = -5$ into the answer you chose. ______________

9. Simplify each of the following. Check with at least three different numbers to see that your answer gives the same value as the original expression.

a. $4x - 4(x - 9) =$ ______________

b. $4 - 4(L - 9) =$ ______________

c. $-17.5 + (5.9f - 12.8) =$ __________

d. $-197 - (-197 - 17m) =$ __________

e. $4(5x - 17) - 5x =$ ______________

f. $-5(6.8n - 13.4) - 16.5 =$ __________

g. $23 - 23(x - 9) =$ ______________

h. $x - (x - 11) =$ ______________

i. $5b - 5(3-b) =$ ______________

j. $27R - 27 - 3(8 + 4x) =$ __________

k. $0.89 - 0.23(0.289d - 1) =$ __________

l. $1987 - (378 - 1987c) =$ __________

Section 2-9 *Solving Equations--A Hands-On Model*

Equations are balance systems. Picture a scale, or balance, where produce is weighed by hanging a known weight on one side of the scale and hanging the produce in a bag or basket on the other side. The desired amount of produce is in the bag when the scale looks balanced. If an additional weight is added on the right, then the same weight of additional produce is added on the left to balance the scale. If produce is taken off the left, then the comparable weight on the right is taken off to keep the balance.

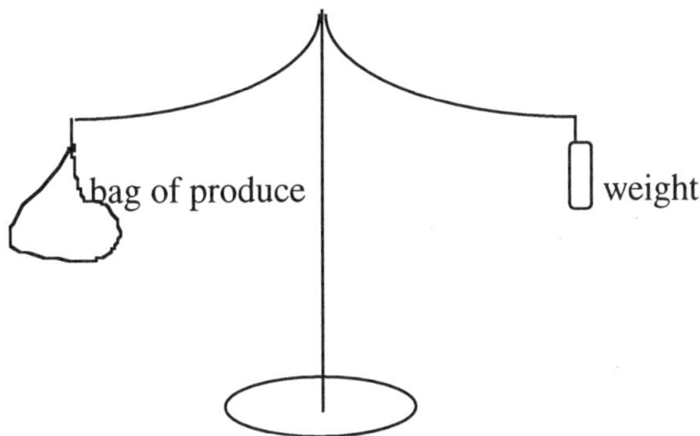

An **algebraic equation** has an equal sign with an algebraic expression on each side of the equal sign. To look at equations as a balance system, we will simulate a small balance scale with the drawing on the next page. On our balance we will place variables and numbers and determine values of the variables that keep a balance. In our hands-on exercises, we will use small squares for x's, large squares for y's and a white bean for the number 1.

Exercise 1-- Balances and Equations (Hands-on activity)

Purpose of the Exercise

- To work with objects to understand equations as a balance system

Materials Needed

- 40 small cardboard squares (about 1/2 in. x 1/2 in.), 40 large cardboard squares (about 1 in. x 1 in.), and 30 white beans per group

Write the letter x on each of the small squares and the letter y on each of the large squares. In each of the following, place the squares and white beans on the left and right sides of the balance as indicated; then determine the number to write on the squares that makes the left and right sides balanced. Use the balance drawn on the next page for the exercises. Each white bean represents the number 1. The first one is shown as an example.

1. a. Place 3 of the squares marked x on the left side of the balance and 6 white beans on the right side of the balance.

b. Pick a number for x that makes the left and right sides balanced and write the number on the back of each of the squares marked x.

The total of 6 on the left balances the 6 on the right.

c. Write the equation represented in part a.
The equation: On the left we have three x's and on the right 6 ones.
3 x's is balanced with 6 beans.
$3x = 6$

d. Write the solution from part b.
The solution: $x = 2$

2. a. Place 4 large squares on one side of the balance and 12 white beans on the other side of the balance.
b. Write a number on each of the large squares that makes the two sides have the same value.
c. Write the equation represented in part a.
equation:

d. Write the solution from part b.
solution:

3. a. Place 4 large squares on one side of the balance and 2 large squares and 6 white beans on the other side of the balance.
b. Write a number on each of the large squares that makes the two sides balanced.
c. Write the equation represented in part a.
equation:

d. Write the solution from part b.
solution:

4. a. Place 3 small squares and 4 white beans on one side of the balance and 7 white beans on the other side of the balance.
b. Write a number on each of the small squares that makes the two sides equal.
c. Write the equation represented in part a.
equation:

d. Write the solution from part b.
solution:

5. a. Place 5 small squares on one side of the balance and 2 small squares and 6 white beans on the other side of the balance.
b. Write a number on each of the small squares that makes the two sides equal.
c. Write the equation represented in part a.
equation:

d. Write the solution from part b.
solution:

6. a. Place 8 small squares and 6 white beans on one side of the balance and 5 small squares and 6 white beans on the other side of the balance.
b. Write a number on each of the small squares that makes the two sides equal.
c. Write the equation represented in part a.
equation:

d. Write the solution from part b.
solution:

7. a. Place 8 large squares and 6 white beans on one side of the balance and 5 large squares and 6 white beans on the other side of the balance.
   b. Write a number on each of the large squares that makes the two sides equal.
   c. Write the equation represented in part a.
      equation:

   d. Write the solution from part b.
      solution:

8. a. Place 15 large squares and 3 white beans on one side of the balance and 5 large squares and 23 white beans on the other side of the balance.
   b. Write a number on each of the small squares that makes the two sides equal.
   c. Write the equation represented in part a.
      equation:

   d. Write the solution from part b.
      solution:

9. a. Place 7 small squares and 7 white beans on one side of the balance and no squares or white beans on the other side.
   b. Write a number on each of the small squares that makes the two sides equal.
   c. Write the equation represented in part a.
      equation:

   d. Write the solution from part b.
      solution:

10. a. Place 14 large squares and 7 white beans on one side of the balance and no squares or white beans on the other side.
    b. Write a number on each of the large squares that makes the two sides equal.
    c. Write the equation represented in part a.
       equation:

    d. Write the solution from part b.
       solution:

11. a. Place 8 large squares and 6 white beans on one side of the balance and 8 large squares and 4 white beans on the other side of the balance.
    b. Write a number on each of the large squares that makes the two sides equal.
    c. Write the equation represented in part a.
       equation:

    d. Explain the difficulty with finding a solution to this problem.

12. a. Place 3 large squares and 6 white beans on one side of the balance and 3 large squares and 6 white beans on the other side of the balance.
    b. Write a number on each of the large squares that makes the two sides equal.
    c. Find another number that makes the two sides equal.
    d. Write the equation represented in part a.
       equation:

    e. Explain why there are many solutions to this equation.

13. Place the appropriate squares and white beans on the balance to represent each of the following equations. In each case, find the value for the variable that "balances" the equation.

a. $4y + 3 = 7$

$y =$ ________

b. $7 = 21x$

$x =$ ________

c. $5y + 5 = y + 1$

$y =$ ________

d. $7x + 3 = x + 9$

$x =$ ________

e. $4 + y = 12$

$y =$ ________

f. $15 = 5 + x$

$x =$ ________

g. $4y = 12$

$y =$ ________

h. $15 = 5x$

$x =$ ________

i. $7y = 0$

$y =$ ________

j. $5x = 10$

$x =$ ________

k. $7 + y = 0$

$y =$ ________

l. $10x = 5$

$x =$ ________

m. $6 + y = 3$

$y =$ ________

n. $11x = 0$

$x =$ ________

o. $6y = 3$

$y =$ ________

p. $11 + x = 0$

$x =$ ________

When solving equations using the balance, squares, and white beans for props, you may have discovered that we can "simplify" more complicated problems by removing duplicates from both sides of the balance.

Example 1: $3x + 2 = 6 + x$ is shown below on the balance.

Without upsetting the balance, we can remove 2 beans from each side and remove a square from each side. This leaves us with $2x = 4$.

X X

x = 2 makes the sides balanced.

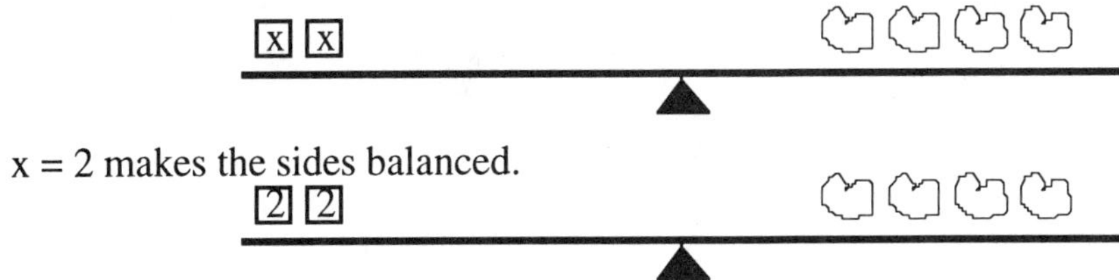

If you had difficulty with some of the more complicated problems in Exercise 1, go back and try them again, removing duplicates; then solve the resulting equation.

Now we will expand the use of the balance system to negative terms. In the balance model, each white bean will represent the number 1, and each red bean will be -1. For the squares, the small white squares will represent x and the small dark colored squares will represent -x.

Similarly the white and dark large squares represent y and -y (read "the opposite of y"). In expanding our use of the balance, we already know we can remove duplicates from both sides of the balance. We can also add duplicates to both sides of the equation without upsetting the balance. The objective in solving an equation is to find a value of the variable that makes the left and right side balanced. To use the balance to solve equations, we will first remove duplicates from both sides. If one side has only squares and the other has only beans, we will then determine the solution. If either side has both squares and beans, then we will put the same thing on both sides of the balance to "cancel" one of the types of objects.

Example 2: $3x + 3 = 2 + x$.

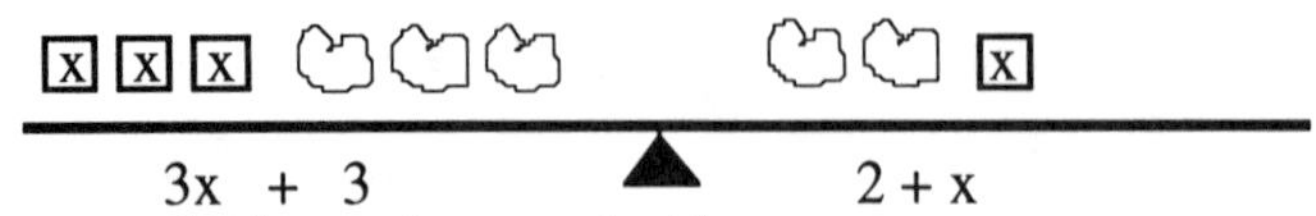

Remove a square and 2 beans from each side.

The resulting equation is $2x + 1 = 0$. At this point we can guess the answer, but suppose we are not good at guessing. We will add a red bean to both sides. Remember from the bean arithmetic, a pair of red and white beans cancel.

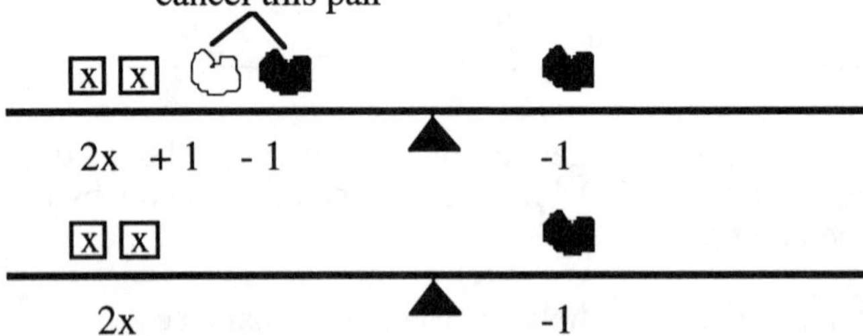

After the red and white pair are canceled, the equation is $2x = -1$ and solution is $x = -0.5$.

Example 3: $4 - 3x = x - 4$.

Or, since subtraction can be written as addition, $4 + (-3x) = x + (-4)$.

-x -x -x x

4 + (-3x) x + (-4)

There are no duplicates, so we will add opposites to cancel. Now add 4 white beans to both sides .

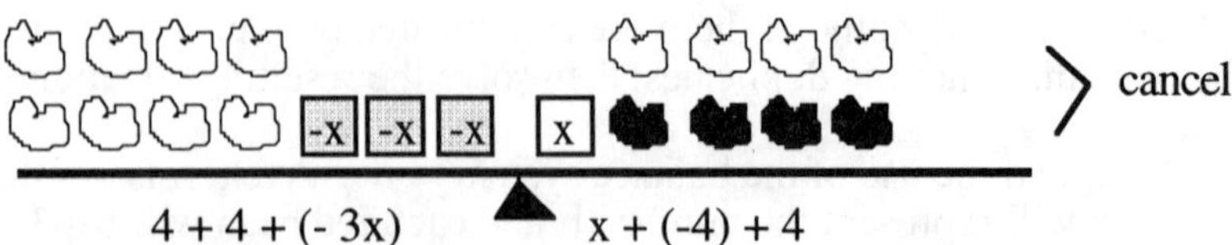

This leaves us with 8 - 3x = x . Now we will add 3 white squares to each side.

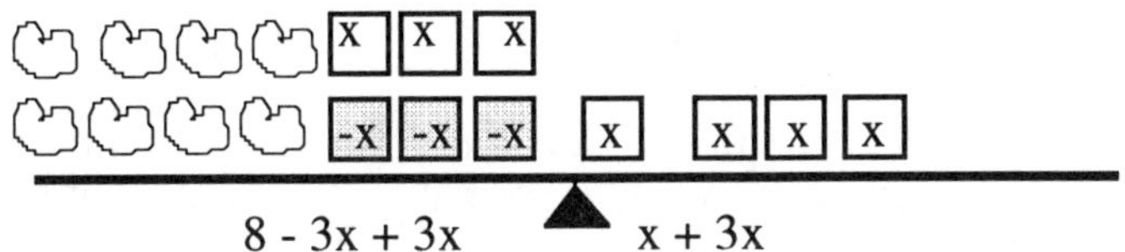

Canceling the positive and negative x's on the left and adding the x's on the right, we get as a final result, 8 = 4x, and x = 2.

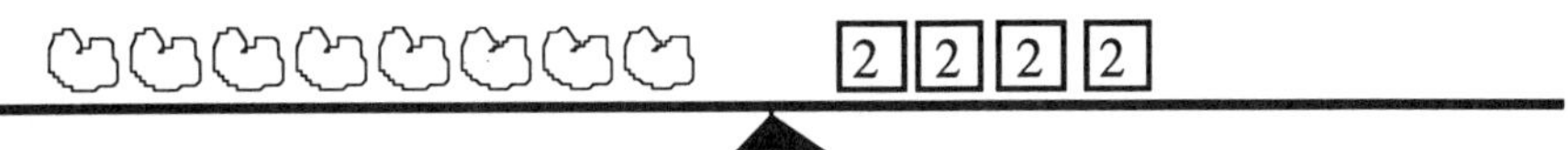

Exercise 2--The Balance and Equations--Negative Terms

Purpose of the Exercise
- To use the balance to discover the rules of equations

Materials Needed
- About 30 of each per group: white beans, red beans, small white squares, small dark squares, large white squares, and large dark squares

In each of the following, represent the equation with the squares and beans on the balance. Remove duplicates from both sides of the balance or add duplicates to both sides of the balance as needed to keep the same balance, ending up with only beans on one side and white squares on the other. On a separate sheet of paper, write the equation that is modeled, each resulting equation, and what you did with the squares and beans to get the resulting equation. Finally, determine a value for the variable that balances the scale or equation. The small squares represent x's, the large squares y's and each white bean is 1. The dark squares and red beans are -x, -y and -1, respectively.

1. 5y - 3 = 2y
2. 4x - 5 = 3 - 4x
3. 7 - 4x = -12
4. 2x - 14 = 5x - 7
5. 6y = 3
6. 5x = 10
7. 5 + x = 10
8. 10y = 5
9. 10 + y = 5
10. 8 + x = 3x + 2
11. 8 + x = 3x - 2
12. 8x = 3x + 5
13. 8x = 4x
14. 8 = 4x
15. y - 7 = 0
16. y + 7 = 0
17. 7y = 0
18. 4x - 5 = x + 1
19. 11x = 11
20. 11 + x = 11

Section 2-10 *Strategy for Solving Equations*

As we discovered working with the balance, we can add the same number of a particular variable and we can also add the same number to both sides of an equation. We can also multiply or divide both sides of an equation by the same number.

In the equation $3x = 6$, it is easy to guess that each of the three x's must have a value of 2 for the equation to be balanced. Now we will formalize a process that will give us that answer without guessing.

Example 1: Solve $3x = 6$.

$\frac{3x}{3} = \frac{6}{3}$ First, divide both sides of the equation by 3.

$x = 2$ Dividing the 3 x's into 3 parts simplifies the left side to x and the division on the right gives 2.

Now, we will summarize the rules of equations.

Rules for Solving Equations

1. The same value can be added to both sides of the equation without changing the balance in the equation.
2. The same value can be subtracted from both sides of the equation without changing the balance in the equation.
3. Both sides of the equation can be multiplied by the same value without changing the balance in the equation. (We would not choose to multiply both sides by zero since that would give us a balance but would leave us with no equation to solve.)
4. Both sides of the equation can be divided by the same **nonzero** value without changing the balance in the equation. (Remember, zero can never be a divisor.)

We could combine the rules above into just two rules. Rule 2 is really a part of Rule 1 since adding a number is the same as subtracting the opposite of that number. Similarly, multiplication and division are inverse operations so we could combine those rules into one rule--however, we need to cover fractions before making that connection.

Now that we have the rules of equations, we need a strategy for what to do first, second, and so on, to solve an equation. We need to be systematic in deciding which rule to use first and how to use it. To complicate the issue even more, unlike the equations we solved using balances, we may have some expressions in parentheses that we need to clear out before proceeding as we did above. Below is a strategy for solving equations. There may be other ways to work through solving an equation, but this gives a systematic approach.

Strategy for Solving Equations

1. Use the distributive law to remove parentheses on each side of the equation.
2. Combine like terms on each side of the equation.
3. Use addition and subtraction rules to get like terms together on the same side of the equation. Combine like terms in the new equation.
4. Use the multiplication and division rules to solve for the variable.
5. Check your solution by substituting it for the variable in the original equation.

When solving an equation for a variable, the objective is to use the rules of equations to transform the equation into the form, "the variable = a number."

Example 2: $-3b + 4 = 25$.

1. and 2. There are no parentheses or like terms on the same side, so steps 1 and 2 are already done.

3. Subtract 4 from both sides of the equation and then combine like terms.

$$-3b + 4 - 4 = 25 - 4$$
$$-3b = 21$$

4. Divide both sides of the equation by -3.

$$\frac{-3b}{-3} = \frac{21}{-3}$$
$$b = -7$$

On the left side of the equation, -3/-3 divides to 1 leaving 1b, or b. On the right side $\frac{21}{-3}$ is -7.

5. Substitute -7 for b into the original equation to see if this balances the equation.

$$-3b + 4 = 25$$
$$-3(-7) + 4 = 25$$
$$21 + 4 = 25$$
$$25 = 25$$

Since both sides of the equation are the same, the solution is correct.

Example 3: Solve $-4(3b - 5) + b = 20 - b$.

Step	Equation	Explanation
1.	$-4(3b) - (-4)(5) + b = 20 - b$	Use distributive law on the left side.
2.	$-12b - (-20) + b = 20 - b$	Multiply within each term.
	$-12b + 20 + b = 20 - b$	
	$-12b + b + 20 = 20 - b$	Commutative law.
	$-11b + 20 = 20 - b$	Combine like terms.
3.	$-11b + 20 + b = 20 - b + b$	Add b to both sides.
	$-10b + 20 = 20$	Combine like terms
	$-10b + 20 - 20 = 20 - 20$	Subtract 20 from both sides.
	$-10b = 0$	Combine like terms.
4.	$\frac{-10b}{-10} = \frac{0}{-10}$	Divide both sides by -10
	$b = 0$	

5. Check the answer:

$$-4(3(0) - 5) + 0 = 20 - 0$$
$$-4(0-5) = 20$$
$$-4(-5) = 20$$
$$20 = 20$$

The answer checks.

Exercise 1--Solving Equations--Choosing the First Step

Purpose of the Exercise

- To practice deciding which first step to take when solving equations

Materials Needed

- Calculator

In each of the following, circle the first step you would take to solve the equation; then solve the equation. Show your work. To check the solution, substitute the answer into the original equation and do the arithmetic. If the solution does not produce a balance, rework the problem. The first one is done as an example.

1. $t + 3.79 = 185.6$

   a. Circle the correct action:
   - Add 3.79 to both sides of the equation.
   - Subtract 3.79 from both sides of the equation. (circled)
   - Add 185.6 to both sides of the equation.
   - Subtract 185.6 from both sides of the equation.

b. Do the arithmetic; show your work.

$t + 3.79 - 3.79 = 185.6 - 3.79$
$t + 0 = 181.81$ (3.79 - 3.79 = 0, and 185.6 - 3.79 = 181.81)
$t = 181.81$

c. Check your work.

The original equation was: $t + 3.79 = 185.6$
Our solution was: $t = 181.81$
Substitute 181.81 for t into the equation: $181.81 + 3.79 = 185.6$
The left side equals the right side, so the answer, $t = 181.81$, is correct.

2. $y - 87.624 = 519$

a. Circle the correct action:
Add 87.624 to both sides of the equation.
Subtract 87.624 from both sides of the equation.
Add 519 to both sides of the equation.
Subtract 519 from both sides of the equation.

b. Do the arithmetic; show your work.

c. Check your work.

The original equation was:

Your solution is:

Substitute the solution into the equation:

Does the left side equal the right side? If not, redo to problem.

3. $3.684y = -6.2628$

a. Circle the correct action:
Subtract 3.684 from both sides of the equation.
Divide both sides by 3.684.
Add -6.2628 to both sides of the equation.
Divide both sides by -6.2628.

b. Do the arithmetic; show your work.

c. Check your work.

The original equation was:

Your solution is:

Substitute the solution into the equation:

Does the left side equal the right side? If not, redo to problem.

4. $c + 17.21 = 83$

a. Circle the correct action:
Add 17.21 to both sides of the equation.
Subtract 17.21 from both sides of the equation.
Add 83 to both sides of the equation.
Subtract 83 from both sides of the equation.

b. Do the arithmetic; show your work.

c. Check your work.

The original equation was:

Your solution is:

Substitute the solution into the equation:

Does the left side equal the right side? If not, redo to problem.

Using a separate sheet of paper, solve each of the following equations for the variable, clearly showing your work. When appropriate, put your solution in decimal form and round to 2 decimal places. Check your work by substituting your solution into the original equation to see if both sides of the equation are equal (or approximately equal if you rounded your answer.) Use a calculator when needed. In Problems 5 - 8, describe your steps in words.

5. $4x - 5 = -23$
6. $-5t + 11 = -7$
7. $-15.9 - 22m = 579$
8. $7(3a - 5) - 22 = 15$
9. $0.57 - 0.57z = 18$
10. $-4.3(2 - 0.5v) + 9 = 15.2$
11. $22y = 22$
12. $-22y = -22$
13. $14.9e - 37 = -37$
14. $18 = 11(27 - t)$
15. $89.3m = -15.8$
16. $89.3 + m = -15.8$
17. $4(-23 + 13r) - r = -5(6 - r)$
18. $17.9 - k = -17.9$
19. $4y - 7 = 18$
20. $5m + 7 + 3m = 28$
21. $18h = 23$
22. $18 + h = 23$
23. $s - 11 = 11$
24. $s + 11 = 11$
25. $11s = 11$
26. $5 + 9p - 4p + 9 = 21$
27. $34.2w = 81$
28. $34.2 + w = 81$
29. $289k = 7856$
30. $5m - 7289 = 9876$
31. $2(19y - 19) - 3y = 2 - 2y$
32. $278.97t = 18.45$
33. $350y - 350 = 0$
34. $178 - 18a = 15$
35. $356 - t = 18.9$
36. $22 - 11t = 0$
37. $11t - 22 = 0$
38. $14.5n - (4n - 14) = 0$
39. $34h + 34 = 92$
40. $78945m = 1097$
41. $87 - 87m = 3(5 - m)$
42. $19m = 19$
43. $19 + m = 19$
44. $19 - m = 19$
45. $76 - 3g = 70$
46. $-17t = 2(-14 + t)$
47. $-z = z - 5$
48. $-2y = 3y - 9$
49. $-w = -12$
50. $k = 2k + 11$
51. $79x = 80x + 786$
52. $34 + 7f - 14 - f = 0$

Chapter 2 *Summary and Test*

The Types of Numbers (page 105)

- **Real numbers** are numbers that can be placed on a number line. All of the numbers that we have worked with are real numbers.
- The **natural,** or **counting,** numbers are 1, 2, 3, 4, . . .
- The **whole numbers** are 0, 1, 2, 3, 4, 5, . . .
- The **integers** are the whole numbers and their opposites.
- A **rational** number is a positive or negative number that has no digits after the decimal point, or has a repeating pattern of digits after the decimal point, or is a terminating decimal (meaning the digits stop eventually.)
- The **irrational numbers** are the real numbers that are not rational.

Rules of Signs (page 115)

Addition

- The sum of two positive numbers is positive.
- The sum of two negative numbers is negative.
- When adding a **positive number and a negative number**, subtract the number with the smaller absolute value from the number with the larger absolute value and keep the sign of the number with the larger absolute value.

Subtraction

- Rewrite subtraction as addition and follow the rules for addition.

Multiplication

- The product of two positive numbers is positive.
- The product of a positive number and negative number is negative.
- The product of two negative numbers is positive.

Division

- A positive number divided by a positive number is a positive number.
- A negative number divided by a positive number is a negative number.
- A positive number divided by a negative number is a negative number.
- A negative number divided by a negative number is a positive number.

Laws of Arithmetic

- The **commutative laws** allow us to change the order. (page 125)
  Commutative law for addition: $a + b = b + a$
  Commutative law for multiplication: $ab = ba$
- The **associative laws** allow us to change the groupings. (page 125)
  Associative law for addition: $(a + b) + c = a + (b + c)$
  Associative law for multiplication: $(ab)c = a(bc)$
- The **distributive law** is: $a(b + c) = ab + ac$ (page 143)

Exponents and Rules of Signs for Exponents (pages 132 and 133)

- In a number of the form, a^n, a is the **base** and n is the **power**, or **exponent**. To simplify such a number, the base is multiplied by itself n times. When a base is negative, the base and the negative sign must be inside parentheses.
  $(-7)^2 = (-7)(-7) = 49$ $\qquad$ $-7^2 = -(7)(7) = -49$
- When the base is positive, the base raised to any power is positive.
- When the **base** in an exponential expression is **negative**,
  and the **power is an even number**, the **result is positive**.
  and the **power is an odd number**, the **result is negative**.

Terminology

- The **absolute value** of a number is the value of the number without regard to sign. (page 113)
- The **minus sign has three meanings**. It is used to indicate subtraction, opposite, or negative. (page 120)
- Two numbers or variables <u>multiplied</u> are called **factors**. (page 135)
- Two numbers or variables <u>added</u> are called **terms**. (page 135)
- **Algebraic expressions** are any combination of variables with arithmetic operations. (page 135)
- An **algebraic equation** has an equal sign with an algebraic expression on each side of the equal sign. (page 149)

The Order of Operations (pages 136, 137)

The basic hierarchy is

1. All powers are computed first.
2. Multiplication and division are done next (from left to right in a problem.)
3. Addition and subtraction are done last.

To change the basic hierarchy

1. Parentheses can be used to change the basic hierarchy. When there are arithmetic operations inside the parentheses, those operations should be done before the operations not in parentheses.
2. When using fraction notation with the numerator on top and the denominator on the bottom, there are implied parentheses around the entire numerator and the entire denominator.

Rules for Solving Equations (page 156)

1. The same value can be added to both sides of the equation without changing the balance in the equation.
2. The same value can be subtracted from both sides of the equation without changing the balance in the equation.
3. Both sides of the equation can be multiplied by the same value without changing the balance in the equation.
4. Both sides of the equation can be divided by the same **nonzero** value without changing the balance in the equation.

Strategy for Solving Equations (page 156)

1. Use the distributive law to remove parentheses on each side of the equation.
2. Combine like terms on each side of the equation.
3. Use addition and subtraction rules to get like terms together on the same side of the equation. Combine like terms in the new equation.
4. Use the multiplication and division rules to solve for the variable.
5. Check your solution by substituting it into the original equation.

Summary Test--Chapter 2

Part 1: Do not use a calculator.

1. Do the arithmetic in the following problems. Show the steps of arithmetic when appropriate.

   a. 87 - 93 = ______________ b. (-5)(-8) = ______________

   c. -15/(-3) = ______________ d. -9 - (-15) = ______________

   e. (9)(-8) = ______________ f. -19 + (-3) = ______________

   g. -7 + 8 -11 + 3 = __________ h. -4(3 - 9) = ______________

2. In each of the following, circle whether the answer is positive or negative.

   a. -897.83 + 763 positive negative b. -(-567)÷(-78) positive negative

   c. -198^2 positive negative d. $(-78)^3$ positive negative

   e. 8.90 - (-7.5) positive negative f. -198 + 215.9 positive negative

3. In each of the following pairs of numbers, insert the sign > or < between the numbers to correctly express the comparison between the two numbers.

   a. 19 -28 b. -89 -67 c. 16.89 9.999

4. In each of the following checkbook transactions, determine the balance after the transaction.

   a. Beginning balance = $83.00
   A check is written for $78.26
   Ending balance = ________

   b. Beginning balance = $215.89
   A deposit of $275 is made.
   Ending balance = __________

   c. The account is overdrawn by $12.98
   A deposit of $25 is made.
   Ending balance = ________

   d. Beginning balance = $14.98
   A check is written for $20
   Ending balance = __________

   e. Beginning balance = $300.10
   A check is written for $450.00
   A deposit is made for $100.00
   Ending balance = ________

   f. The account is overdrawn by $72.00
   A check is written for $12.00
   A deposit of $95.00 is made.
   Ending balance = __________

5. Evaluate the following expressions involving absolute values.

a. $|-3| - (-2)$ b. $4 - |-3|$

c. $4||$ d. $4|-|$

6. In each of the following, add, subtract, and combine like terms as indicated.

a. $5x + 3y - 7 - x + y$ b. $(7t - 11a) + (4t - 5)$

c. Add $-5x + 3$ to $-5y + 3x$ d. $14 + 3x - y - 4 - 4x$

7. Solve each of the following equations for the variable.

a. $5x - 7 = -3x + 2$ b. $-5y - 2 = y - 8$

c. $3(2x - 4) + 1 = 2(1 - x) - 1$ d. $-14x + 14 = 3x - 3$

Part 2: Use a calculator as needed.

8. Compute the following arithmetic problems.

a. $(-22.5)(17.4) - 12.98/17.3$ b. $\dfrac{-14(22)+4.9(-22)}{14.8\cdot 3^2-15}$

c. $-5(15.2 + 8) - 278.93$ d. $-11.2 + 23 + 84 - 17 - 4$

9. Evaluate each of the following if $k = -22.3$, $m = 18$, $n = 59.8$, and $p = -0.78$.

a. $4kp =$ __________ b. $-3 \div k =$ __________ c. $2p^2n =$ __________

d. $n/p =$ __________ e. $3m - n =$ __________ f. $3m - k =$ __________

10. The beginning balance in a checkbook is \$756.89. The following transactions are made.

A check is written for \$82.75.
A check is written for \$328.96
A deposit of \$80.00 is made.
A check is written for \$415.98

a. What is the balance in the account after the four transactions? (In case there is not enough money in the account, assume the bank will carry a negative balance.)

b. How much money would have to be deposited before a check for \$298.67 would clear the bank?

Challenge Questions

11. At 7:00 a.m. the temperature was 64° F, but it was 82° F by 1:00 p.m. What was the rise in temperature from 7:00 to 1:00?

12. At 3:00 a.m. the temperature was - 4° F. The temperature rose 15 degrees by noon. What was the temperature at noon? (Hint: Put the number on a number line.)

13. At the beginning of the month, Heather owed $237.53. Ten days later, she had made enough money to pay the debt and have $38.73 left over. How much money did she make?

14. At the beginning of the month, Heather owed $237.53. Ten days later, she had made enough money to pay $120.45 of the debt. How much did she still owe?

15. Steve dived from a diving board that was 10 feet above the water. At his lowest point, he was 6.5 feet below the surface of the water. How far did Steve fall from the board to his lowest point on the dive?

Chapter 3 Factors and Geometric Shapes

Key Topics

- Geometric shapes and formulas
- The Meaning of π
- The Pythagorean theorem
- Factoring and factors
- Greatest common factor and least common multiple

Materials Needed for the exercises (one set per group)

- Calculator
- Ruler with inch and centimeter markings
- Graph paper
- 25 squares of the same size (1/2 in. by 1/2 in. is a good size)
- 40 cubes of the same size (1/2 in. by 1/2 in. by 1/2 in. is a good size)
- 3 circular objects of different sizes such as lids
- String
- Large sheet of paper such as a sheet of newsprint
- Scissors
- Meter stick or metric measuring tape
- Yardstick or measuring tape

Section 3-1 *Linear Measurements and Units*

A **linear unit** is a particular length used for measuring the length of a line segment. The units inch, foot, yard, mile, centimeter, meter, and kilometer are called linear units because they are used to measure distances or lengths. We use linear units to measure our height, the distance across a room, or the width of a doorway. We measure such lengths in inches, feet, centimeters, or meters. For greater lengths, such as the distance between two cities or the radius of Earth, we use larger units, such as miles or kilometers.

Historically, a foot was the approximate length of a man's foot. An inch and a yard can also be approximated with body measurements. To approximate an inch, measure the length from the middle joint to the top of your thumb.

A seamstress or tailor may estimate a yard of fabric by stretching the fabric between the nose and the stretched out arm length. Try measuring 1 yard of rope or string in this manner. Check your measurement with a yardstick or tape measure to see how close to 1 yard your stretch is.

A foot, inch, or yard were not very standard when they were first used as measurements--the size of the unit varied from person to person depending on the size of the foot of whoever was doing the measuring. As society's needs for consistent, accurate units grew, the lengths of an inch, foot, and yard were standardized to the lengths we now have on our rulers, tape measures, and yardsticks. The English system, as it is historically called, uses inches, feet, yards, and miles to measure distances. The United States is one of the few countries that still uses this measurement system.

Most countries, including England, now use the metric system of measurement instead of the English system. In the United States, the metric system is becoming more prevalent. Some of the units of linear measurement in the metric system are centimeters, meters, and kilometers.

Exercise 1--Measuring Lengths (Group)

Purpose of the Exercise
- To learn about standard and nonstandard measurements

Materials Needed
- Yardsticks, meter sticks, or measuring tapes

Each group will do one of the following, and then report the results to the class.

1. Measure the length of the classroom from front to back, side to side, and corner to corner in foot-lengths. To do this, have one group member walk from one wall to the other at each step placing heel of front foot against toe of back foot. On the board, make a drawing to represent the floor of the room and label the measurements on the drawing. For example, if the width of the room is found to be 20 foot-lengths, the label should read, "20 of Laura's feet" (if Laura paced off the room).

2. Using a tape measure or a yardstick, measure the length of the classroom in feet from front to back, side to side, and corner to corner. On the board, make a drawing to represent the floor of the room and label the measurements on the drawing. When writing the labels include units, for example, "20 feet," not just "20."

3. Using a tape measure or a meter stick, measure the length of the classroom in meters from front to back, side to side, and corner to corner. On the board, make a drawing to represent the floor of the room and label the measurements on the drawing. When writing the labels include units, for example, "20 meters," not just "20."

4. Using a tape measure or a yardstick, measure the length of the classroom in yards from front to back, side to side, and corner to corner. On the board, make a drawing to represent the floor of the room and label the measurements on the drawing. When writing the labels include units, for example, "20 yards."

5. Using a tape measure or yardstick, measure the length of a table or desk in inches from front to back, side to side, and corner to corner. On the board, make a drawing to represent the top of the desk and label the measurements on the drawing. When writing the labels include units, for example, "20 inches."

6. Using a tape measure or meter stick, measure the length of a table or desk in centimeters from front to back, side to side, and corner to corner. On the board, make a drawing to represent the top of the desk and label the measurements on the drawing. When writing the labels include units, for example, "20 centimeters."

7. Measure each group member's height in inches, then in centimeters. On the board, make a drawing to represent each person and label the measurements both in inches and in centimeters. When writing the labels, include units.

8. Using a tape measure or yardstick, measure the width and height of a doorway in inches, as well as the diagonal distance from corner to corner. On the board, make a drawing to represent the doorway, and label the measurements on the drawing. When writing the labels, be sure to include units.

Any group that finishes early should carry out the measurements again using a different unit. When all the results have been recorded on the board, each group will report what they did and their results.

9. The entire class lists the kinds of things they measured and discuss what is meant by "linear" measurements. In the discussion, name at least five things that will not fit in the classroom that would be measured in yards, miles, or kilometers.

Section 3-2 *Rectangles--Shapes with Dimensions*

As we have learned, a **linear unit** is a particular length used for measuring the length of a line segment--for example, a foot (ft), an inch (in.), a meter (m), a mile (mi), or a kilometer (km). These units are called linear because they measure lengths along a line. Lines are said to be one-dimensional because they have length, but no width or height.

A rectangle is two-dimensional because it is measured in two directions--vertically and horizontally. We call the two **dimensions** of a rectangle **length** and **width**. A square is a special type of rectangle for which the length and width are equal. A rectangle, as shown below, is a four-sided closed figure where the opposite sides are parallel and the angles are all 90° (as when a vertical line meets a horizontal line).

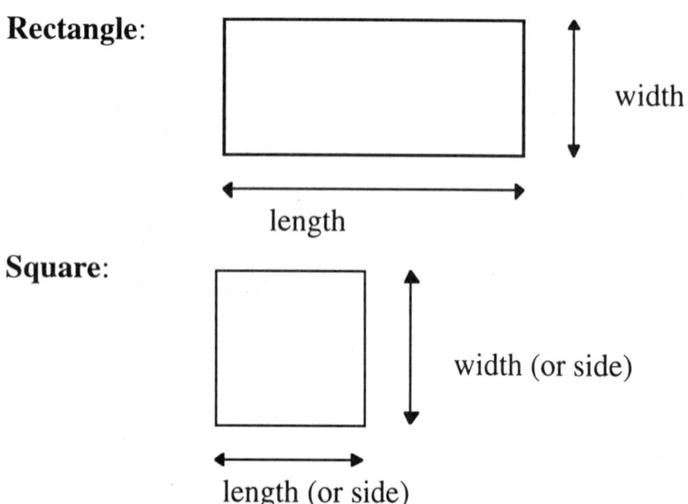

When working with rectangles, we use linear units to measure lengths of sides. We use **square units** in measuring what we call the **area** of a rectangle. In the following exercise, you will be constructing rectangles from squares and drawing conclusions about what is called the perimeter and the area of a rectangle. Before jumping in, however, here are some vocabulary terms that you will need.

Vocabulary for Rectangles

1. A nonsquare rectangle has two longer sides and two shorter sides. We commonly call the longer sides the length of the rectangle and the shorter sides the width. The **length** and **width** are the **dimensions** of the rectangle. (The definitions of the length and width of a rectangle are interchangeable--the longer sides can be the width, and the shorter sides the length.)
2. If the rectangle is a **square**, all four sides have the same measurement. In a square, the length equals the width.
3. The space inside the rectangle is the **area** of the rectangle. Since we measure that space by the number of squares of a particular size it takes to fill the space, area is measured in **square units.**
4. A **square unit** is two-dimensional and has the shape of a square--as in square inch (in.^{2}), square foot (ft^2), square mile (mi^2).
5. The length around the entire rectangle is called the **perimeter** of the rectangle. There are four sides to the rectangle, so the perimeter is just the sum of the lengths of the four sides.

Exercise 1--Finding the Area and Perimeter of a Rectangle

Purpose of Exercise
- To better understand dimensions, rectangles, square units, area, and perimeter

Materials Needed
- Graph paper
- Straight edge or ruler

Vocabulary
When we draw a square that is 1 unit length by 1 unit length, this square is called a **square unit**, abbreviated "U^2" or "sq. U." If the linear unit is a hand, we might write, "H^2" or "sq. H." If the unit is an inch we would write "$in.^2$" or "sq. in."

In each of the following, on graph paper, draw the rectangle described and carry out each of the following steps.
- Count the number of rows of squares and the number of squares in each row of the rectangle.
- Count and record the number of squares in the entire rectangle.
- Count and record the number of linear units around the entire rectangle.
- Using the side of one square for your unit, record the length and width of the rectangle. Include proper units with your answer.
- Using the side of one square for your unit, record the perimeter and area of the rectangle. Include proper units with your answer.

The first problem is done as an example.

1. Draw 4 squares side by side and then 4 other squares directly below the first 4 squares.

Number of rows of squares = 2 Number of squares in each row = 4

Total number of squares = 8 Number of units around rectangle = 12

length = 2U width = 4U perimeter = 12 U area = 8 U^2

Or, since the length and width are interchangeable, we could have said

length = 4U width = 2U perimeter = 12 U area = 8 U^2

2. Draw 3 squares side by side.

Number of rows of squares = _____ Number of squares in each row = _______

Total number of squares = ________ Number of units around rectangle = _____

length = _______ width = _______ perimeter = _______ area = _______

3. Draw 1 square that is 1 unit long each direction.

Number of rows of squares = _____ Number of squares in each row = _______

Total number of squares = ________ Number of units around rectangle = _____

length = _______ width = _______ perimeter = _______ area = _______

4. Draw 3 squares side by side connected to 3 squares directly below the first 3.

   Number of rows of squares = _____ Number of squares in each row = _______

   Total number of squares = ________ Number of units around rectangle = ____

   length = _______ width = _______ perimeter = _______ area = _______

5. Draw a rectangle that is 4 units by 3 units.

   Number of rows of squares = _____ Number of squares in each row = _______

   Total number of squares = ________ Number of units around rectangle = ____

   length = _______ width = _______ perimeter = _______ area = _______

6. Experiment with other drawings, putting together squares to form a rectangle.

   Number of rows of squares = _____ Number of squares in each row = _______

   Total number of squares = ________ Number of units around rectangle = ____

   length = _______ width = _______ perimeter = _______ area = _______

   Number of rows of squares = _____ Number of squares in each row = _______

   Total number of squares = ________ Number of units around rectangle = ____

   length = _______ width = _______ perimeter = _______ area = _______

   Number of rows of squares = _____ Number of squares in each row = _______

   Total number of squares = ________ Number of units around rectangle = ____

   length = _______ width = _______ perimeter = _______ area = _______

7. Summarize what you discovered. If A stands for area, P for perimeter, L for length and W for width of a rectangle, write the formulas for area and perimeter.

   A = ____________________ P = ______________________

Exercise 2--Dividing a Square Unit into Parts

Purpose of the Exercise

- To study the area of a rectangle when it involves fractions

Materials Needed

- A ruler or straight edge

In the following exercises you will be working with fractions. You do not need to know how to multiply or add fractions to answer the questions. All you need to know is that if something is divided into 2 equal parts, each part is 1/2 of the whole. If something is divided into 3 equal parts, each part is 1/3 of the whole. Make the observations visually, not by multiplying fractions.

1. Shown below is one square unit. Divide the square into left and right halves by drawing a vertical line down the middle of the square. Shade the area to the right of the middle line.

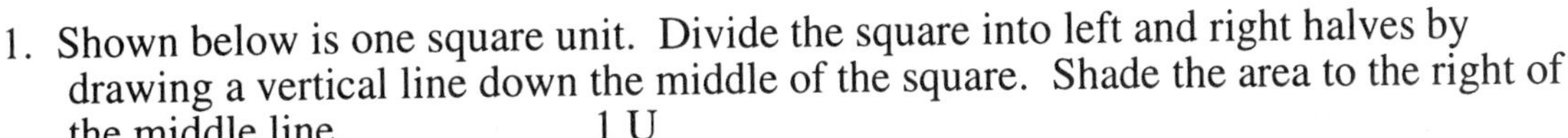

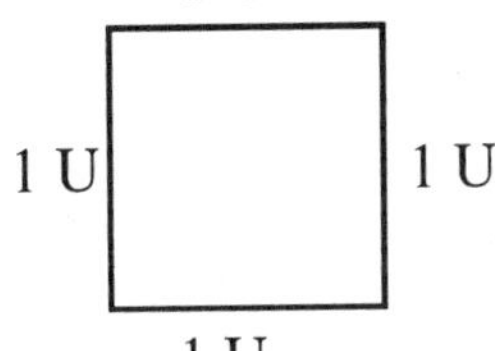

   a. On the drawing, label each of the dimensions of the **shaded area**. (Remember that each of the four sides of the original square is one unit, or 1 U.)

   b. Record the dimensions, perimeter, and area of the **entire square**. Include proper units with your answer.

   length = _______ width = _______ perimeter = _______ area = _______

   c. By observing how the shaded region compares to the entire region, determine and record the dimensions, perimeter, and area of the **shaded** region. Include proper units with your answer.

   length = _______ width = _______ perimeter = _______ area = _______

2. Shown below is one square unit. Draw a vertical line through the middle of the square and draw a horizontal line through the middle of the square. Shade the small square in the upper-left corner of the large square.

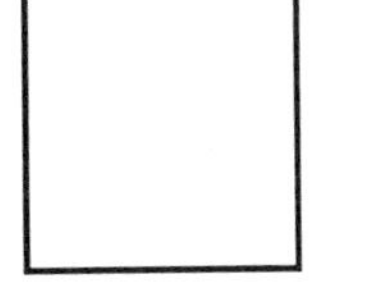

   a. On the drawing, label each of the dimensions of the shaded area. (Remember that each of the four sides of the original square is one unit.)

   b. By observing how the shaded region compares to the entire region, determine and record the dimensions, perimeter, and area of the **shaded** region. Include proper units with your answer.

   length = _______ width = _______ perimeter = _______ area = _______

Exercise 3--Using the Formulas You Have Discovered

Purpose of the Exercise

- To practice using the area and perimeter formulas

Materials Needed

- Graph paper

In each of the following, use graph paper to draw rectangles with the given dimensions. Compute the perimeter and area of the rectangle using the formulas you discovered.

1. length = 5 U, width = 2 U perimeter = ________ area = ________
2. length = 2 U, width = 3 U perimeter = ________ area = ________
3. length = 1.5 U, width = 2 U perimeter = ________ area = ________
4. length = 2.5 U, width = 2 U perimeter = ________ area = ________
5. length = 0.5 U, width = 4 U perimeter = ________ area = ________
6. length = 6 U, width = 4 U perimeter = ________ area = ________
7. length = 6 U, width = 1 U perimeter = ________ area = ________
8. length = 3 U, width = 5 U perimeter = ________ area = ________
9. length = 5 U, width = 0.7 U perimeter = ________ area = ________
10. length = 4 U, width = 0.2 U perimeter = ________ area = ________
11. length = 7 U, width = 6 U perimeter = ________ area = ________
12. length = 2.5 U, width = 3 U perimeter = ________ area = ________
13. length = 1 U, width = 0.5 U perimeter = ________ area = ________
14. length = 2 U, width = 5 U perimeter = ________ area = ________
15. length = 0.5 U, width = 2 U perimeter = ________ area = ________
16. length = 1.5 U, width = 2 U perimeter = ________ area = ________

Section 3-3 *Applications of Linear Units, Rectangles, Area, and Perimeter*

As we have already discovered, linear units are units like inches, yards, miles, meters, or kilometers that can be used to measure lengths. Square units are squares that are a measure of area. A square inch is a square in which the length and width are both 1 inch. A square foot is a square in which the length and width are both 1 foot.

Examples of linear measurements in real life

The dimensions (length and width) of a room
The height of a building
The distance between two buildings
The length of fence around a plot of land
The amount of lace to go around the edge of a table cloth
The distance from Earth to the Sun
The length of a stick
The length of a coastline
The distance from Tokyo to Singapore
The number of miles traveled
The thickness of a book
The width of a videotape
The length of a videotape
The dimensions of a countertop
The width of a window
The length of baseboard needed to go around a room

Examples of square measurements in real life

The floor space or area of a room
The square feet of counter space
The amount of land in a city
The land area of a country
Wall space
The amount of carpet needed to cover a floor

The length, width, and perimeter of a rectangle are measured in **linear units** because they measure lengths or distances. The abbreviation for linear measures do not have a square (a superscript 2) on the symbol.

To save some writing, we are going to **abbreviate area as A, length as L, width as W, and perimeter as P**.

Formulas

Area of a rectangle: **A = L • W** (The dot between the L and the W is a symbol for multiplication. The dot can also be omitted. The formula, A = LW is the same formula.)

Perimeter of a rectangle: **P = 2L + 2W**. (Or P = L + L + W + W)

Exercise 1--Recognizing Areas and Perimeters in Real Life

Purpose of the Exercise

- To distinguish between linear and square units
- To determine whether to use the area or the perimeter formula

In each of the following questions, determine whether the quantity mentioned is an area, a perimeter, or a dimension (length or width) of a rectangle. Determine whether the quantity would be described in linear or square units. Circle the correct answers.

1. Amount of fencing needed to enclose a farm

   Type of measurement: area perimeter dimension

   Type of unit: linear unit square unit

2. Amount of floor space in a building

   Type of measurement: area perimeter dimension

   Type of unit: linear unit square unit

3. Height of the Empire State Building

   Type of measurement: area perimeter dimension

   Type of unit: linear unit square unit

4. Decorative stripe around a picture frame

   Type of measurement: area perimeter dimension

   Type of unit: linear unit square unit

5. Length of a cake pan

   Type of measurement: area perimeter dimension

   Type of unit: linear unit square unit

Exercise 2--Rectangles, Area, and Perimeter--Applications and Drill

Purpose of the Exercise

- To determine, in context, when to use the area and perimeter formulas
- To practice using the area and perimeter formulas

In each of the following problems, determine what formula to use; then compute the quantities. Include proper units with each answer. Use a calculator as needed.

1. A plot of land is 3 miles by 7 miles. Find the land area and how much fencing it would take to fence in the land.

   Formula for land area ________ Formula for length of fencing ___________

   Land area = ________________ Length of fencing = ____________________

2. A scarf is 32 inches by 8 inches. How many square inches of fabric are in the scarf and how much trim would it take to go around the scarf?

   Formula for no. of square inches ________ Formula for amount of trim _______

   Sq. inches = ________________ How much trim? = ______________________

3. Suppose a country is approximately 179 mi by 253 mi. How many square miles are in the country and how long is its entire border?

   Formula to compute no. of square miles _______ Formula for length of border _____

   Number of square miles = _____________ Length of border = _____________

4. A wall is 14 ft long by 8 ft high. Suppose you want to paint the wall and also to put trim across the top of the wall. What is the area of the wall and how much trim would it take?

   Formula for area ___________ Formula for length of trim _______________

   Area = ___________________ Length of trim = _____________________

5. For a room that is 8 feet by 11 feet:
   a. Formula for computing square feet of ceiling tiles ___________________
   b. How many square feet of ceiling tiles are needed to redo the ceiling? __________
   c. Formula for computing the length of baseboard around the room _____________
   d. How many feet of baseboard are needed to go around the room? ____________

6. For a one-story house that is 50 ft by 35 ft:
   a. Formula for computing the amount of living space ____________________
   b. How much living space is there in the house (in square feet)? _____________
   c. Formula for length of ribbon around the house ______________________
   d. If you were going to put a ribbon around the entire house, how much ribbon would it take? ___________________

7. Find the area and perimeter in each of the following rectangles.
   a. L = 17.3 km, W = 14.9 km   A = ___________ P = ___________
   b. L = 5280 ft, W = 10 ft   A = ___________ P = ___________
   c. L = 8.5 in, W = 11 in   A = ___________ P = ___________
   d. L = 12 ft, W = 14 ft   A = ___________ P = ___________
   e. L = 92 cm, W =81 cm   A = ___________ P = ___________
   f. L = 100 cm, W = 100 cm   A = ___________ P = ___________
   g. L = 2 mi, W = 2 mi   A = ___________ P = ___________

Abbreviations

Centimeter: cm	Meter: m
Mile: mi	Foot: ft
Inch: in.	Yard: yd

Table 3-1
100 centimeters = 1 meter
12 inches = 1 foot
3 feet = 1 yard
5280 feet = 1 mile

Exercise 3--Units and Conversions

Purpose of the Exercise

- Comparisons between linear units and comparisons between the related square units

Materials Needed

- Calculator

1. Compute the following areas and perimeters.

   a. L = 5280 ft, W = 5280 ft A = __________ P = __________

   b. L = 1 m, W = 1 m A = __________ P = __________

   c. L = 3 ft, W = 3 ft A = __________ P = __________

   d. L = 1 yd, W = 1 yd A = __________ P = __________

   e. L = 12 in., W = 12 in. A = __________ P = __________

   f. L = 1 ft, W = 1 ft A = __________ P = __________

   g. L = 100 cm, W = 100 cm A = __________ P = __________

   h. L = 1 m, W = 1 m A = __________ P = __________

2. Compare the areas in parts a through f above. Review the information in Table 3-1 and then answer the following questions.

   a. 1 meter = __________ centimeters

   b. 1 square meter = __________ square centimeters

   c. 1 mile = __________ feet

   d. 1 square mile = __________ square feet

   e. 1 yard = __________ feet

   f. 1 square yard = __________ square feet

   g. 1 foot = __________ inches

   h. 1 square foot = __________ square inches

Section 3-4 *Comparison between Linear Units and Square Units*

We know that there are 3 feet in 1 yard, so sometimes we make the mistake of thinking that there are 3 square feet in 1 square yard. The relationship between linear units such as feet and yards is not the same as the relationship between their square units--square feet and square yards. In this section we will learn how the relationship between two linear units determines the relationship between the corresponding square units.

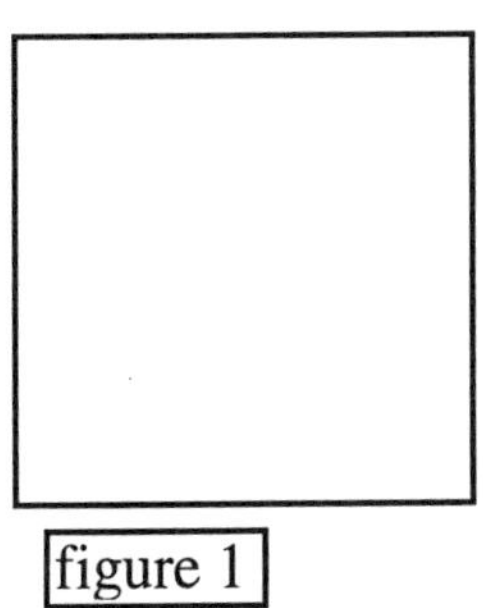

figure 1

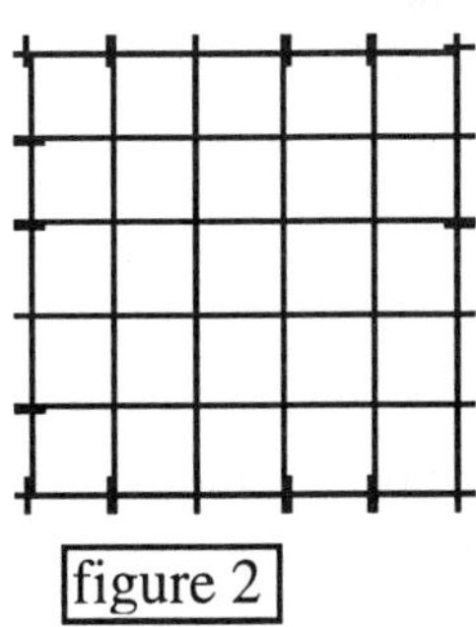

figure 2

If we treat the length of the square in figure 1 as one unit (1 U), and divide the square both vertically and horizontally into 5 equal parts as in figure 2, we can make the following observations.

- The length of a side of each of the small squares is 1/5 U.
- The area of the entire (large square) is 1 square unit, or 1 U^2.
- The grid cuts the larger square into 25 smaller squares.
- 1 of the small squares makes up 1/25 of the large square.
- 2 of the small squares makes up 2/25 of the large square.
- 2 large squares would be 50 of the small squares.

In the exercises below, you will work with fractions, but only visually. You do not need to know how to do arithmetic with fractions to do the exercises. You do need to know that when something is divided into 5 equal parts, for example that one of the parts is 1/5 of the whole, two parts are 2/5, and so on.

Exercise 1--A Square Divided into Smaller Squares

Purpose of Exercise
- To improve visual understanding of units, fractions, and comparisons of square and linear units

Materials Needed
- 6-inch or 12-inch ruler

1 in.

1 in.

- Divide the square on the left with **3 vertical lines** at the 1/4-, 1/2-, and 3/4-inch marks.
- Divide the square with **3 horizontal lines** at the 1/4-, 1/2- and 3/4-inch marks.

Answer the following questions by observing the diagram and your ruler. Trace the diagram onto another sheet of paper as many times as needed to answer the questions. Include proper units where appropriate. If the answer does not represent a length or area, do not put a unit with your answer.

1. This grid cuts the square inch into how many small squares? ________________
2. What are the dimensions of each of the small squares? ________________
3. What fraction of a square inch is 1 of the small squares? ______________
4. What fraction of a square inch is 3 of the small squares? ________________
5. What fraction of a square inch is 4 of the small squares? _______________
6. What is the area of a rectangle that is 2 inches by 4 inches? _____________
7. What fraction of a square inch is 16 of the small squares? _______________
8. How many of the small squares would fit in 2 square inches? ____________
9. How many of the small squares would fit in 10 square inches? ___________
10. How many 1/4 inches make up 1 inch? ___________________________
11. How many 1/4-inch squares make up 1 square inch? ________________
12. How many 1/4 inches make up 2 inches? ___________________________
13. How many 1/4-inch squares make up 2 square inches? ________________
14. How many 1/4 inches make up 3 inches? ___________________________
15. How many 1/4-inch squares make up 3 square inches? ________________

Exercise 2--Inches, Yards, Meters, Square Inches, Square Yards, Square Meters

Purpose of the Exercise

- To compare inches and feet and square inches and square feet
- To compare inches and yards, square inches, and square yards
- To compare feet and yards, square feet and square yards
- To compare centimeters and meters, square centimeters and square meters

Materials Needed

- Meter stick, yardstick, 12-inch ruler (or use the yardstick)
- Board space or large sheets of newsprint taped together for a large drawing surface

Work in four groups and follow the instructions for your group.

Group 1

First: Observe markings on the meter stick to answer the following questions.

a. How many centimeters are in a meter? ______________________________
b. 1 centimeter is how much of a meter? ______________________________
c. 5 centimeters is how much of a meter? ______________________________
d. 17 centimeters is how much of a meter? _____________________________
e. 99 centimeters is how much of a meter? _____________________________
f. How many centimeters would it take to equal 2 meters? ________________
g. How many centimeters would it take to equal 1/2 meter? _______________

Second:

a. Using your meter stick, draw a square on the board that is 1 meter by 1 meter.
b. Make marks for the first 3 centimeters on each of the 4 sides of the square meter.
c. Connect the marks on opposite sides to begin to divide the square meter into a grid of square centimeters.

Third: By observing the square meter divided into a grid, answer the following questions:

a. How many square centimeters are in a square meter? ____________________
b. 1 square centimeter is how much of a square meter? ____________________
c. 5 square centimeters is how much of a square meter? ____________________
d. 17 square centimeters is how much of a square meter? ____________________
e. 99 square centimeters is how much of a square meter? ____________________
f. How many square centimeters would it take to equal 2 square meters? __________
g. How many square centimeters would it take to equal 1/2 square meter? __________

Group 2

First: Answer the following questions by observing markings on a yardstick.

a. How many inches are in a yard? ____________________
b. 1 inch is how much of a yard? ____________________
c. 5 inches is how much of a yard? ____________________
d. 17 inches is how much of a yard? ____________________
e. 35 inches is how much of a yard? ____________________
f. How many inches would it take to equal 2 yards? ____________________
g. How many inches would it take to equal 1/2 yard? ____________________

Second:

a. Use your yardstick to draw a square on the board that is 1 yard by 1 yard.
b. Make marks for the first 3 inches along each of the 4 sides.
c. Connect the marks on opposite sides to begin to divide the square yard into a grid of square inches.

Third: By observing the square yard cut into a grid, answer the following questions:

a. How many square inches are in a square yard? ____________________
b. 1 square inch is how much of a square yard? ____________________
c. 5 square inches is how much of a square yard? ____________________
d. 17 square inches is how much of a square yard? ____________________
e. 35 square inches is how much of a square yard? ____________________
f. How many square inches would it take to equal 2 square yards? __________
g. How many square inches would it take to equal 1/2 square yard? __________

Group 3

First: Answer the following questions by observing markings on a yardstick.

a. How many feet are in a yard? ____________________
b. 1 foot is how much of a yard? ____________________
c. 5 feet is how much of a yard? ____________________
d. 17 feet is how much of a yard? ____________________
e. 35 feet is how much of a yard? ____________________
f. How many feet would it take to equal 2 yards? ____________________
g. How many feet would it take to equal 1/2 yard? ____________________

Second:

a. Use your yardstick to draw a square on the board that is 1 yard by 1 yard.
b. Make marks for each foot on each of the 4 sides.
c. Connect the marks on opposite sides to divide the square yard into a grid of square feet.

Third: By observing the square yard divided into a grid, answer the following questions:

a. How many square feet are in a square yard? ____________________
b. 1 square foot is how much of a square yard? ____________________
c. 5 square feet is how much of a square yard? ____________________
d. 17 square feet is how much of a square yard? ____________________
e. 35 square feet is how much of a square yard? ____________________
f. How many square feet would it take to equal 2 square yards? __________
g. How many square feet would it take to equal 1/2 square yard? __________

Group 4

First: Observe markings on a 12-inch ruler to answer the following questions.

a. How many inches are in a foot? ____________________

b. 1 inch is how much of a foot? ____________________

c. 5 inches is how much of a foot? ____________________

d. 17 inches is how much of a foot? ____________________

e. 35 inches is how much of a foot? ____________________

f. How many inches would it take to equal 2 feet? ____________________

g. How many inches would it take to equal 1/2 foot? ____________________

Second:

a. Use your ruler to draw a square on the board that is 1 foot by 1 foot.

b. Make marks for each inch on each of the 4 sides.

c. Connect the marks on opposite sides to divide the square foot into a grid of square inches.

d. If time is short, make only the first few vertical and horizontal lines but notice how the grid will look.

Third: By observing the square foot divided into a grid, answer the following questions:

a. How many square inches are in a square foot? ____________________

b. 1 square inch is how much of a square foot? ____________________

c. 5 square inches is how much of a square foot? ____________________

d. 17 square inches is how much of a square foot? ____________________

e. 35 square inches is how much of a square foot? ____________________

f. How many square inches would it take to equal 2 square feet? __________

g. How many square inches would it take to equal 1/2 square foot? __________

Exercise 3--Challenge Questions

Purpose of the Exercise

- To apply your knowledge and skills to tougher questions and concepts

Materials Needed

- Calculator

Information: 1 mile = 5280 feet 1 mile = 1760 yards

1. 1 square mile = __________ square feet
2. 1 square mile = __________ sq. yards
3. 1 foot = __________ miles
4. 1 square foot = __________ sq. miles
5. 1 yard = __________ miles
6. 1 square yard = __________ sq. miles
7. 2900 feet = __________ miles
8. 2900 square feet = __________ sq. miles
9. 2900 yards = __________ miles
10. 2900 square yards = __________ sq. miles
11. 10 miles = __________ feet
12. 10 miles = __________ yards
13. 10 sq. miles = __________ sq. feet
14. 10 square miles = __________ sq. yards

Questions 15-20 apply to a rectangular plot of land that is 2 miles wide and 3 miles long. Include appropriate units with your answers.

15. The dimensions and perimeter of the property in miles are:
L = _______ W = _______ P = _______

16. The dimensions and perimeter of the property in feet are:
L = _______ W = _______ P = _______

17. The dimensions and perimeter of the property in yards are:

L = _______ W = _______ P = _______

18. The area of the property in square miles is: ________________________________
19. The area of the property in square yards is: ________________________________

20. The area of the property in square feet is: ________________________________

21. 5280 ft = 1 mi

a. 1 ft = ____________ mi | b. 3 mi = ____________ ft

c. 5 ft = ____________ mi | d. 15 ft = ____________ mi

e. 1 mi^2 = __________ ft^2 | f. 1 ft^2 = ____________ mi^2

g. 5 mi^2 = __________ ft^2 | h. 5479 mi^2 = __________ ft^2

22. 1000 m = 1 km

a. 1 m = ____________ km | b. 3 km = ____________ m

c. 3 m = ____________ km | d. 15 km = ____________ m

e. 1 km^2 = __________ m^2 | f. 1 m^2 = ____________ km^2

g. 5 km^2 = __________ m^2 | h. 5 m^2 = ____________ km^2

23. 4 T's make up 1 R where T and R are linear units. In the drawing, 1 R is the length of the side of the large square and 1 T is the length of the side of a small square.

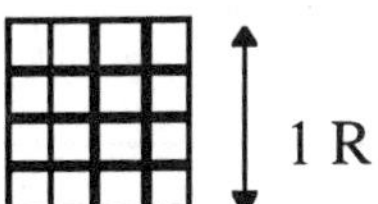

1 R ———
1 T —

a. 1 R = __________ T | b. 2 R = ____________ T

c. 1 T = __________ R | d. 3 T = ____________ R

e. 1 R^2 = __________ T^2 | f. 5 R^2 = ____________ T^2

g. 1 T^2 = __________ R^2 | h. 7 T^2 = ____________ R^2

24. 47 T's make up 1 R where T and R are linear units.

a. 1 R = __________ T | b. 5 R = ____________ T

c. 1 R^2 = __________ T^2 | d. 5 R^2 = ____________ T^2

e. 1 T = __________ R | f. 7 T = ____________ R

g. 1 T^2 = __________ R^2 | h. 25 T^2 ____________ R^2

Section 3-5 *Squares as Shapes and Squares as Units*

As you recall from Section 1-3, a square is a rectangle in which the length equals the width. We use the word "square" in several different contexts.

L = 5 U
W = 3 U
A = 15 U^2

In the above rectangle, L =5 U and W = 3 U, so the area A = 15 square units or 15 U^2. The U^2 is a "square unit" for measuring the area.

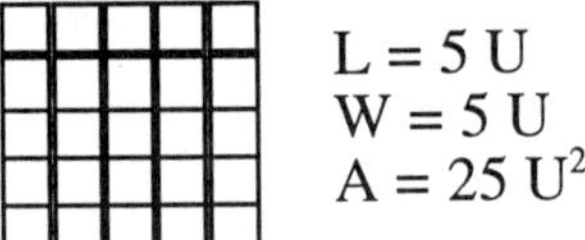

L = 5 U
W = 5 U
A = 25 U^2

Not only is the shape a "5-unit square," meaning it is a square that measures 5 U by 5 U, but U^2 is a "square unit" for measuring the area.

The formulas for area and perimeter of a square come directly from the formula for area and perimeter of a rectangle. Since the length and width of a square are the same, it is redundant to use the letters L and W to represent the same length. Instead, we use the one letter "s" for the length of the side of a square. The area formula for a rectangle, $A = L \cdot W$, becomes $A = s \cdot s$ which we write as s^2 and read as "s squared" or "s raised to the second power." The perimeter of a square is the length around the square. Since a square has four sides, P = 4s.

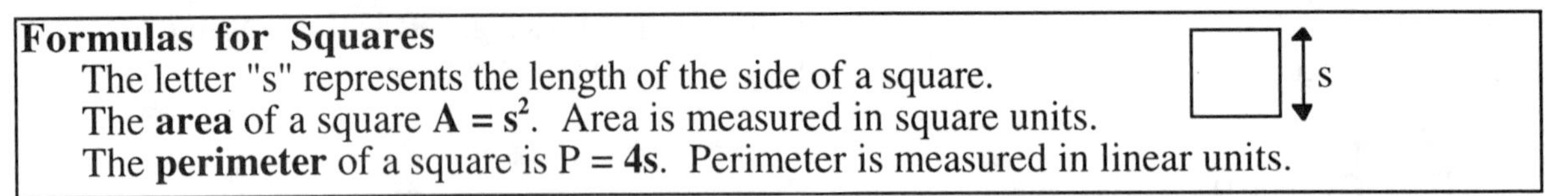

Formulas for Squares

The letter "s" represents the length of the side of a square.
The **area** of a square $\mathbf{A = s^2}$. Area is measured in square units.
The **perimeter** of a square is P = **4s**. Perimeter is measured in linear units.

In the formula, $A = s^2$, we used the symbol s^2 to mean $s \cdot s$, or s times s. We have also used similar symbols, U^2 or m^2 or in.^{2} to mean square units, square meters, or square inches, respectfully. It is important to notice the distinction between the two symbols.

In the formula, $A = s^2$, s is a variable, not a unit. If the value of s is 9, s^2 means 9^2 or $9 \cdot 9$,which is 81.

9 in.

$A = 9^2$ square inches
= 81 in.^{2}

If you measure the square, you will notice that it is not actually 9 inches in length and width. Instead, it has been drawn "to scale" to fit the space.

The above square is called a **9-inch square** since it is 9 inches by 9 inches. A "9-inch square," however, is not to be confused with a rectangle that has an area of "9 square inches." Such a rectangle might have L = 9 in. and W = 1 in.

Example 1: A room with an area of **80 ft^2** might be **8 ft by 10 ft,** or **5 ft by 16 ft,** or **20 ft by 4 ft.** A building with an **80-foot square** foundation is a square that is **80 feet by 80 feet.**

Example 2: A 15-inch square is a square that is 15 inches by 15 inches.

Area = 225 square inches

A 15 square-inch rectangle is a rectangle with an area of 15 square inches. There are two rectangles described below that could fit the description:

L = 15 in., W = 1 in., A = 15 in.^{2}

L = 5 in., W = 3 in., A = 15 in.^{2}

Exercise 1--Symbols for Square Units and Numbers or Variables Squared

Purpose of the Exercise

- To distinguish between symbols for square units and symbols for numbers squared
- To practice using the area and perimeter formulas for squares

Materials Needed

- Calculator

In each of the following, state whether the "squared" symbol is the name of a square unit or whether it indicates a number to be squared.

1. 45 ft^2 ____________________ 2. s^2 ____________________

3. 14^2 ____________________ 4. 14 km^2 ____________________

5. 7 m^2 ____________________ 6. 7^2 m ____________________

7. 9 in.^{2} ____________________ 8. 9^2 in. ____________________

Using the formulas $A = s^2$ and $P = 4s$, compute the area and perimeter of each of the following squares. Include proper units with your answer. Use a calculator as needed.

9. s = 82 mi, A = ____________ 10. s = 17.2 in., A = ____________
   P = ____________ P = ____________

11. s = 1.25 cm, A = ____________ 12. s = 24 ft, A = ____________
   P = ____________ P = ____________

13. s = 287 km, A = ____________ 14. s = 96.8 yd, A = ____________
   P = ____________ P = ____________

15. s = 489 mi, A = ____________ 16. s = 189.5 yd, A = ____________
   P = ____________ P = ____________

17. s = 5.36 m, A = ____________ 18. s = 111 km, A = ____________
   P = ____________ P = ____________

Exercise 2-- 5 Square Inches or 5-Inch Square

Purpose of the Exercise

- To compare and understand terminology such as 4-inch square vs. 4 square inches

Materials Needed

- 6-inch or 12-inch ruler
- Calculator (optional)

In Problems 1-6, draw the square or rectangle described. There may be more than one way to draw the rectangles--one diagram will be enough. Give the area and dimensions of each shape.

1. A 4-inch square

A = ______ L = ________ W = ______

2. A rectangle with an area of 4 square inches.

A = ______ L = ________ W = ______

3. A 7-inch square

A = ______ L = ________ W = ______

4. A rectangle with an area of 7 square inches.

A = ______ L = ________ W = ______

5. A 2-inch square

A = ______ L = ________ W = ______

6. A rectangle with an area of 2 square inches.

A = ______ L = ________ W = ______

In Problems 7-18, record the area and dimensions of each shape.

7. An 8-inch square

A = ______ L = ________ W = ______

8. A rectangle with an area of 8 square inches.

A = ______ L = ________ W = ______

9. A 12-inch square

A = ______ L = ________ W = ______

10. A rectangle with an area of 12 square inches.

A = ______ L = ________ W = ______

11. A 14-inch square

A = ______ L = ________ W = ______

12. A rectangle with an area of 14 square inches.

A = ______ L = ________ W = ______

13. A 81-inch square

A = ______ L = ________ W = ______

14. A rectangle with an area of 81 square inches.

A = ______ L = ________ W = ______

15. A 289-inch square

A = ______ L = ________ W = ______

16. A rectangle with an area of 289 square inches.

A = ______ L = ________ W = ______

17. A 36-inch square

A = ______ L = ________ W = ______

18. A rectangle with an area of 36 square inches.

A = ______ L = ________ W = ______

In previous exercises in the chapter, when using the formulas for area and perimeter, we substituted the numbers for the letters, did the arithmetic and then determined the unit, square or linear, that was appropriate to include with the answer. In the following examples, we will include the units with the numbers that we substitute into the formula and notice that when we include the unit in our computations from the beginning, we "multiply" two linear units to get a square unit or add 2 linear units to get a linear unit. Looking at the units included in the computations in this way, notice that the areas still turn out to be measured in square units and the perimeters are still measured in linear units.

Example 3

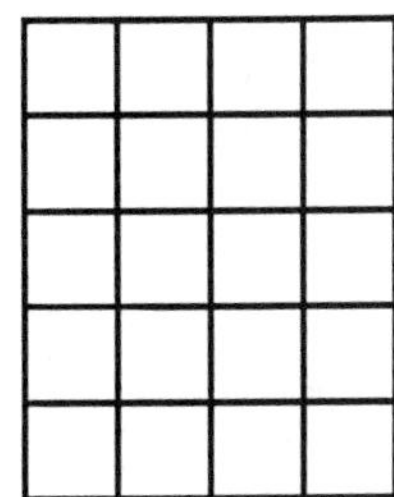

5 in.

4 in.

L = 4 in.

W = 5 in.

$A = LW = (4 \text{ in.})(5 \text{ in.}) = 20 \text{ in.}^2$

$P = 2L + 2W = 2(4 \text{ in.}) + 2(5 \text{ in.}) = 8 \text{ in.} + 10 \text{ in.} = 18 \text{ in.}$

Example 4

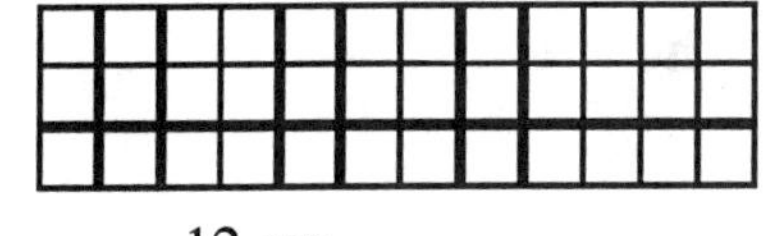

3 cm

12 cm

L = 12 cm

W = 3 cm

$A = (12 \text{ cm})(3 \text{ cm}) = 36 \text{ cm}^2$

$P = 2(12 \text{ cm}) + 2(3 \text{ cm}) = 30 \text{ cm}$

Example 5

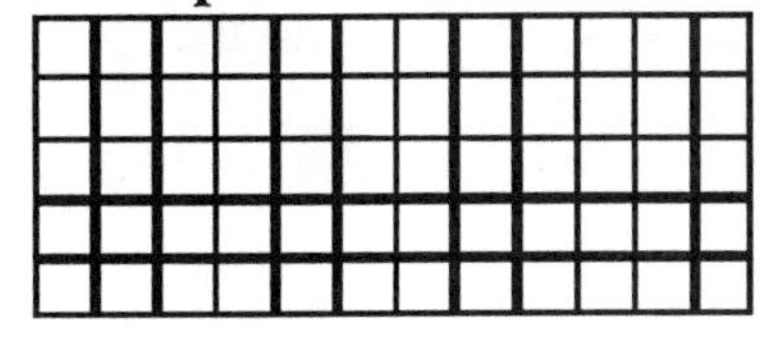

5 cm

12 cm

L = 12 cm

W = 5 cm

$A = (12 \text{ cm})(5 \text{ cm}) = 60 \text{ cm}^2$

$P = 2(12 \text{ cm}) + 2(5 \text{ cm}) = 24 \text{ cm} + 10 \text{ cm} = 34 \text{ cm}$

Example 6

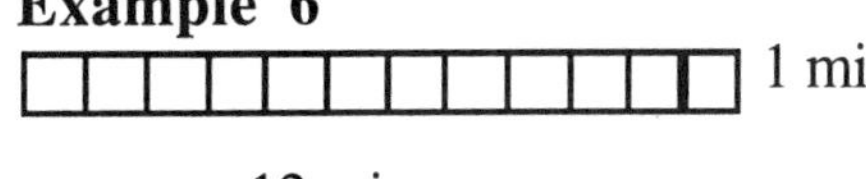

1 mi

12 mi

L = 12 mi

W = 1 mi

$A = (12 \text{ mi})(1 \text{ mi}) = 12 \text{ mi}^2$

$P = 2(12 \text{ mi}) + 2(1 \text{ mi}) = 24 \text{ mi} + 2 \text{ mi} = 26 \text{ mi}$

Example 7

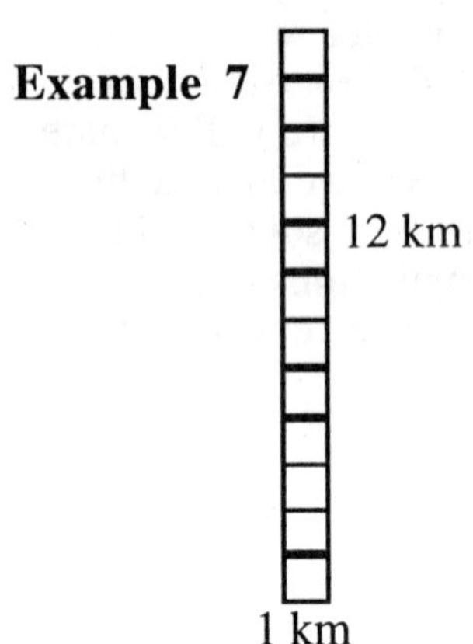

L = 1 km
W = 12 km
A = (1 mi)(12 mi) 12 km^2
P = 2(1 mi) + 2(12 mi) = 2 mi + 24 mi 26 km

Tricky Discussion Question In the following rectangle, L = 4 and W = 4.

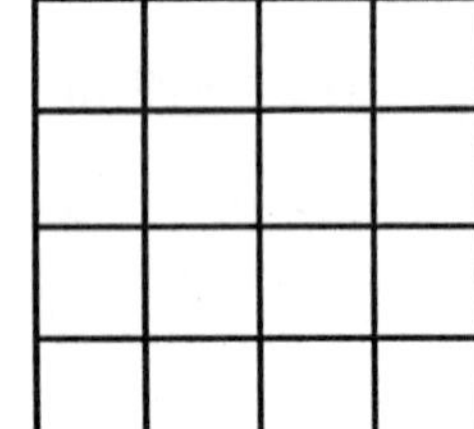

P = 2L + 2W = 2(4) + 2(4) = 16
A = LW = (4)(4) = 16

Does the area equal the perimeter in the square on the left?

Exercise 3--Perimeter and Area--Including Units in the Computations

Purpose of the Exercise

- To practice doing computations including units
- To practice getting information from a drawing to compute perimeter and area

In each of the following, determine the length and width from the drawing and compute the area and perimeter. As in the examples above, show the setup, including the units, and then do the computations.

1. 3 ft; 5 ft

L = ____________________

W = ____________________

P = ____________________ = ____________
setup answer

A = ____________________ = ____________
setup answer

2. 14.2 mi; 18.9 mi

L = ____________________

W = ____________________

P = ____________________ = ____________
setup answer

A = ____________________ = ____________
setup answer

3.

122 in.

54 in.

L = ____________________

W = ____________________

P = ____________________ = ____________
setup answer

A = ____________________ = ____________
setup answer

4.

122 cm.

54 cm.

L = ____________________

W = ____________________

P = ____________________ = ____________
setup answer

A = ____________________ = ____________
setup answer

5.

1.9 m

1.5 m

L = ____________________

W = ____________________

P = ____________________ = ____________
setup answer

A = ____________________ = ____________
setup answer

6.

17 yd

12 yd

L = ____________________

W = ____________________

P = ____________________ = ____________
setup answer

A = ____________________ = ____________
setup answer

7.

67 mi

29 mi

L = ____________________

W = ____________________

P = ____________________ = ____________
setup answer

A = ____________________ = ____________
setup answer

Section 3-6 *Factors and Prime Numbers*

As we have learned, the area of a rectangle consists of a number of square units and can be computed with the formula A = LW. In this section we will be using those same concepts in a little different way to learn how to "factor" a number. But first, let's cover some of the vocabulary needed to discuss the concepts.

Vocabulary

As we already know, the **whole numbers** are 0, 1, 2, 3, 4, 5, 6, 7, and so on.

If two whole numbers are multiplied together to get a third number, the first two numbers are called **factors** of the third number, and the third number is called a **multiple** of the other numbers.

For example; (4)(3) = 12 , (2)(6) = 12

Since 4, 3, 2, and 6 are all factors of 12, 12 is a multiple of 4, 3, 2, and 6.

A whole number is said to be **divisible (can be divided)** by each of its factors.

A whole number that is divisible by two is an **even number**.

A whole numbers that is not divisible by two is an **odd number.**

A **prime** number is a whole number larger than 1 in which the only factors are the number itself and 1. A prime number is said to be **nonfactorable**.

A **composite** number is a whole number that will not only factor as itself times 1 but will also factor in at least one other way. A composite number is **factorable.**

Example 1: 3 is a prime number because the only way to write 3 as the product of two whole numbers is 3 = (3)(1). (There was only one way to make a rectangle using 3 squares.) The only factors of 3 are 3 and 1.

Example 2: 4 is a composite number because not only does 4 = (4)(1) but also 4 = (2)(2). The factors of 4 are 4, 1, and 2. 4 is a multiple of 2.
5 is a prime number because the only factors of 5 are 5 and 1.

Example 3: 16 is a composite number because 16 can be factored in several ways:
16 = (4)(4)   16 = (8)(2)   16 = (16)(1)
Since the factors of 16 are 4, 8, 2, 16, and 1, we say that 16 is divisible by 4, 8 and 2. A number is always divisible by itself and by 1, so we would not always say that 16 is divisible by 1, for example.

Example 4: 30 is a composite number because 30 can be factored as
30 = (15)(2) 30 = (5)(6) and 30 = (3)(10)
Since the factors of 30 are 15, 2, 5, 6, 3, and 10, 30, and 1, we say that 30 is divisible by 15, 2, 5, 6, 3, and 10.

In the following exercise, you will use squares to build rectangles and the A = LW formula to learn how to factor a number into two factors.

Exercise 1--Squares, Rectangles, and Factors

Purpose of the Exercise

- To review the formula, A = LW, for area of a rectangle
- To learn about factors of numbers
- To learn about prime numbers

Materials Needed

- 25 cardboard or ceramic squares of the same size (for each group). Each square is 1 U by 1 U.

For each of the following, use the number of squares indicated and see how many rectangles you can make with that number of squares. Each square is 1 U by 1 U. Write the dimensions of all of the rectangles possible. If two rectangles are formed by switching the length and the width, just write that rectangle once. For each rectangle, write factors of the number of squares that form the rectangle. Include appropriate units with the answers. Remember that lengths and widths are linear units, the unit symbol will just be U, and that an area is the number of squares. The unit for area is square units, abbreviated U^2. In some cases you may have only one or two rectangles. In those cases, leave the other spaces blank. The first one is done as an example.

1. 2 squares

   rectangle 1 length = 2 U width = 1 U area = $2\ U^2$ 2 = (2)(1)

   rectangle 2 length = _____ width = _____ area = ______

   rectangle 3 length = _____ width = _____ area = ______

   rectangle 4 length = _____ width = _____ area = ______

   Is 2 a prime number or a composite number? prime

2. 3 squares

   rectangle 1 length = _____ width = _____ area = ______ 3 = ______

   rectangle 2 length = _____ width = _____ area = ______ 3 = ______

   rectangle 3 length = _____ width = _____ area = ______ 3 = ______

   rectangle 4 length = _____ width = _____ area = ______ 3 = ______

   Is 3 a prime number or a composite number? ____________

3. 5 squares

   rectangle 1 length = _____ width = _____ area = ______ 5 = ______

   rectangle 2 length = _____ width = _____ area = ______ 5 = ______

   rectangle 3 length = _____ width = _____ area = ______ 5 = ______

   rectangle 4 length = _____ width = _____ area = ______ 5 = ______

   Is 5 a prime number or a composite number? ____________

4. 6 squares

   rectangle 1 length = _____ width = _____ area = ______ 6 = ______

   rectangle 2 length = _____ width = _____ area = ______ 6 = ______

   rectangle 3 length = _____ width = _____ area = ______ 6 = ______

   rectangle 4 length = _____ width = _____ area = ______ 6 = ______

   Is 6 a prime number or a composite number? ____________

5. 7 squares

   rectangle 1 length = _____ width = _____ area = ______ 7 = ______

   rectangle 2 length = _____ width = _____ area = ______ 7 = ______

   rectangle 3 length = _____ width = _____ area = ______ 7 = ______

   rectangle 4 length = _____ width = _____ area = ______ 7 = ______

   Is 7 a prime number or a composite number? ____________

6. 8 squares

rectangle 1	length = _____	width = _____	area = ______	8 = ______
rectangle 2	length = _____	width = _____	area = ______	8 = ______
rectangle 3	length = _____	width = _____	area = ______	8 = ______
rectangle 4	length = _____	width = _____	area = ______	8 = ______

Is 8 a prime number or a composite number? ______________

7. 9 squares

rectangle 1	length = _____	width = _____	area = ______	9 = ______
rectangle 2	length = _____	width = _____	area = ______	9 = ______
rectangle 3	length = _____	width = _____	area = ______	9 = ______
rectangle 4	length = _____	width = _____	area = ______	9 = ______

Is 9 a prime number or a composite number? ______________

8. 10 squares

rectangle 1	length = _____	width = _____	area = ______	10 = ______
rectangle 2	length = _____	width = _____	area = ______	10 = ______
rectangle 3	length = _____	width = _____	area = ______	10 = ______
rectangle 4	length = _____	width = _____	area = ______	10 = ______

Is 10 a prime number or a composite number? ______________

9. 11 squares

rectangle 1	length = _____	width = _____	area = ______	11 = ______
rectangle 2	length = _____	width = _____	area = ______	11 = ______
rectangle 3	length = _____	width = _____	area = ______	11 = ______
rectangle 4	length = _____	width = _____	area = ______	11 = ______

Is 11 a prime number or a composite number? ______________

10. 12 squares

rectangle 1	length = _____	width = _____	area = ______	12 = ______
rectangle 2	length = _____	width = _____	area = ______	12 = ______
rectangle 3	length = _____	width = _____	area = ______	12 = ______
rectangle 4	length = _____	width = _____	area = ______	12 = ______

Is 12 a prime number or a composite number? ______________

11. 13 squares

rectangle 1	length = _____	width = _____	area = ______	13 = ______
rectangle 2	length = _____	width = _____	area = ______	13 = ______
rectangle 3	length = _____	width = _____	area = ______	13 = ______
rectangle 4	length = _____	width = _____	area = ______	13 = ______

Is 13 a prime number or a composite number? ______________

12. 14 squares

rectangle 1 length = _____ width = _____ area = ______ 14 = ______

rectangle 2 length = _____ width = _____ area = ______ 14 = ______

rectangle 3 length = _____ width = _____ area = ______ 14 = ______

rectangle 4 length = _____ width = _____ area = ______ 14 = ______

Is 14 a prime number or a composite number? ____________

13. 15 squares

rectangle 1 length = _____ width = _____ area = ______ 15 = ______

rectangle 2 length = _____ width = _____ area = ______ 15 = ______

rectangle 3 length = _____ width = _____ area = ______ 15 = ______

rectangle 4 length = _____ width = _____ area = ______ 15 = ______

Is 15 a prime number or a composite number? ____________

14. 16 squares

rectangle 1 length = _____ width = _____ area = ______ 16 = ______

rectangle 2 length = _____ width = _____ area = ______ 16 = ______

rectangle 3 length = _____ width = _____ area = ______ 16 = ______

rectangle 4 length = _____ width = _____ area = ______ 16 = ______

Is 16 a prime number or a composite number? ____________

15. 17 squares

rectangle 1 length = _____ width = _____ area = ______ 17 = ______

rectangle 2 length = _____ width = _____ area = ______ 17 = ______

rectangle 3 length = _____ width = _____ area = ______ 17 = ______

rectangle 4 length = _____ width = _____ area = ______ 17 = ______

Is 17 a prime number or a composite number? ____________

16. 18 squares

rectangle 1 length = _____ width = _____ area = ______ 18 = ______

rectangle 2 length = _____ width = _____ area = ______ 18 = ______

rectangle 3 length = _____ width = _____ area = ______ 18 = ______

rectangle 4 length = _____ width = _____ area = ______ 18 = ______

Is 18 a prime number or a composite number? ____________

17. 19 squares

rectangle 1 length = _____ width = _____ area = ______ 19 = ______

rectangle 2 length = _____ width = _____ area = ______ 19 = ______

rectangle 3 length = _____ width = _____ area = ______ 19 = ______

rectangle 4 length = _____ width = _____ area = ______ 19 = ______

Is 19 a prime number or a composite number? ____________

18. 20 squares

rectangle 1 length = _____ width = _____ area = ______ 20 = ______

rectangle 2 length = _____ width = _____ area = ______ 20 = ______

rectangle 3 length = _____ width = _____ area = ______ 20 = ______

rectangle 4 length = _____ width = _____ area = ______ 20 = ______

Is 20 a prime number or a composite number? ______________

19. 21 squares

rectangle 1 length = _____ width = _____ area = ______ 21 = ______

rectangle 2 length = _____ width = _____ area = ______ 21 = ______

rectangle 3 length = _____ width = _____ area = ______ 21 = ______

rectangle 4 length = _____ width = _____ area = ______ 21 = ______

Is 21 a prime number or a composite number? ______________

20. 22 squares

rectangle 1 length = _____ width = _____ area = ______ 22 = ______

rectangle 2 length = _____ width = _____ area = ______ 22 = ______

rectangle 3 length = _____ width = _____ area = ______ 22 = ______

rectangle 4 length = _____ width = _____ area = ______ 22 = ______

Is 22 a prime number or a composite number? ______________

21. 23 squares

rectangle 1 length = _____ width = _____ area = ______ 23 = ______

rectangle 2 length = _____ width = _____ area = ______ 23 = ______

rectangle 3 length = _____ width = _____ area = ______ 23 = ______

rectangle 4 length = _____ width = _____ area = ______ 23 = ______

Is 23 a prime number or a composite number? ______________

22. 24 squares

rectangle 1 length = _____ width = _____ area = ______ 23 = ______

rectangle 2 length = _____ width = _____ area = ______ 23 = ______

rectangle 3 length = _____ width = _____ area = ______ 23 = ______

rectangle 4 length = _____ width = _____ area = ______ 23 = ______

Is 23 a prime number or a composite number? ______________

23. In which of Problems 1-22 could you form only one rectangle? ________________

24. In which problems could you form two or more rectangles? ____________________

25. List all of the prime numbers from 2 to 25. ______________________________

26. List all of the composite numbers between 2 and 25. _______________________

27. List all of the even numbers between 1 and 25. ____________________________

28. List all of the odd numbers between 1 and 25. _____________________________

29. List all of the multiples of 3 between 1 and 25. ________________________

30. List all of the multiples of 5 between 1 and 25. ________________________

31. How many different rectangles could you make with 100 squares? Explain how you decided.

Factoring a Number Completely

To factor a number completely, we first find two factors that multiplied together equal the number. We then continue to factor each nonprime factor until all of the factors are prime numbers. For example, suppose we want to factor 24 completely. We first find two numbers that multiplied together give 24. Using the squares as aids, let's arrange the 24 squares in 4 rows of 6 squares each as shown below.

24 = (4)(6)

Next, we factor each of the nonprime factors. Since 4 = (2)(2) and 6 = (2)(3),

$$\begin{aligned} 24 &= (4)(6) \\ &= (2)(2)(2)(3) \end{aligned}$$

Since 2 and 3 are prime numbers, **24 = 2 • 2 • 2 • 3** shows the complete, or **prime, factorization** of 24.

Another way to write this process is in a "tree" form.

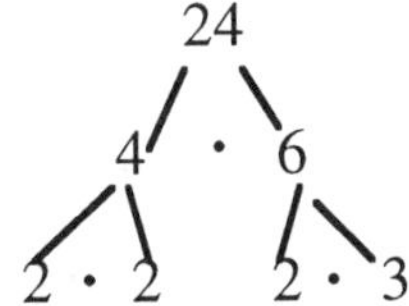

4 and 6 multiply to give 24.

2, 2, 2 and 3 are prime factors of 24.

The complete (or prime) factorization comes from writing 24 as the product of all the numbers at the ends of the "tree." **24 = (2)(2)(2)(3)**

When factoring 24, we started with 24 = (4)(6). We could just as easily have correctly started with 24 = (3)(8). The steps below show how the tree would have led us to the same results.

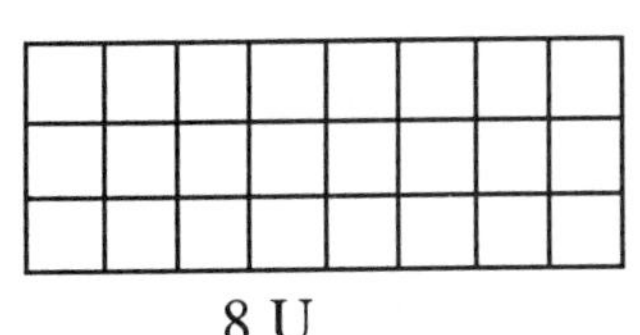

3 U

$$\begin{aligned} 24 &= (3)(8) \\ &= (3)(2)(4) \\ &= (3)(2)(2)(2) \end{aligned}$$

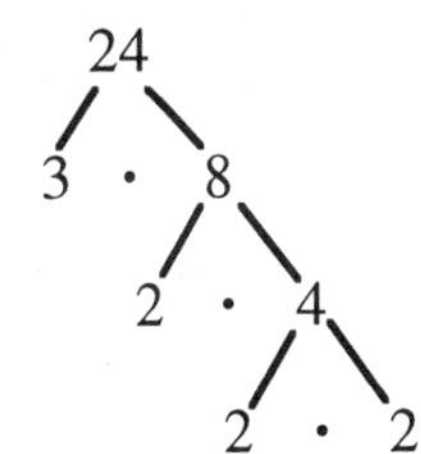

The ends, or leaves, of the tree are 3, 2, 2, and 2. So 24 = (3)(2)(2)(2)

Exercise 2--Factoring

Purpose of the Exercise

- To develop and practice factoring skills

Materials Needed

- Squares
- Calculator

Factor each of the following numbers into prime numbers. Form squares into rectangles as needed as an aid to first factoring the number into two factors. If the number is prime, state that. Use your calculator as needed to multiply the factors and check your work.

1. 33 = ______________ 2. 45 = ______________

3. 16 = ______________ 4. 41 = ______________

5. 82 = ______________ 6. 44 = ______________

7. 18 = ______________ 8. 19 = ______________

9. 25 = ______________ 10. 27 = ______________

11. 9 = ______________ 12. 28 = ______________

13. 26 = ______________ 14. 35 = ______________

15. 36 = ______________ 16. 38 = ______________

17. 50 = ______________ 18. 100 = ______________

19. 75 = ______________ 20. 24 = ______________

21. 32 = ______________ 22. 31 = ______________

23. 40 = ______________ 24. 55 = ______________

25. 60 = ______________ 26. 63 = ______________

The Greatest Common Factor and Least Common Multiple

As mentioned before, 24 is a multiple of 2 as well as 3, 4, 6, and 8 because 2, 3, 4, 6, and 8 are all factors of 24. In Chapter 5, when adding fractions, we will need to find the smallest number that is a multiple of two numbers.

Vocabulary

A **common multiple** of two or more natural numbers is a number that is a multiple of all of the numbers.

For example, 12, 24, and 36 are all multiples of both 3 and 4 so they are common multiples of 3 and 4.

A **least common multiple** of two or more natural numbers is the smallest number that is a multiple of all of the numbers. The least common multiple is abbreviated, **LCM**.

For example, 12 is the LCM of 3 and 4 because it is the smallest multiple of both 3 and 4.

A **common factor** of two or more natural numbers is a number that is a factor of all of the numbers.

For example, 2, 6, and 3 are common factors of 18 and 24 because 2, 6, and 3 are factors of 18 and 24.

The **greatest common factor** of two or more natural numbers is the largest factor of all of the numbers. The greatest common factor is abbreviated **GCF**.

For example, 6 is the GCF of 18 and 24 because 6 is the largest factor of both 18 and 24.

Example 5: Find several common multiples, the least common multiple, and the greatest common factor of 5 and 3.

Common multiples: 15, 30, 45, 60 LCM = 15

5 and 3 are both prime, 5 = 5 x 1 and 3 = 3 x 1 so 1 is the largest number that is factor of both.

GCF = 1

Example 6: Find several common multiples, the least common multiple, and the greatest common factor of 6 and 9.

Common multiples: 18, 36, 54, 72
LCM = 18
6 = (3)(2) and 9 = (3)(3)
GCF = 3

Finding the least common multiple and the greatest common factor of numbers that have many factors can be a bit more complicated than guessing, so it is good to have a systematic way to find the least common multiple that works no matter how large the numbers. A common multiple has to contain all of the factors of each of the numbers, so the least common multiple is the most efficient way to include all of the factors.

To Find a Least Common Multiple

1. Prime factor each number. (Write each number as a product of prime factors.)
2. If a factor appears more than once in one of the numbers, write it in exponential form.
3. The least common multiple is the product of each factor raised to the highest power that appears in any one of the numbers.

Example 7: Find the least common multiple of 20 and 18.

1. 20 = (2)(2)(5)
   18 = (3)(3)(2)
2. $20 = (2^2)(5)$
   $18 = (3^2)(2)$
3. $LCM = (2^2)(5)(3^2) = (4)(5)(9) = (20)(9) = 180$

Find a Greatest Common Factor

1. Prime factor each number.
2. If a factor appears more than once in one of the numbers, write it in exponential form.
3. The greatest common factor is the product of the factors that appear in all numbers. If a factor is raised to a power, use the lowest power that appears in any one number.

Example 8: Find the least greatest common factor of 20 and 18.

1. 20 = (2)(2)(5)
   18 = (3)(3)(2)
2. $20 = (2^2)(5)$
   $18 = (3^2)(2)$
3. GCF = 2 (2 is the only factor that appears in both numbers.)

Example 9: Find the least common multiple and greatest common factor of 24 and 36.

$24 = (2)(2)(2)(3) = (2^3)(3)$
$36 = (2)(2)(3)(3) = (2^2)(3^2)$

2^2 is a factor of 36 and 2^3 is a factor of 24, so 2^3 (the higher power) is a factor of the least common multiple. Similarly, 3 is a factor of 24 and 3^2 is a factor of 36, so the higher power, 3^2, is a factor of the least common multiple.

$LCM = (2^3)(3^2) = 72$
$GCF = (2^2)(3) = 12$

Example 10: Find the least common multiple and greatest common factor of 60 and 150.

$60 = (6)(10) = (2)(3)(2)(5) = (2^2)(3)(5)$
$150 = (15)(10) = (3)(5)(2)(5) = (2)(3)(5^2)$
$LCM = (2^2)(3)(5^2) = 300$
$GCF = (2)(3)(5) = 30$

Example 11: Find the least common multiple and greatest common factor of 20, 30 and 40.

$20 = (2)(10) = (2)(2)(5) = (2^2)(5)$
$30 = (3)(10) = (3)(2)(5)$
$40 = (4)(10) = (2)(2)(2)(5) = (2^3)(5)$
$LCM = (2^3)(5)(3) = 120$
$GCF = (2)(5) = 10$

Exercise 3--Least Common Multiple and Greatest Common Factor

Purpose of the Exercise

- To practice finding the LCM and GCF

Prime factor each number in the following problems, writing duplicate factors in exponential form. Then find the least common multiple and the greatest common factor.

1. 18 and 27

   18 = ____________________
   27 = ____________________
   GCF = ____________________
   LCM = ____________________

2. 33 and 44

   33 = ____________________
   44 = ____________________
   GCF = ____________________
   LCM = ____________________

3. 9 and 12

   9 = ____________________
   12 = ____________________
   GCF = ____________________
   LCM = ____________________

4. 15 and 35

   15 = ____________________
   35 = ____________________
   GCF = ____________________
   LCM = ____________________

5. 27 and 16

   27 = ____________________
   16 = ____________________
   GCF = ____________________
   LCM = ____________________

6. 13 and 26

   13 = ____________________
   26 = ____________________
   GCF = ____________________
   LCM = ____________________

7. 28, 42, and 35

   28 = ____________________
   42 = ____________________
   35 = ____________________
   GCF = ____________________
   LCM = ____________________

8. 21, 14, and 15

   21 = ____________________
   14 = ____________________
   15 = ____________________
   GCF = ____________________
   LCM = ____________________

Section 3-7 *Rectangular Boxes, Cubes, Volume, and Surface Area*

A rectangle is a **two-dimensional** shape because it has **two dimensions--length and width**. Although triangles and circles are not described in terms of length and width, they take up space in two directions, up-down and left-right. They are flat, like a piece of paper--except they have no real thickness. For this reason we refer to circles, triangles, and other flat shapes as two-dimensional.

rectangle triangle circle irregular 2-dimensional shape

A **three-dimensional** object is an object like a box or ball. In addition to length and width, they have thickness, and are not flat. The three dimensions of a rectangular box are called **length, width, and height.** Other objects such as rocks, furniture, and dishes are three-dimensional because they take up space in three directions, up-down, left-right, and front-back. By comparison, two-dimensional objects do not have thickness, so they do not take up space in one of the directions.

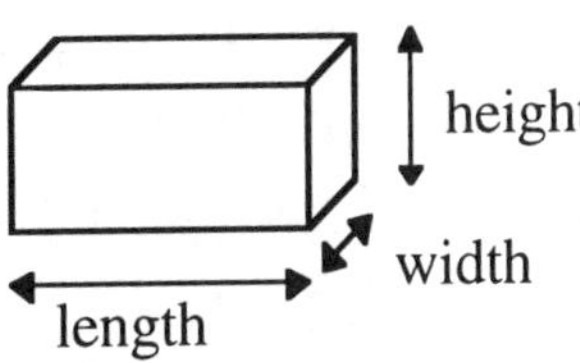

Vocabulary

A **solid rectangular box** (a special case of a shape called prism) is a three-dimensional box where each of its six sides is a rectangle and the sides join in right (90°) angles.

The dimensions of a rectangular box are **length**, **width,** and **height**.

A **cube** is a rectangular box for which each of its six sides are squares. In a cube, the length, width, and height are all equal.

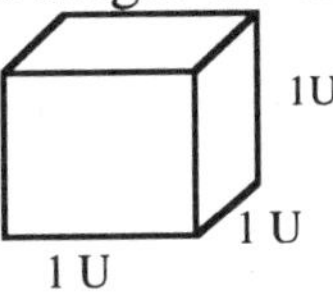

Volume is the amount of space within a three-dimensional structure. It is measured in the number of cubes of a particular size that will fill the space.

Volume is measured in **cubic units** (or cubes). In a cubic foot, for example, the length, width, and height each measure 1 foot. The unit cubic feet is abbreviated cu ft or ft^3. Cubic yards is abbreviated , cu yd or yd^3.

The **surface area** of a rectangular box is the total of the areas of its six sides--top and bottom, left and right, front and back.

Exercise 1--Volume and Surface Area

Purpose of the Exercise

- To get familiar with three-dimensional rectangular shapes, volumes, and surface area

Materials Needed

- 40 cubes

In each problem below, shape as many solid rectangular boxes possible using all of the cubes indicated. Assume the length of a side of one cube is one unit (abbreviated 1 U.) Record the dimensions and volume of each solid rectangular box. If one box is the same as another box turned in a different position, write its dimensions only once.

For each solid rectangular box, record the dimensions and area of the front and back sides, left and right sides, and top and bottom sides. Add the areas of the six sides--front, back, left, right, top, and bottom--to compute the total surface area. Write appropriate units by the answers. The first one is done as an example.

1. 2 cubes

box 1 length = 1 U width = 1U height 2 U

length x width x height = (1 U)(1 U)(2 U) = 2 U^3 number of cubes = 2

dimensions of front/back = LxH = 1U x 2U area of front/back = 2 U^2

dimensions of left/right = WxH = 1U x 2U area of left/right = 2 U^2

dimensions of top/bottom = LxW = 1U x 1W area of top/bottom = 1 U^2

total surface area = front+back+left+right+top+bottom = 2 + 2 + 2 +2 + 1 + 1 = 10 U^2

Since 2 is a prime number, only one box can be made from 2 cubes.

2. 3 cubes

box 1 length = ______ width = _______ height _____

length x width x height = ____________________ number of cubes = ____

dimensions of front/back = _________ area of front/back = ___________

dimensions of left/right = _________ area of left/right = ___________

dimensions of top/bottom = ________ area of top/bottom = ___________

total surface area = ______________________________

3. 9 cubes

box 1 length = ______ width = _______ height _____

length x width x height = ____________________ number of cubes = ____

dimensions of front/back = _________ area of front/back = ___________

dimensions of left/right = _________ area of left/right = ___________

dimensions of top/bottom = ________ area of top/bottom = ___________

total surface area = ______________________________

box 2 length = ______ width = _______ height _____

length x width x height = ____________________ number of cubes = ____

dimensions of front/back = _________ area of front/back = ___________

dimensions of left/right = _________ area of left/right = ___________

dimensions of top/bottom = ________ area of top/bottom = ___________

total surface area = ______________________________

4. 12 cubes (There are four possibilities. Record information for three of them.)

box 1 length = ______ width = ________ height ______

length x width x height = ______________________ number of cubes = ____

dimensions of front/back = __________ area of front/back = ____________

dimensions of left/right = ___________ area of left/right = _____________

dimensions of top/bottom = _________ area of top/bottom = ____________

total surface area = _________________________________

box 2 length = ______ width = ________ height ______

length x width x height = ______________________ number of cubes = ____

dimensions of front/back = __________ area of front/back = ____________

dimensions of left/right = ___________ area of left/right = _____________

dimensions of top/bottom = _________ area of top/bottom = ____________

total surface area = _________________________________

box 3 length = ______ width = ________ height ______

length x width x height = ______________________ number of cubes = ____

dimensions of front/back = __________ area of front/back = ____________

dimensions of left/right = ___________ area of left/right = _____________

dimensions of top/bottom = _________ area of top/bottom = ____________

total surface area = _________________________________

5. 13 cubes

box 1 length = ______ width = ________ height ______

length x width x height = ______________________ number of cubes = ____

dimensions of front/back = __________ area of front/back = ____________

dimensions of left/right = ___________ area of left/right = _____________

dimensions of top/bottom = _________ area of top/bottom = ____________

total surface area = _________________________________

6. 16 cubes (There are four possibilities. Record information for three of them.)

box 1 length = ______ width = ________ height ______

length x width x height = ______________________ number of cubes = ____

dimensions of front/back = __________ area of front/back = ____________

dimensions of left/right = ___________ area of left/right = _____________

dimensions of top/bottom = _________ area of top/bottom = ____________

total surface area = _________________________________

box 2 length = ______ width = ________ height ______

length x width x height = ______________________ number of cubes = ____

dimensions of front/back = __________ area of front/back = ____________

dimensions of left/right = ___________ area of left/right = _____________

dimensions of top/bottom = _________ area of top/bottom = ____________

total surface area = _________________________________

box 3 length = ______ width = ________ height _____

length x width x height = ______________________ number of cubes = ____

dimensions of front/back = __________ area of front/back = ____________

dimensions of left/right = ___________ area of left/right = ____________

dimensions of top/bottom = _________ area of top/bottom = ____________

total surface area = _________________________________

7. 24 cubes (There are six possible boxes. Record information about four of them.)

box 1 length = ______ width = ________ height _____

length x width x height = ______________________ number of cubes = ____

dimensions of front/back = __________ area of front/back = ____________

dimensions of left/right = ___________ area of left/right = ____________

dimensions of top/bottom = _________ area of top/bottom = ____________

total surface area = _________________________________

box 2 length = ______ width = ________ height _____

length x width x height = ______________________ number of cubes = ____

dimensions of front/back = __________ area of front/back = ____________

dimensions of left/right = ___________ area of left/right = ____________

dimensions of top/bottom = _________ area of top/bottom = ____________

total surface area = _________________________________

box 3 length = ______ width = ________ height _____

length x width x height = ______________________ number of cubes = ____

dimensions of front/back = __________ area of front/back = ____________

dimensions of left/right = ___________ area of left/right = ____________

dimensions of top/bottom = _________ area of top/bottom = ____________

total surface area = _________________________________

box 4 length = ______ width = ________ height _____

length x width x height = ______________________ number of cubes = ____

dimensions of front/back = __________ area of front/back = ____________

dimensions of left/right = ___________ area of left/right = ____________

dimensions of top/bottom = _________ area of top/bottom = ____________

total surface area = _________________________________

8. 36 cubes (There are eight possible boxes. Record information about four of them.)

box 1 length = ______ width = ________ height _____

length x width x height = ______________________ number of cubes = ____

dimensions of front/back = __________ area of front/back = ____________

dimensions of left/right = ___________ area of left/right = ____________

dimensions of top/bottom = _________ area of top/bottom = ____________

total surface area = _________________________________

box 2 length = ______ width = _______ height _____

length x width x height = ____________________ number of cubes = ____

dimensions of front/back = _________ area of front/back = ___________

dimensions of left/right = __________ area of left/right = ____________

dimensions of top/bottom = ________ area of top/bottom = ___________

total surface area = ______________________________

box 3 length = ______ width = _______ height _____

length x width x height = ____________________ number of cubes = ____

dimensions of front/back = _________ area of front/back = ___________

dimensions of left/right = __________ area of left/right = ____________

dimensions of top/bottom = ________ area of top/bottom = ___________

total surface area = ______________________________

box 4 length = ______ width = _______ height _____

length x width x height = ____________________ number of cubes = ____

dimensions of front/back = _________ area of front/back = ___________

dimensions of left/right = __________ area of left/right = ____________

dimensions of top/bottom = ________ area of top/bottom = ___________

total surface area = ______________________________

9. 39 cubes

box 1 length = ______ width = _______ height _____

length x width x height = ____________________ number of cubes = ____

dimensions of front/back = _________ area of front/back = ___________

dimensions of left/right = __________ area of left/right = ____________

dimensions of top/bottom = ________ area of top/bottom = ___________

total surface area = ______________________________

box 2 length = ______ width = _______ height _____

length x width x height = ____________________ number of cubes = ____

dimensions of front/back = _________ area of front/back = ___________

dimensions of left/right = __________ area of left/right = ____________

dimensions of top/bottom = ________ area of top/bottom = ___________

total surface area = ______________________________

10. Observations:

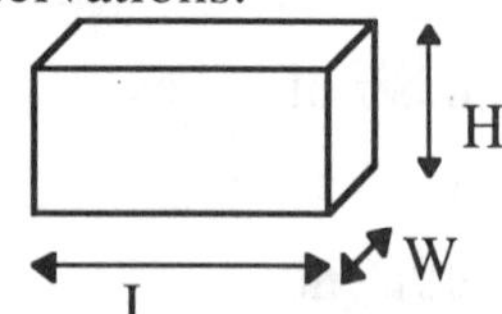

a. If the dimensions of a rectangular box are L, W, and H (length, width, and height), write a formula for the volume, V, of the box. V = ________________

b. If the dimensions for a rectangular box are L, W, and H, write a formula for the area of each of the six sides of the box.

area of front side = ____________ area of back side = ________________

area of left side = ______________ area of right side = _________________

area of top = __________________ area of bottom = __________________

c. State whether the following quantities are measured in linear units, square units, or cubic units.

length ___________ width ____________ height ____________

surface area ____________ volume _________________

Formulas

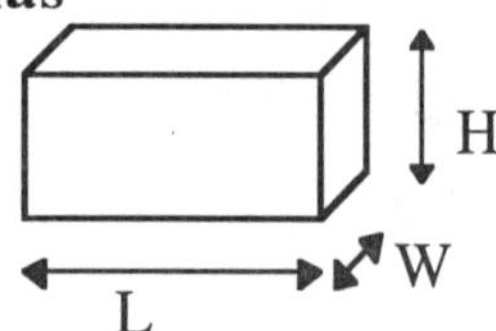

The **volume of a rectangular box** is **V = L · W · H** (or V = LWH).
The **surface area of a rectangular box** is **SA = 2LW + 2LH + 2HW.**

Units

The **dimensions** of a rectangular box, length, width and height, are measured in **linear units**. The **volume** is measured in **cubic units.**
Area is measured in **square units**, so **surface area** is measured in **square units.**

Exercise 2--Rectangular Boxes, Cubes, and Volumes

Purpose of the Exercise

- To better understand dimensions and volumes for rectangular boxes

Materials Needed

- Calculator
- Cubes

In each of the following, determine the dimensions of the box; then use the formulas you developed in Exercise 1 to compute the volume and surface area of the box. Assume that the length of a side of the cube is 1 U. Include correct units with your answers. (**V** = volume, and **SA** = surface area.)

1. 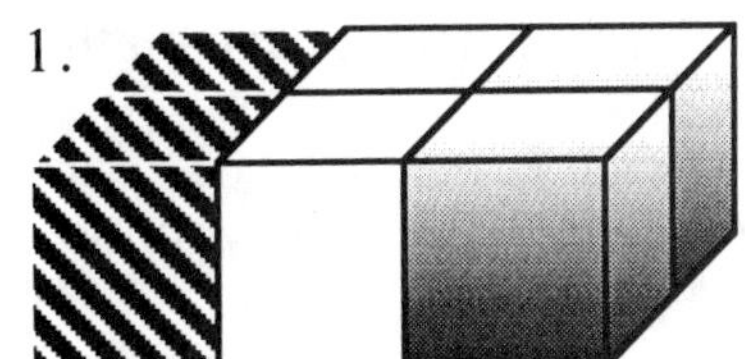

L = ______ W = ______

H = ______

V = ______ SA = ______

2. If we add another layer of cubes to those in Problem 1, we get the following box.

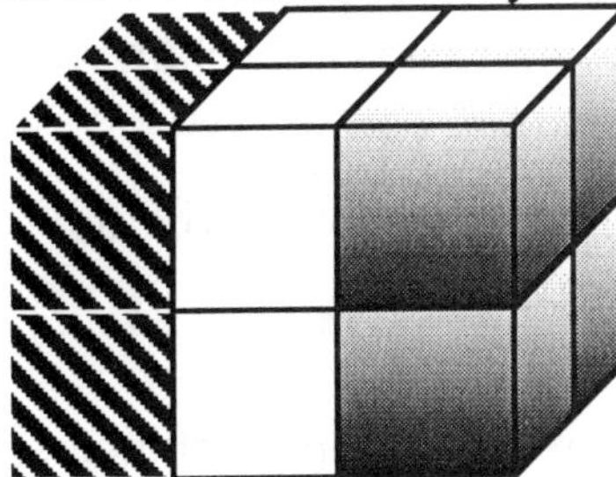

L = ______ W = ______

H = ______

V = ______ SA = ______

3. If we place the two layers from Problem 2 side by side (instead of stacked), we also get a box that is double the size of the box in Problem 1.

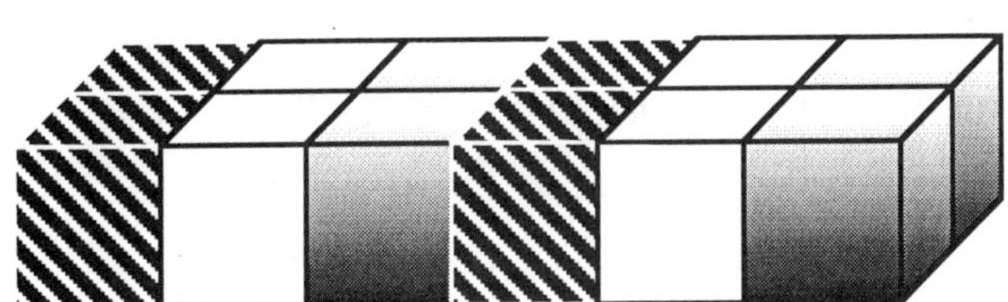

L = ______

W = ______

H = ______

V = ______

SA = ______

4. Compare the original box in Problem 1 with the boxes in Problems 2 and 3, in which one of the dimensions of the original box doubled. What affect did doubling one dimension have on the volume of the box?

5. Now let's see what happens to the dimensions and volume of the box when we stack three layers of cubes like those in Problem 1.

L = ______ W = ______

H = ______

V = ______ SA = ______

6. The original box had 6 cubes ($V = 6\ U^3$). What effect did tripling the height have on the volume of the box?

7. If the volume of a box is $10\ U^3$ and the height is then doubled, what will be the volume of the new box?

8. If the volume of a box is $10\ U^3$ and the width is doubled, what will be the volume of the new box?

9. If the volume of a box is $12\ U^3$ and the length is multiplied by 4, what is the volume of the new box?

10. If the volume of a box is 7^3 and the length, width, and height are all doubled, will the volume of the new box be 14^3? Justify your answer.

As with rectangles, the dimensions of rectangular boxes do not have to be natural numbers. When the dimensions are not natural numbers, the area and volume are not necessarily perfect squares or cubes of a natural number. When the dimensions are not natural numbers or are large numbers, you may want to use a calculator to do the calculations.

Example 1: A box with dimensions, L = 12.3 U, W = 10.2 U, and H = 14.7 U, has a volume of $V = (12.3\ U)(10.2\ U)(14.7\ U) = 125.46\ U^3$.

On a scientific calculator, the above multiplication is easy. Here are the key strokes:
12.3 [x] 10.2 [x] 14.7 [=]

Exercise 3--Computing Volumes and Surface Areas Using a Calculator

Purpose of the Exercise
- To reinforce use of the volume formula for rectangular boxes

Materials Needed
- Calculator

Find the volume and surface area of each of the boxes with the given dimensions. Include the appropriate units with your answers. Use a calculator as needed for computations.

1. L = 2.5 mi, W = 4.9 mi, H = 6.7 mi V = ________ SA = ________
2. L = 1.7 in., W = 2.4 in., H = 6 in. V = ________ SA = ________
3. L = 75 ft, W = 49 ft, H = 10.2 ft V = ________ SA = ________
4. L = 16 yd, W = 4 yd, H = 1.9 yd V = ________ SA = ________
5. L = 95 ft, W = 18 ft, H = 12 ft V = ________ SA = ________
6. L = 1 m, W = 2.3 m, H = 4.2 m V = ________ SA = ________
7. L = 18.7 ft, W = 29.3 ft, H = 2.9 ft V = ________ SA = ________
8. L = 48 cm, W = 135 cm, H = 95 cm V = ________ SA = ________

Projects to be done outside of class. Pick your favorite.

1. Call a business that sells furnaces or air conditioners to find out how cubic units are used to help determine your heating and cooling needs. Write a report describing what cubic units are used.
2. Look at a gas company bill and find out what cubic units are used in computing your gas bill. Write a report of your findings. Include in the report what unit is used and how you are billed.
3. Look at specifications for a car. How many different items are described in cubic units? Write a report about your findings.
4. What kinds of linear units are used in describing refrigerators? What kinds of cubic units are used in describing refrigerators? Write a report about your findings.
5. In a library, look up the history of measurements and how they have evolved. Write a report of your findings.
6. Check with the local fire department to find out how square or cubic units affect the occupancy limits of a building or room. Write a report about your findings.
7. Interview someone (other than a mathematics teacher) who uses mathematics in his or her job. Ask the person to give you an example of the kind of problem that arises. Write a report giving the name of the person you interviewed, their company, and position. Describe the math problem and how it is used in his/her business.
8. Find out how a person who pours concrete determines the amount of concrete to mix. Write a report about your findings.
9. Make up your own project that uses square units or cubic units. Write a report describing your project and conclusions.

For any of the following projects you choose, gather the information requested and do the math required (if any). Write a report with your conclusions explaining where or how you got the information and clearly showing all mathematical computations

10. Visit a store that sells carpet. Pick out your favorite carpet and carpet pad. Measure your living room and figure out how much it would cost to buy the carpet, pad, and other necessary supplies if you do the installation. How much would it cost if you hire someone to install the carpet? Write a report about your findings.
11. Measure your kitchen counter. At a store that sells ceramic tile, pick out your favorite tile and estimate how much it would cost to cover your counter. Include in your estimate putty, sealant, partial tiles at the edges and cost of renting tile cutters or other costs you might incur. Write a report showing all of your measurements, mathematical computations, and estimated costs. Explain why the costs are estimates and may not be exact.
12. Call a pool company. If you were to build a backyard pool, decide what size you would build and how much water it would take to fill your pool.
13. Measure all the walls of a room in your home. Estimate how much plasterboard (dry wall) would be needed to cover the walls. If you are not familiar with how this material is sold, check with a lumber yard. Measure the ceiling of the room. How much plasterboard would cover the ceiling?
14. Measure the walls of a room in your home. Estimate how much paint you would need to give all the walls one coat.
15. Measure the windows in your house. Also measure the distance from the window sill to the floor. Determine what size drapes you should buy assuming that you want them to reach the floor. Check at a store and see what choices fit your windows.

Section 3-8 *Circles and π*

Many of the geometric shapes that we study today, like triangles, circles, and rectangles, occur so naturally that they have been studied for at least 2500 years. Greek society from the days of Plato, Aristotle, and Pythagoras was intrigued by geometry. They discovered most of the formulas that we still find useful in mathematics and life today. One of their fascinations was with the circle. A circle can be created with no more than a rope, a stake, and one person. No doubt, the Greek mathematicians drew many circles in the sand by staking down one end of a rope and walking at rope's length around the stake, marking the circular path. The ability to draw many circles easily, no doubt, helped them study circles. The dot in the center below represents the stake and the circle is the path.

The Greeks observed that since the length of the rope did not change, the distance from the center of a circle to the rim of the circle is the same no matter which point on the rim you use. The length from the center to the circle is called the radius. The Greek mathematicians also observed that the length across the circle through the center is twice the radius.

Vocabulary

The distance from the center of the circle to the rim is called the **radius** of the circle.
The distance across the circle through the center is called the **diameter** of the circle.
The diameter crosses the widest part of the circle.

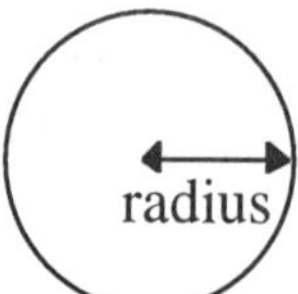

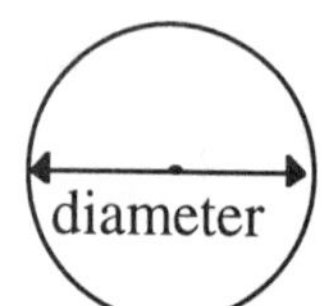

The length around the outside of a circle is called the **circumference** of the circle.

One of the really fascinating discoveries of the Greek mathematicians was the number called "pi" or π. The number has a Greek letter for a name in honor of the Greek mathematicians who discovered it. In the following exercise, we will discover for ourselves what the Greeks found in about 300 BC.

Exercise 1--Diameter, Circumference, and π

Purpose of the Exercise

- To understand the number π

Materials Needed

- Three circular objects of different sizes, such as lids
- String (long enough to go around the largest circular object)
- Large piece of paper
- Ruler with centimeter markings
- Calculator

Part 1

- Stretch the string across the diameter of the circle. You may have to move it around and try several positions to find the widest part.
- Mark the length of the diameter on a paper. Call this unit a diameter (1 d).
- Draw a length of 4 d on the paper (or board.) Be as precise as possible, showing the 1-d mark, 2-d mark, 3-d mark, and 4-d mark.

- Divide the fourth diameter (between 3-d and 4-d) into eighths. Make marks at 1/8, 2/8 (1/4), and so on.
- Your drawing will look something like this:

0 d 1 d 2 d 3 d 4 d

Part 2

- Stretch your string around the rim of your circular object and mark the length of the circumference. Be as accurate as possible.
- Measure this length along the ruler segment that you made with the unit, d.
- When measuring along your ruler, about how long is this segment?
- Measure the diameter and circumference in centimeters.
- Using a calculator, divide the circumference by the diameter (as measured in centimeters.) Record your answers below.
- Find the π button on your calculator and see what it gives you for the approximation for π. Record the number below--including all of the digits shown. If you have difficulty getting the calculator to display a value close to 3.1416 for π, consult your instructor for help.

$\frac{\text{circumference}}{\text{diameter}}$ = _______ (first circle)________ (second circle) _________ (third circle)

$\pi \approx$ _______________. (The symbol ≈ means "approximately equal to.")

Part 3 Repeat parts 1 and 2 with your other circular objects.

What Is π?

You will notice that for each of your circles, the circumference of the circle is slightly more than 3 diameters. Accurate measurements of the circumference of a circle always turn out to be about 3.14 diameters--no matter how large the circle is. The exact number of diameters in a circumference is the number π. π is not a rational number, so when we use π in computations, we round it to a particular decimal place. We sometimes use 3.14, 3.1416 or $\frac{22}{7}$ to approximate π, but the number your calculator gives for is more accurate than any of these estimates. However it is still not exactly π.

From observing the relationship between circumference and diameter of many circles, we find that the circumference is the diameter times π. We also know that the radius is the length from the center to the rim of the circle and the diameter is the length from rim to rim through the center. This makes the diameter twice the radius of a circle. This leads us to the following formulas for circles. In the formulas, d is the diameter, r is the radius, and C is the circumference of a circle.

Formulas

$d = 2r$ $r = \frac{1}{2}d$

$C = \pi d$ $C = 2\pi r$

Circumference, diameter, and radius are lengths, so they are measured in linear units.

Why are the two formulas for the circumference the same?

The Area of a Circle

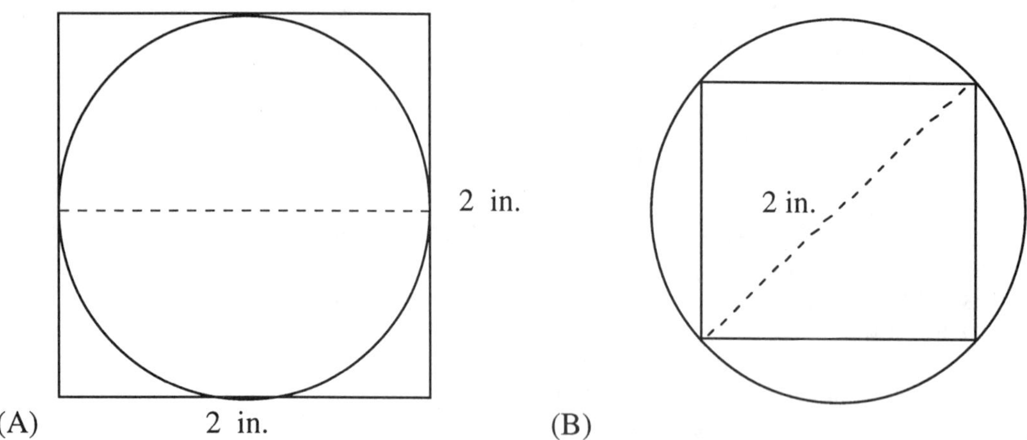

Consider the squares and the circle drawn above. Both circles are the same size, with diameter of 2 in. **Measure the diameter of each circle to verify the length.**

In (A), the circle is drawn within the square. The length of a side of the square is the same as the diameter of the circle. The diameter of the circle is 2 in., so the radius of the circle is 1 in. The square is said to be **circumscribed** around the circle. The formula for area of a square is $A = s^2$, so the area of the square = 2 in. x 2 in. = 4 in^2. Since the circle is inside the square, the area of the circle must be less than 4 in^2.

In (B), a square is drawn inside the circle, with a radius of 1 in. (diameter of 2 in.) This square is said to be **inscribed** in the circle. The line drawn from corner to corner of the square is a **diagonal** of the square. Notice that the diagonal of the inscribed square is also the diameter of the circle. Since the diameter of the circle is 2 in., the sides of the square are smaller than 2 in. **Measure the sides of the square to verify their lengths.** Record the length and compute the area of the square in (B). The sign "≈" means "is approximately equal to."

In (B), length ≈______________ Area of inscribed square ≈ ________________

In (A), the area of the circumscribed square = 4 in^2.

The actual area of the circle with radius 1 appears to be less than the area of the circumscribed circle but greater than the area of the inscribed square. The actual area of the circle is about 3.14 in^2. Does this fall within the areas of the inscribed and circumscribed squares?

> **Formula**
> The area of a circle of radius r is **Area = times the radius squared.** Written symbolically, $\mathbf{A = \pi r^2}$
> As always, area is measured in square units; therefore the area of a circle is measured in square units.

Use the formula for the above circle of radius 1 in. to verify that $A = \pi$ in^2.

When using π in a formula, we sometimes leave the answer for circumference or diameter with "π" as part of the answer; at other times we compute the answer using a calculator to get a decimal approximation. The following examples show the key strokes on a scientific calculator to compute circumference and area. They also cover how to find the radius if you know the diameter or to get the diameter from the radius.

Example 1: In a circle of radius 5 cm, find the diameter, circumference, and the area of the circle.

$C = 2\pi r = (2)(\pi)(5\text{ cm})$. To get a decimal approximation for C, do the following key strokes:

2 [x] π [x] 5 [=] $C \approx 31.416$ cm (This is rounded to 3 decimal places.)

Try the key strokes. If you do not get something close to 31.416, get help from your instructor. The answer will vary some due to round-off errors.

$A = \pi r^2$. $A = (\pi)(5\text{ cm})^2 = 25\pi\text{ cm}^2$. (Remember, area is measured in square units.) Using the calculator, this gives $A \approx 73.5398\text{ cm}^2$ (rounded to four decimal places). The key strokes would be

π [x] 5 [x] 5 [=] or 25 [x] π [=]

Note: For circumference, you could have used the formula C = d instead and substituted the diameter into the formula instead of the radius. Either formula should give the same answer. Check it out.

Example 2: The diameter of a circle is 14.9 in. Find the radius, circumference, and area of the circle. First, let's find the radius. The formula is, $r = \frac{1}{2}d$. An easy way to take 1/2 of a number is to divide that number by 2, so r = (14.9 divided by 2).

Calculator key strokes: 14.9 [x] 2 [=]

You should get 7.45; so r = 7.45 in.

$C = 2\pi r = (2)(\pi)(7.45)$ in. ≈ 46.8 in. Check it out.

The area, $A = \pi r^2 = (\pi)(7.45\text{ in.})(7.45\text{ in.}) \approx 174.366\text{ in.}^2$

Exercise 2--Circumference and Area of a Circle

Purpose of Exercise

- To practice using the formulas for circumference and area of circle

Materials Needed

- Calculator

In each of the following circles, compute the radius or diameter, whichever is not given. Write the formula you are using to compute area and circumference, show the substitutions and use a calculator to find the answers correct to two decimal places. Include the appropriate unit with each of your answers. The letter C represents circumference and A represents area. The first one is done as an example.

1.

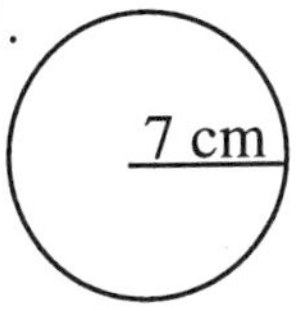

r = 7 cm d = 14 cm Formula: $C = 2\pi r$

$C = (2)(\pi)(7\text{ cm}) \approx 43.98$ cm $A = \pi r^2 = (\pi)(7\text{ cm})^2 \approx 153.94\text{ cm}^2$

2.

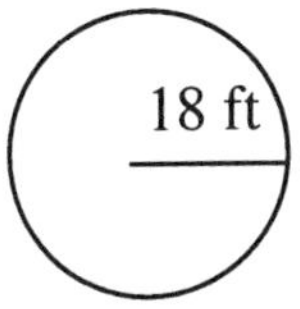

r = ________ d = ________ Formula: C = ________

C = ________________ A = ____________________

3. 81 m

r = ________ d = ________ Formula: C = ________

C = ________ A = ________

4. 4.3 yd

r = ________ d = ________ Formula: C = ________

C = ________ A = ________

5. 12.5 in

r = ________ d = ________ Formula: C = ________

C = ________ A = ________

6. 22 km

r = ________ d = ________ Formula: C = ________

C = ________ A = ________

7. 13.75 mi

r = ________ d = ________ Formula: C = ________

C = ________ A = ________

8. 82 ft

r = ________ d = ________ Formula: C = ________

C = ________ A = ________

9. 3.5 yd

r = ________ d = ________ Formula: C = ________

C = ________ A = ________

10. 13 m

r = ________ d = ________ Formula: C = ________

C = ________ A = ________

11. 215 yd

r = ________ d = ________ Formula: C = ________

C = ________ A = ________

12. 19.1 m

r = ________ d = ________ Formula: C = ________

C = ________ A = ________

Section 3-9 *Geometric Shapes--Triangle, Parallelogram, Trapezoid*

In previous sections, we have studied rectangles and circles (two-dimensional geometric shapes). In this section we will expand our study to other two-dimensional shapes, such as triangles, parallelograms, and trapezoids. We will define these shapes and work with formulas for the area and perimeter of the shapes. We will also work with less regular shapes and compute or estimate area and perimeter when the formulas may be more complex or may combine several formulas.

Vocabulary

A **right angle** (or 90° angle) is formed when two lines are perpendicular.
An **acute angle** is an angle smaller than 90°.
An **obtuse** angle is an angle larger than 90°.

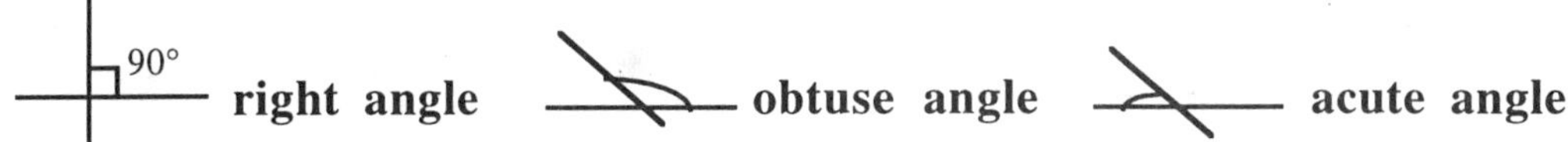

The **perimeter** of any closed geometric shape is the length around the outside of the shape. If the shape has distinct sides, the perimeter is the sum of the lengths of the sides.

A **triangle** is a three-sided closed geometric shape. The measures of the angles of any triangle add up to 180°.

A **right triangle** is a triangle in which one of the angles is a right angle.

The **hypotenuse** (abbreviated hyp. in the diagrams below) of a right triangle is the side opposite the right angle.

The **legs** of a right triangle are the other two sides.

An **obtuse** triangle is one in which one of the angles is obtuse.

An **acute** triangle is one in which all the angles are acute.

A triangle has three **vertices**. Each of the corners of the triangle is a **vertex**.

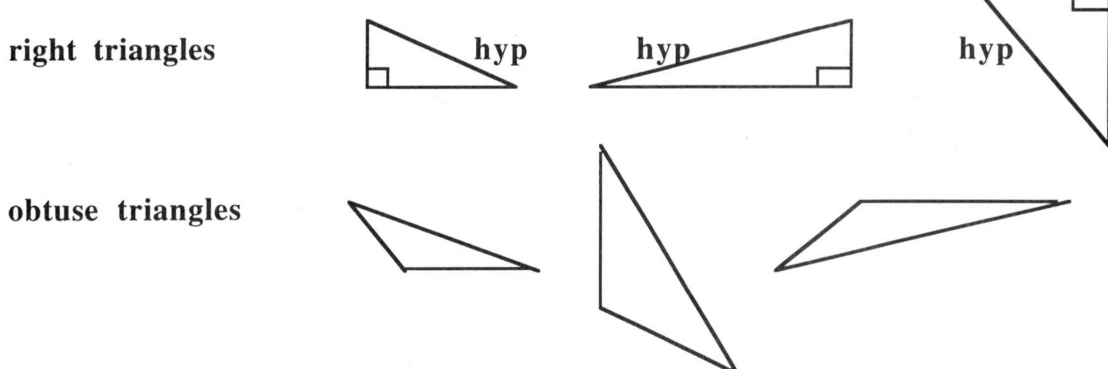

The formula for the area of a triangle refers to base and height of the triangle. In any triangle, the base can be any one of the three sides of the triangle. The height then is relative to the base chosen. The height is the distance measured along a line perpendicular to the base from the opposite vertex. Sometimes the base has to be extended to draw the height.

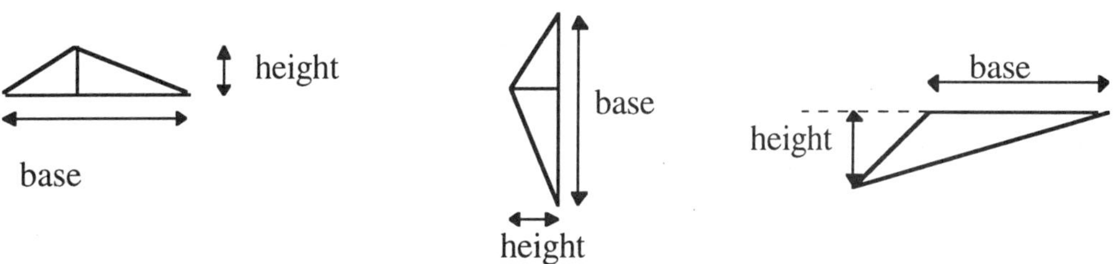

Formulas for the Triangle

$P = a + b + c$ — P = perimeter; a, b, and c are the lengths of the sides.

$A = \frac{1}{2} b \cdot h$ — A = area, b= base, h = height

The sum of the measures of the angles of a triangle is 180°.

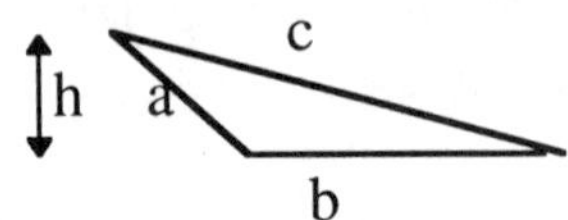

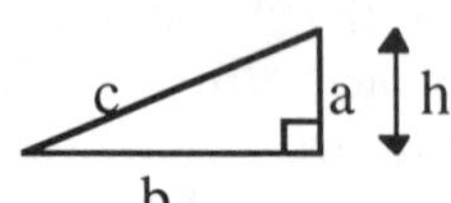

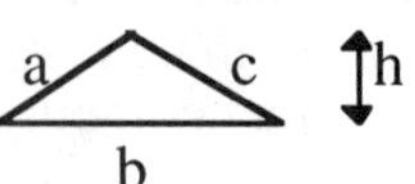

Vocabulary

A **line segment** is part of a straight line. A line segment has a particular length.

A **quadrilateral** is a four-sided closed geometric shape in which each of the sides is a line segment. The sum of the measures of the angles of any type of quadrilateral is 360°.

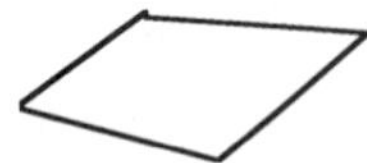

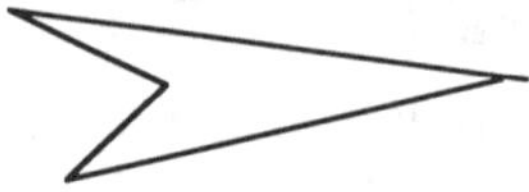

A **vertex** of a quadrilateral is the point where two sides meet. There are four **vertices** in a quadrilateral.

A **parallelogram** is a quadrilateral in which both sets of opposite sides are parallel.

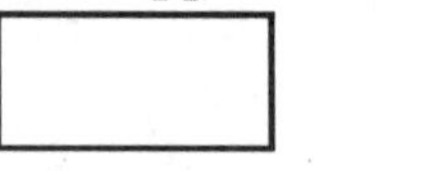

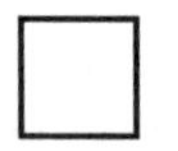

A **rectangle** is a parallelogram in which the angles are all **right** angles.

A **square** is a rectangle in which all sides are equal.

A **trapezoid** is a quadrilateral in which two of the sides are parallel.

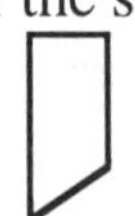

Area and Perimeter Formulas for a Parallelogram

$$A = b \bullet h \quad \text{and} \quad P = 2a + 2b$$

Letters a and b represent the two different lengths of sides of the parallelogram, h represents the height, A the area, and P the perimeter. **Remember that lengths are measured in linear units and area in square units.**

parallelogram

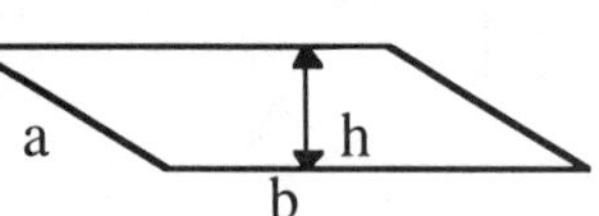

Area and Perimeter Formulas for a Trapezoid

$$P = a + b + c + d \qquad A = \frac{1}{2} h (b + d)$$

where a, b, c, d are the four sides, b and d are the parallel sides, and h is the distance between the parallel sides.

trapezoid

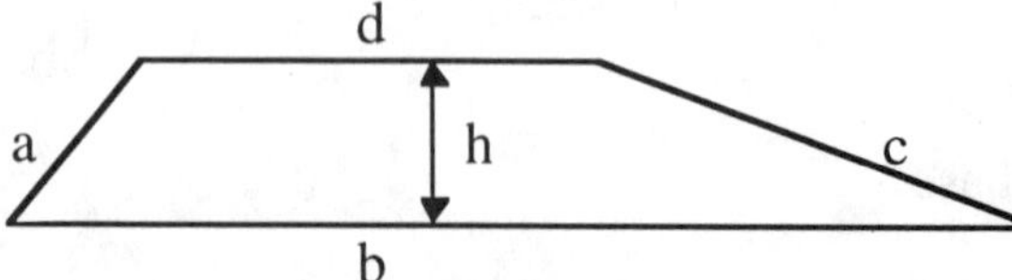

Exercise 1--Verifying the Formulas for Trapezoids and Parallelograms

Purpose of the Exercise

- To understand the formulas for area and perimeter of a trapezoid and a parallelogram

Materials Needed

- Graph paper
- Ruler or straight edge
- Scissors

Instructions

- On graph paper, draw two different parallelograms and two different trapezoids.
- Estimate the dimensions of these figures by counting the number of graph paper squares that they cover. One box on the graph paper equals one unit.
- Use the appropriate formula to compute the area of the geometric figure.
- For each parallelogram, cut off a triangle, and reposition it so that the shape looks like a rectangle (see the diagram below). Record the length and width of the rectangle. Compute the area of the rectangle and compare the answer to the area of the parallelogram. The two answers should be the same.
- For each trapezoid, partition the trapezoid into a parallelogram and a triangle. (See the diagram below.) Compute the area of the parallelogram and of the triangle. Add the two areas and compare with the area of the trapezoid. The two answers should be the same.
- Record the data below.

Parallelogram

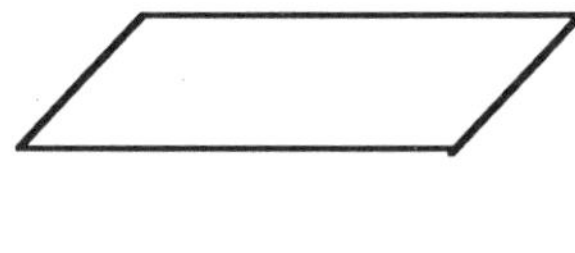

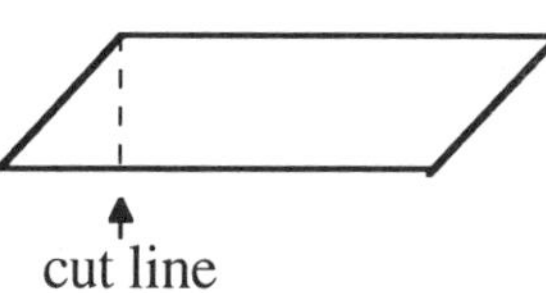

cut line

The parallelogram reshaped to a rectangle.

Trapezoid

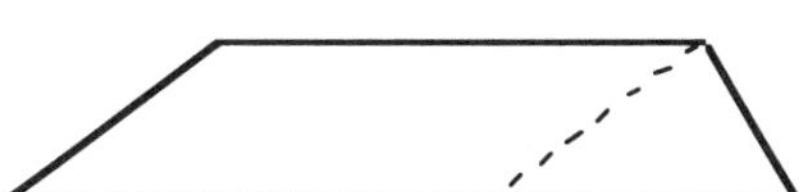

1. Parallelogram 1: a = ______ b = _____ h = _____ A = _______ P = _______

   Rectangle (after cut and move) : L = ______ W = _______ A = ___________

2. Parallelogram 2: a = ______ b = _____ h = _____ A = _______ P = _______

   Rectangle (after cut and move) : L = ______ W = _______ A = ___________

3. Trapezoid 1: a = _____ b = _____ c = _____ d = ____ A = _____ P = ______

   Area of triangle = ____________ Area of rectangle = ____________

   Sum of the 2 areas = ____________

4. Trapezoid 2: a = _____ b = _____ c = _____ d = ____ A = _____ P = ______

   Area of triangle = ____________ Area of rectangle = ____________

   Sum of the 2 areas = ____________

Exercise 2--Rectangles, Circles, Parallelograms, and Trapezoids

Purpose of the Exercise

- To practice using formulas to find perimeter and areas of various geometric shapes

Materials Needed

- Calculator

For each of the following, name the shape and compute the area (A), perimeter (P), and circumference (C) as indicated. (If the shape has more than one name, list the most specific shape. For example if the figure is both a rectangle and a square, list square as the name.) Use a calculator as needed and include proper units in your answer.

1.

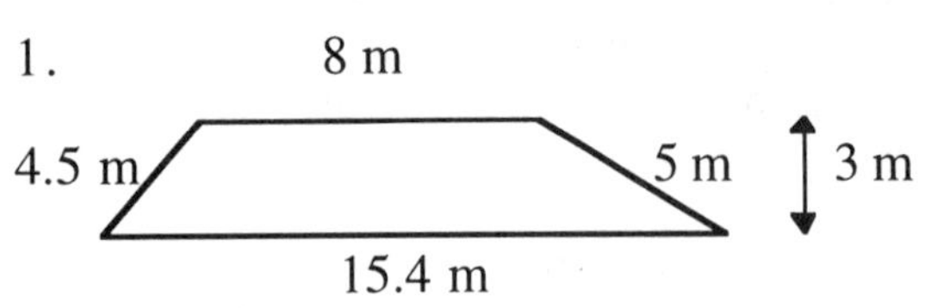

Name of shape ____________________

A = ______________________________

P = ______________________________

2.

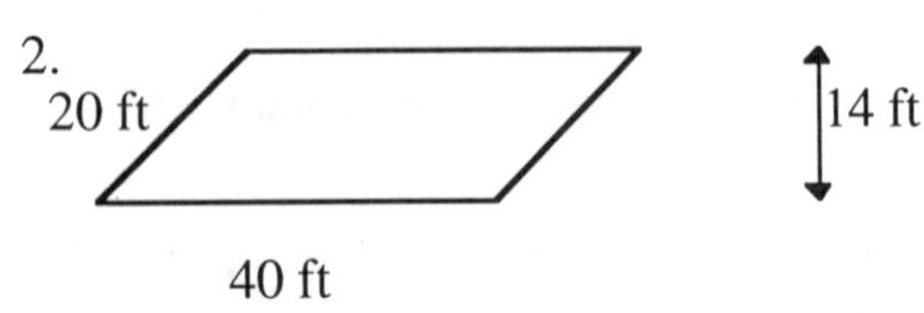

Name of shape ____________________

A = ______________________________

P = ______________________________

3.

5 mi

13.2 mi

Name of shape ____________________

A = ______________________________

P = ______________________________

4.

73 yd

Name of shape ____________________

A = ______________________________

C = ______________________________

5.

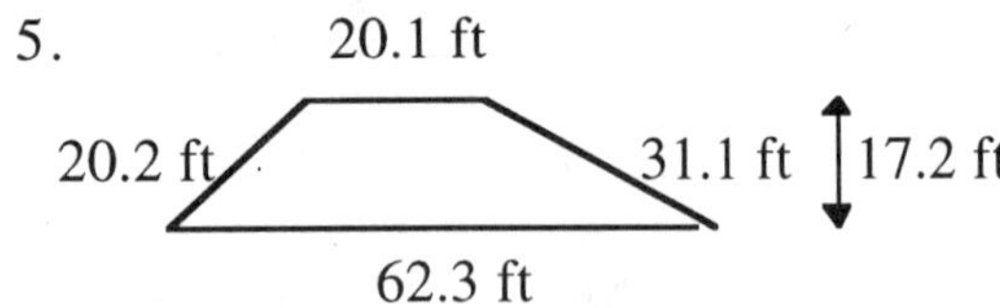

Name of shape ____________________

A = ______________________________

P = ______________________________

6.

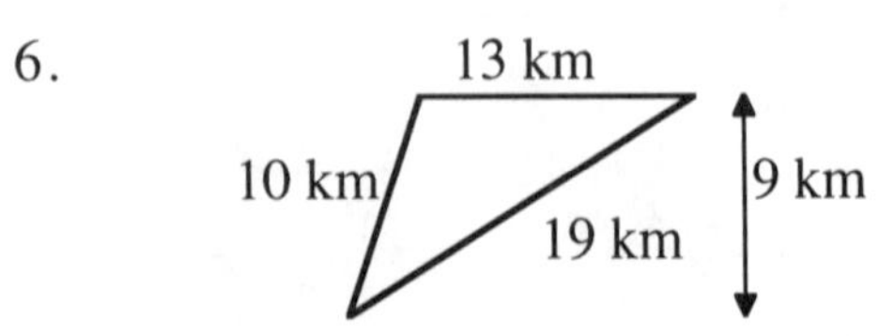

Name of shape ____________________

A = ______________________________

P = ______________________________

7. 195.3 in. 195.3 in.

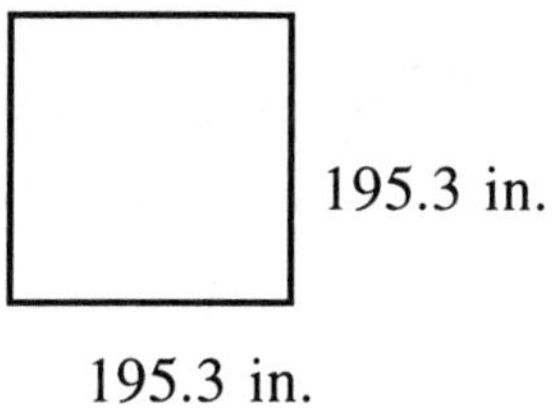

Name of shape ______________

A = ______________

P = ______________

8. 22 m 71 m

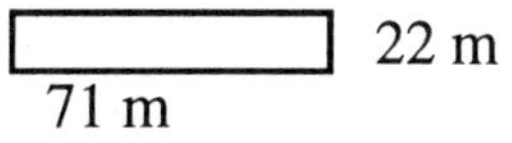

Name of shape ______________

A = ______________

P = ______________

9. 11.9 in. 13.8 in. 7.0 in.

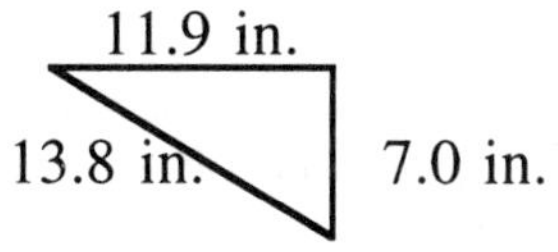

Name of shape ______________

A = ______________

P = ______________

10. 23 yd 11 yd 7 yd 6 yd 29 yd

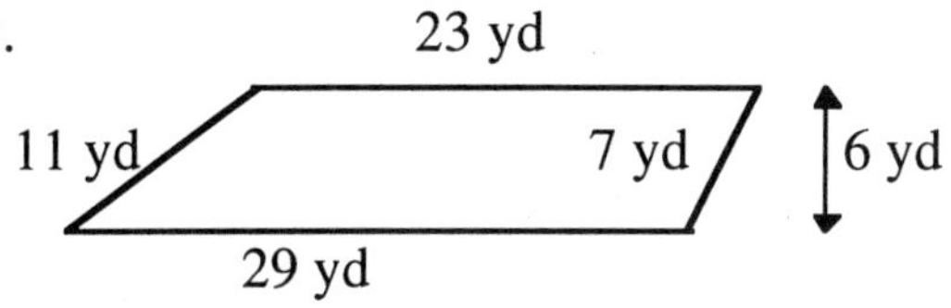

Name of shape ______________

A = ______________

P = ______________

11. 19.4 m

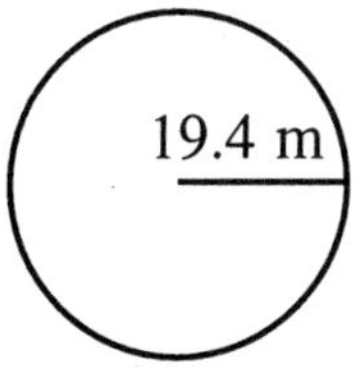

Name of shape ______________

A = ______________

C = ______________

12. 10.5 m. 14.1 m 31.2 m 19.1 m

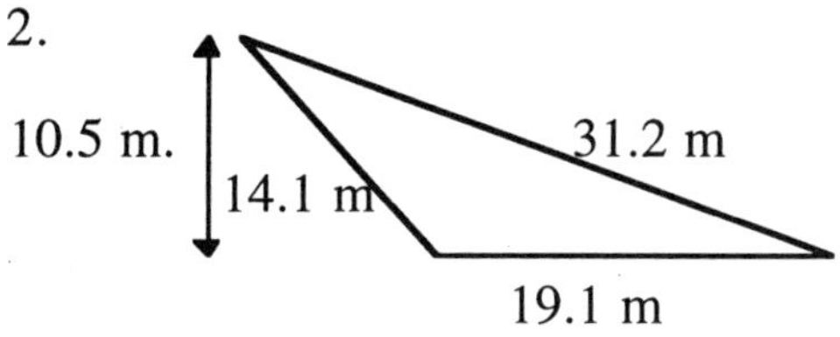

Name of shape ______________

A = ______________

P = ______________

13. 0.2 mi 0.2 mi

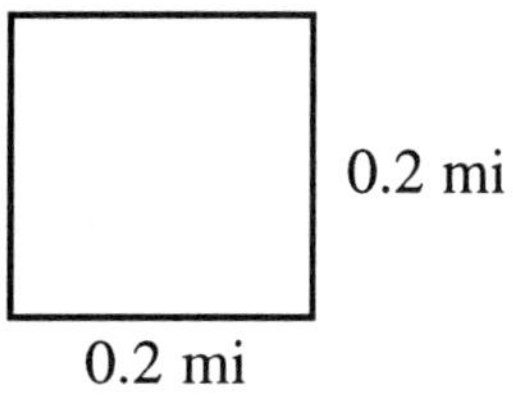

Name of shape ______________

A = ______________

P = ______________

14. 6.5 km 10 km 21 km

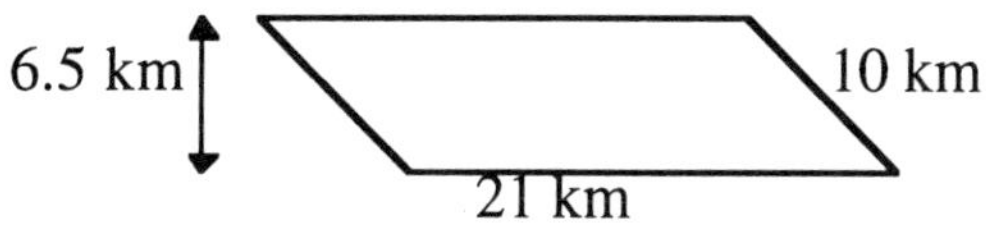

Name of shape ______________

A = ______________

P = ______________

Other Geometric Shapes

Many times we need to compute or estimate the area of shapes that are not a circle, rectangle, parallelogram, trapezoid, or triangle. Sometimes we can get an accurate answer by breaking the area into two or more shapes to use common formulas. Other times the shapes are so irregular that we just estimate by drawing other shapes that have about the same area.

Exercise 3--Area and Perimeter of Other Geometric Shapes

Purpose of the Exercise

- To practice computing area and volume of composite shapes and irregular shapes

Materials Needed

- Calculator and ruler

In Problems 1-4, partition the shape into as many parts as needed to compute the area and perimeter of composite parts. Add the parts to compute area and perimeter of the entire shape.

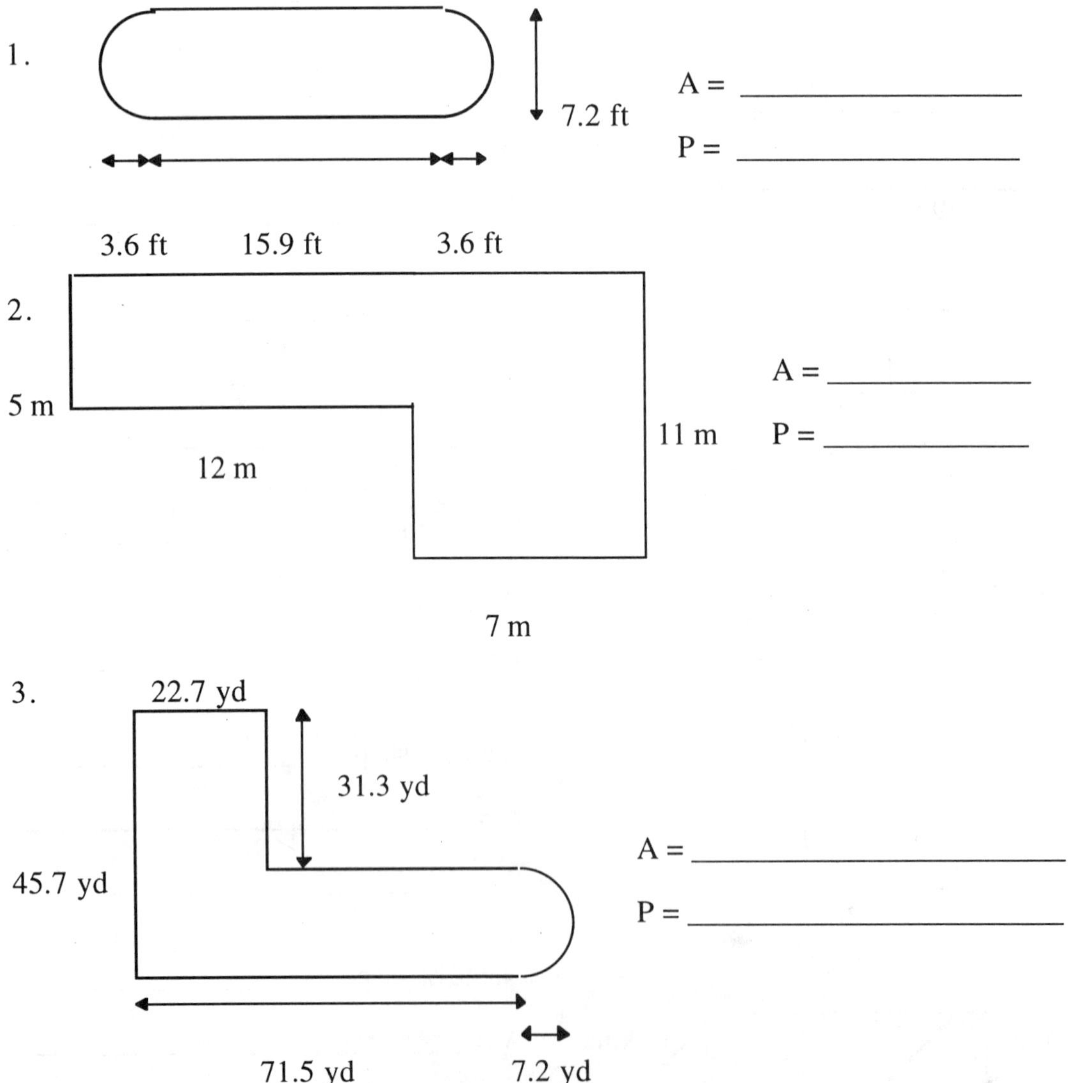

4.

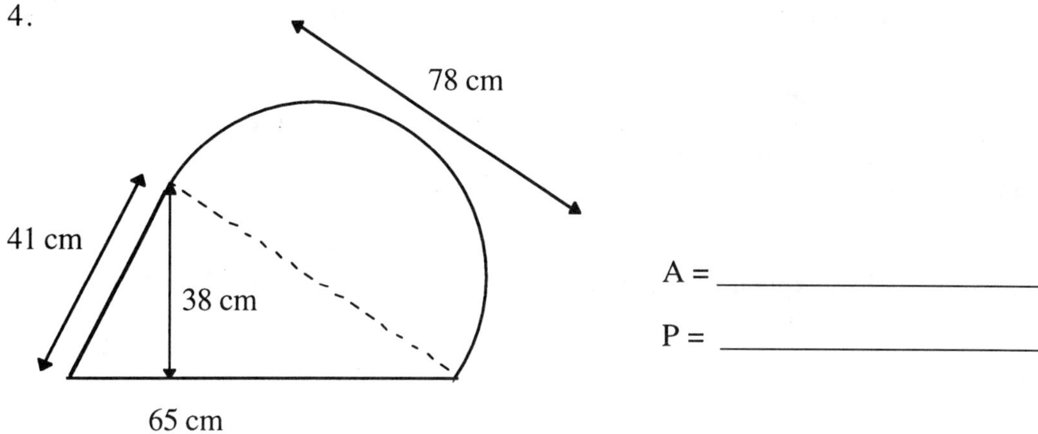

A = ____________________

P = ____________________

In each of the following, estimate the area and perimeter of the shape. You may use the grid for the estimate, or you may draw shapes that seem to have the same area around the pieces. If you are working in a group, have some group members use one approach and others use another approach; then compare answers.

5.

6.

7.

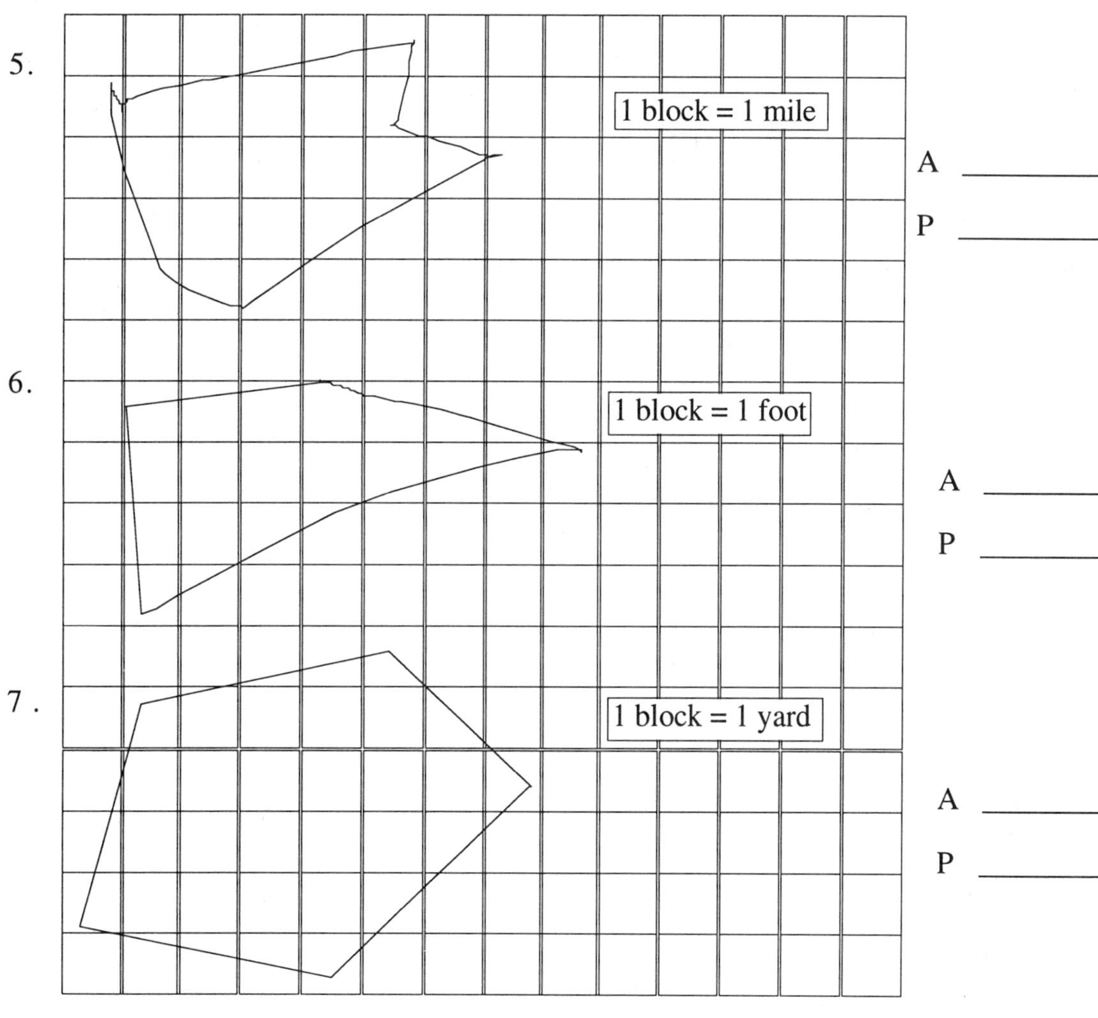

5. A ________ P ________

6. A ________ P ________

7. A ________ P ________

Exercise 4--What's Wrong with This Picture?

Purpose of the Exercise

- To catch the common mistakes made in finding area, perimeter, circumference, volume, and surface area problems

Materials Needed

- Calculator

In each of the following, explain what is wrong with the answer(s) or problem. In some cases, there may be more than one error. Find all of the errors. **A** = area, **P** = perimeter, **C** = circumference, **V** = volume, **SA** = surface area. **L**, **W**, and **H** represent length, width, and height, respectively.

1. 25.6 ft

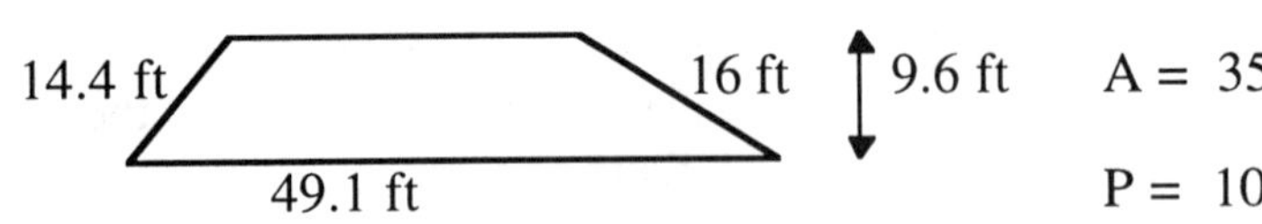

14.4 ft, 16 ft, 9.6 ft, 49.1 ft

A = 358.35 ft

P = 105.1 ft

What's wrong? ____

2.

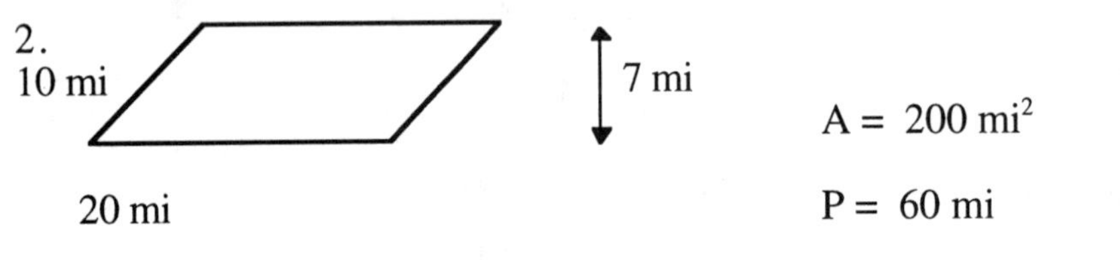

10 mi, 7 mi

A = 200 mi^2

20 mi

P = 60 mi

What's wrong? ____

3.

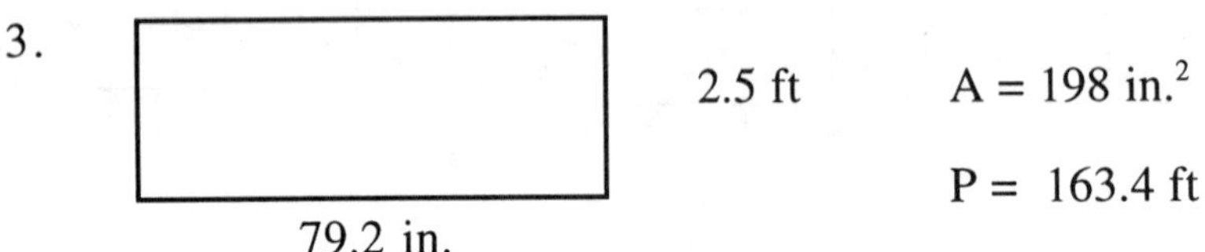

2.5 ft

A = 198 $in.^2$

P = 163.4 ft

79.2 in.

What's wrong? ____

4. 123 yd

A = 47529.15526 yd^2

C = 772.8 yd

What's wrong? ____

5.

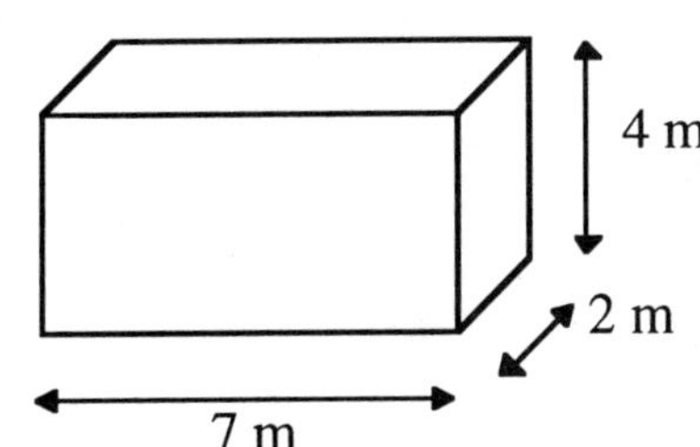

V = 56 m^2

SA = 100 m

What's wrong? ______________________________

6.

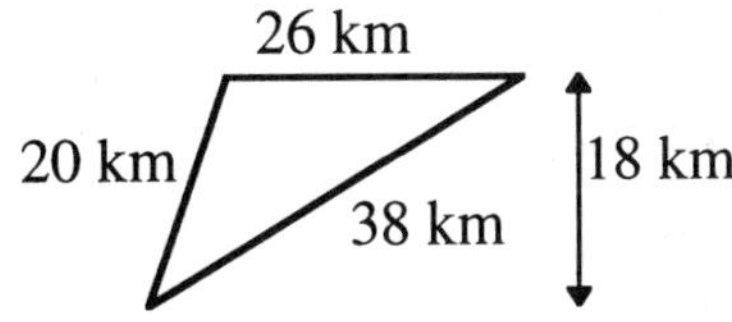

A = 468 km^2

P = 84 km^2

What's wrong? ______________________________

7.

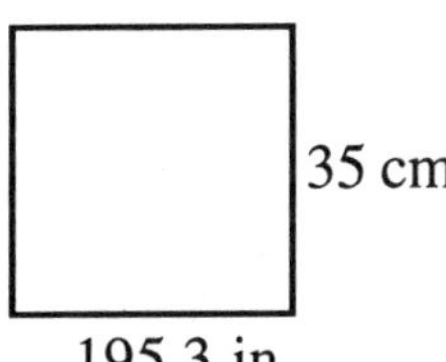

A = 6835.5 cm^2

P = 230.3 cm.

What's wrong? ______________________________

8.

A = 47529.15526

C = 772.8

What's wrong? ______________________________

9.

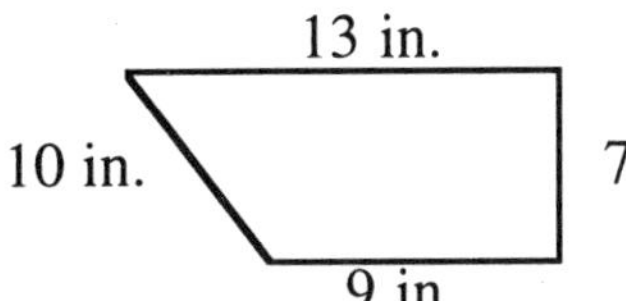

A = 91 in.^{2}

P = 129 in.

What's wrong? ______________________________

10.

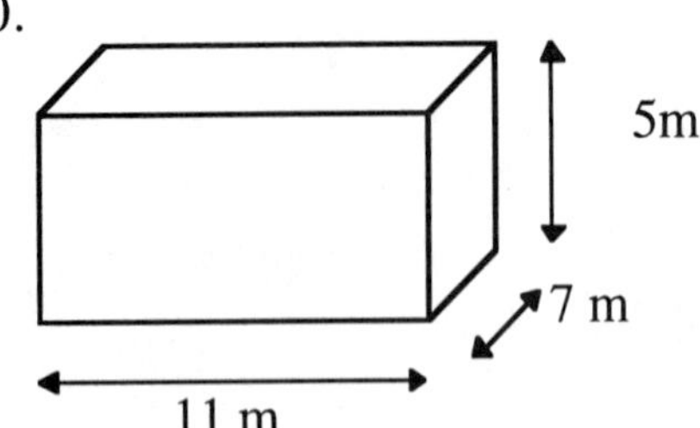

$V = 385\text{ m}^3$

$SA = 334\text{ m}^3$

What's wrong? ______________________________

11.

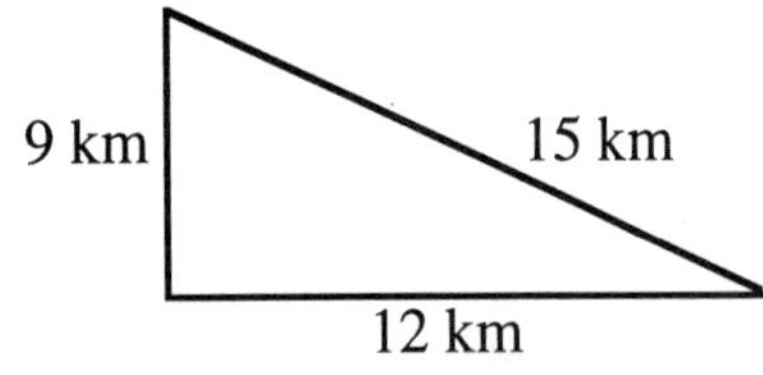

$A = 108\text{ km}^2$

$P = 36\text{ km}$

What's wrong? ______________________________

12. In the following, explain why the first estimate of the area is obviously too small and the second estimate is too large.

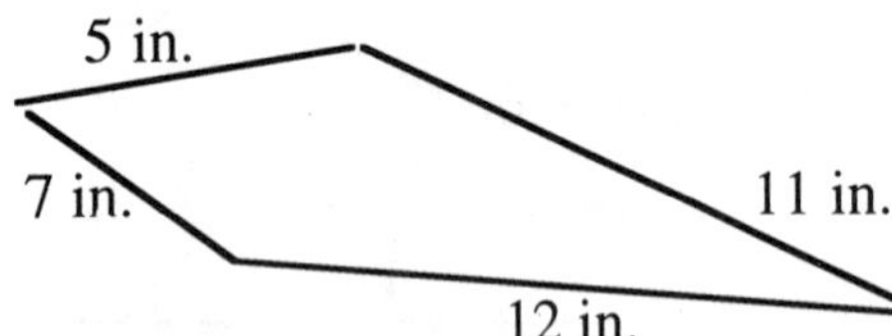

$A \approx 35\text{ in.}^2$

Explanation ______________________________

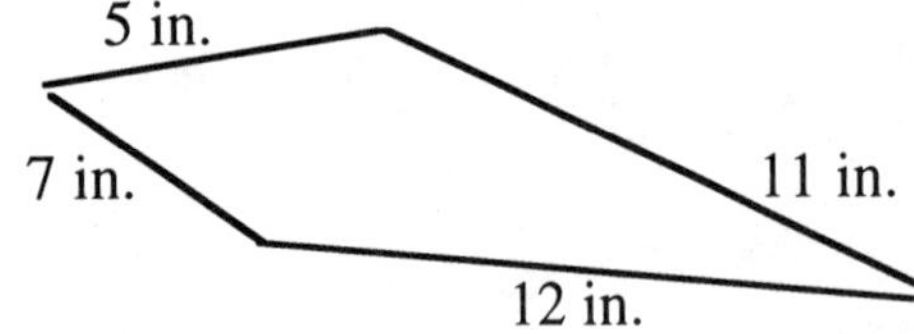

$A \approx 225\text{ in.}^2$

Explanation ______________________________

Section 3-10 *The Pythagorean Theorem*

In ancient Greece (about 500 BC), a student of Pythagoras discovered a particular relationship between the three sides of a **right** triangle. If one square is drawn, using one leg as a side of the square and another square is drawn using the other leg as a side, then the area of these two squares together is equivalent to the area of the square that has the hypotenuse as a side. Since the area of a square of side a is a^2, the area of a square of side b is b^2, and the area of the square of side c (the hypotenuse) is c^2, then $a^2 + b^2 = c^2$.

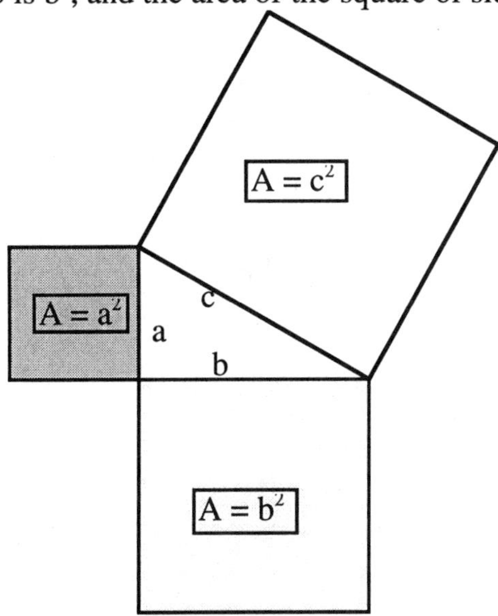

Formula for the Pythagorean Theorem

$$a^2 + b^2 = c^2$$

where a and b are the lengths of the sides of the two legs and c is the length of the hypotenuse of a right triangle.

The Pythagorean theorem has been used many times to confirm that a triangle is a **right** triangle. It was also used in primitive societies in construction to get a right angle at the corner of rooms or buildings.

One particular type of right triangle that has been useful is the "three-four-five" triangle. A three-four-five triangle is one in which the two legs are three units and four units long, and the hypotenuse is five units long. Observe the triangle below with the squares of the sides and hypotenuse shown.

The left side of the Pythagorean Theorem, $a^2 + b^2$ is

$$(3)^2 + (4)^2 = 9 + 16 = 25$$

The right side of the Pythagorean Theorem, c^2 is

$$(5)^2 = 25$$

The left side = the right side, so the triangle with these dimensions is a right triangle.

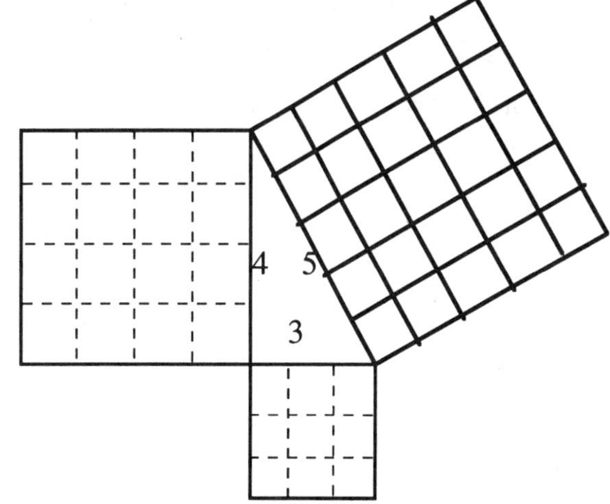

Exercise 1--The Pythagorean Theorem

Purpose of the Exercise

- To determine from the lengths of the sides whether or not a triangle is a right triangle

Materials Needed

- Calculator
- 6- or 12-inch ruler

In each of the following exercises, determine whether or not the triangle is a right triangle by determining whether the squares of the two sides, a and b, equal the square of the third side, c. Fill in the blanks with the answers. When the units are small enough, use your ruler and a separate sheet of paper to draw the rectangles indicated to compare results.

1. a = 8 in., b = 6 in., c = 10 in.

   $a^2 + b^2 =$ ______ $c^2 =$ _____

   Right triangle? yes no

2. a = 8 ft, b = 6 ft, c = 10 ft

   $a^2 + b^2 =$ ______ $c^2 =$ _____

   Right triangle? yes no

3. a = 1.3 cm, b = 2.1 cm, c = 3 cm

   $a^2 + b^2 =$ ______ $c^2 =$ _____

   Right triangle? yes no

4. a = 6.1 in, b = 9.2 in, c = 10 in

   $a^2 + b^2 =$ ______ $c^2 =$ _____

   Right triangle? yes no

5. a = 2 in, b = 2 in, c = 3 in

   $a^2 + b^2 =$ ______ $c^2 =$ _____

   Right triangle? yes no

6. a = 40 m, b = 30 m, c = 50 m

   $a^2 + b^2 =$ ______ $c^2 =$ _____

   Right triangle? yes no

7. a = 3 in, b = 3 in, c = 4 in

   $a^2 + b^2 =$ ______ $c^2 =$ _____

   Right triangle? yes no

8. a = 4.5 cm, b = 6 cm, c = 7.5 cm

   $a^2 + b^2 =$ ______ $c^2 =$ _____

   Right triangle? yes no

9. a = 1.5 in, b = 2 in, c = 2.5 in

   $a^2 + b^2 =$ ______ $c^2 =$ _____

   Right triangle? yes no

10. a = 1 in, b = 1 in, c = 2 in

    $a^2 + b^2 =$ ______ $c^2 =$ _____

    Right triangle? yes no

11. a = 1.3 in, b = 2.1 in, c = 3 in

    $a^2 + b^2 =$ ______ $c^2 =$ _____

    Right triangle? yes no

12. a = 2 in, b = 3 in, c = 4 in

    $a^2 + b^2 =$ ______ $c^2 =$ _____

    Right triangle? yes no

13. a = 1 in, b = 2 in, c = 3 in

    $a^2 + b^2 =$ ______ $c^2 =$ _____

    Right triangle? yes no

14. a = 9 in, b = 12 in, c = 15 in

    $a^2 + b^2 =$ ______ $c^2 =$ _____

    Right triangle? yes no

The Pythagorean Theorem and Square Roots

The Pythagorean theorem can be used to compute the hypotenuse of a right triangle if the lengths of the other two sides are known. First we will take a look at a variation of the Pythagorean theorem.

As stated previously, the Pythagorean theorem is given by the formula $c^2 = a^2 + b^2$. Another way to state this same formula is $c = \sqrt{a^2 + b^2}$. This is read, "**c equals the square root of a^2 plus b^2.**"

To understand this new version of the Pythagorean theorem and the meaning of square roots, we will draw some right triangles for which we know the length of the two legs. We will measure the length of the hypotenuse, and then compute the length by using the Pythagorean theorem and a calculator.

First, we will learn how to use a calculator to compute a square root. The symbol for square root is $\sqrt{\ }$. Look for this symbol on your calculator. To find the square root of 3 (written as $\sqrt{3}$) first key in a 3; then press the square root key. Since keying in the square root symbol varies from calculator to calculator, experiment on your calculator to see how to get the answer. If you have difficulties, see your instructor. The calculator should give the approximate value for $\sqrt{3}$ as 1.732050808. (The number of digits varies with different calculators.)

Example 1: Now we will construct a triangle that has $\sqrt{5}$ units as the length of its hypotenuse. The unit we will use will be 1 inch. We will draw a triangle that has legs of length 1 inch and 2 inches. Using a square to make sure the angles meet in right angles, first draw the two legs. Then draw in the hypotenuse.

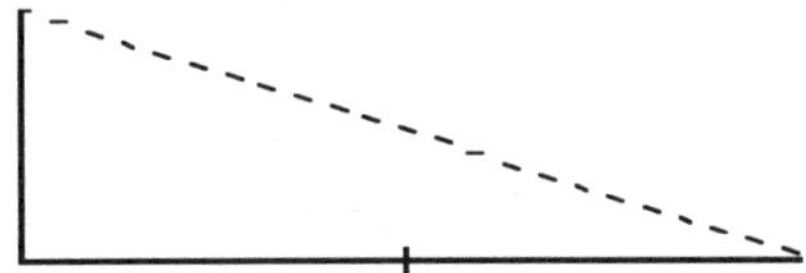

By the Pythagorean theorem, the length of the hypotenuse should be $\sqrt{1^2 + 2^2}$ or $\sqrt{5}$.

To compute $\sqrt{1^2 + 2^2}$, we compute 1^2 then 2^2 and then add the answer to get $1 + 4 = 5$.

Next, we compute the square root of the answer.

The length of the hypotenuse should be approximately 2.236 inches (or about 2.2)

Measure the hypotenuse above with a ruler and see if this looks like a reasonable answer.

Exercise 2-- Finding the Length of the Hypotenuse

Purpose of the Exercise

- To practice using the Pythagorean theorem and finding square roots with a calculator

Materials Needed

- Graph paper
- Calculator
- Straight edge

Instructions

- In each of the following problems, the two lengths given are the legs of a right triangle.
- Draw each triangle on graph paper. Use the length of one square as 1 unit (1 U) and use the graph paper and the straight edge to accurately draw a right angle.
- Measure the hypotenuse by laying another sheet of graph paper across it. Estimate any decimal parts visually.
- Compute the length of the hypotenuse using the Pythagorean theorem.

- Compute the area and perimeter of each of the triangles, using the value you got from the Pythagorean theorem for the hypotenuse. Include appropriate units with your answers.
- In the following problems, a and b are the lengths of the legs of the right triangle and c is the length of the hypotenuse. P is the perimeter and A is the area.

1. a = 1 U, b = 1 U

   c ≈ ________________ (from measuring)

   c ≈ ________________ (from formula)

   A ≈ ________________

   P ≈ ________________

2. a = 1 U, b = 2 U

   c ≈ ________________ (from measuring)

   c ≈ ________________ (from formula)

   A ≈ ________________

   P ≈ ________________

3. a = 2 U, b = 3 U

   c ≈ ________________ (from measuring)

   c ≈ ________________ (from formula)

   A ≈ ________________

   P ≈ ________________

4. a = 3 U, b = 3 U

   c ≈ ________________ (from measuring)

   c ≈ ________________ (from formula)

   A ≈ ________________

   P ≈ ________________

5. a = 3 U, b = 4 U

   c ≈ ________________ (from measuring)

   c ≈ ________________ (from formula)

   A ≈ ________________

   P ≈ ________________

6. a = 3 U, b = 5 U

   c ≈ ________________ (from measuring)

   c ≈ ________________ (from formula)

   A ≈ ________________

   P ≈ ________________

6. Challenge question: In each of the following situations, Clint and Danny start at the same place but Clint drives 25 miles from the starting position while Danny drives 17 miles. For each of the following situations, draw a diagram to show Clint's and Danny's starting position and path. Determine how far apart Danny and Clint are after the drive.

   a. Clint and Danny both drive due East.

   b. Clint drives due north and Danny drives due South.

   c. Clint drives due north and Danny drives due West. (Compute the distance along the shortest path between Clint and Danny.)

Section 3-11 *Spheres, Cylinders, and Cones*

All three-dimensional structures have volume and surface area. In rectangular boxes, we defined volume as the number of cubic units that fit in the box, and surface area as the number of square units that would cover all six sides of the box. Below are formulas for the volume and surface area of other three-dimensional geometric shapes.

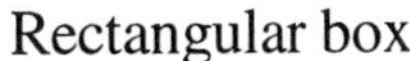

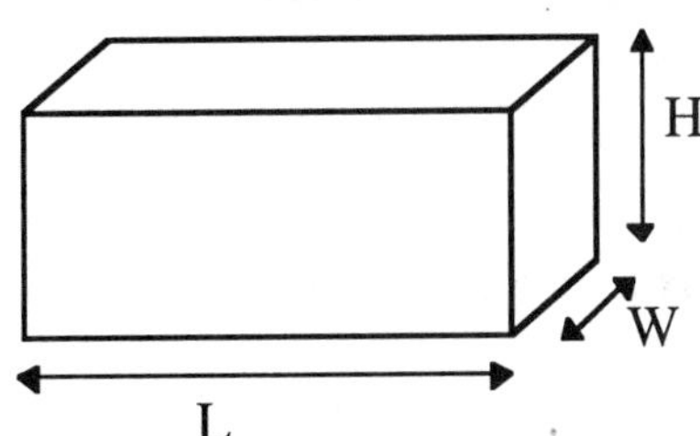

Cube

s
(all sides have the same length.)

Right circular cylinder

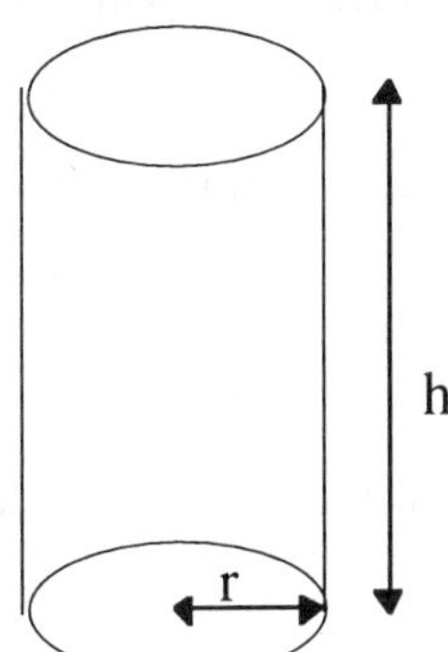

Right circular cone

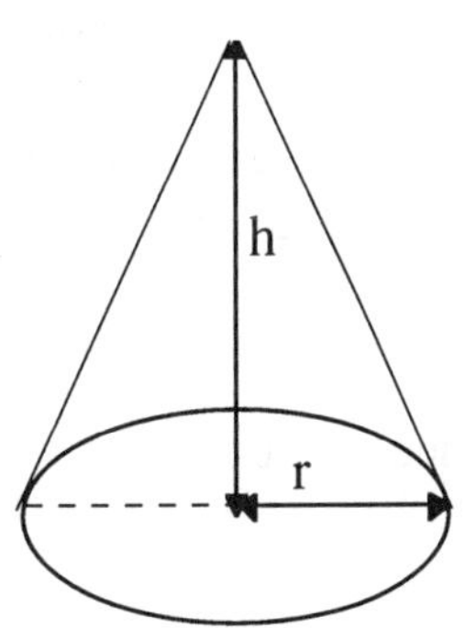

Sphere

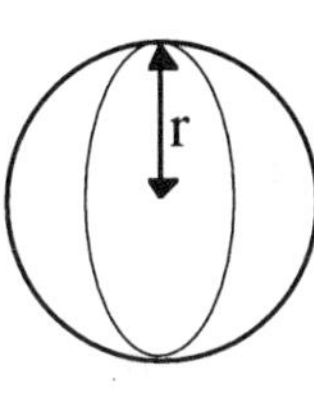

Volume and Surface Area Formulas

Volume of a rectangular box: $\mathbf{V = LWH}$

Volume of a cube: $\mathbf{V = s^3}$

Surface area of a rectangular box: $\mathbf{SA = 2L+2W+2H}$

Surface area of a cube: $\mathbf{SA = 6s^2}$

Volume of a right circular cylinder: $\mathbf{V = \ r^2h}$

Surface area of a right circular cylinder (top and bottom included): $\mathbf{SA = 2\ rh + 2\ r^2}$

Surface area of a right circular cylinder (without top and bottom): $\mathbf{SA = 2\ rh}$

Volume of a sphere: $\mathbf{V = \frac{4}{3}\ r^3}$ Surface area of a sphere: $\mathbf{SA = 4\ r^2}$

Volume of a right circular cone: $\mathbf{V = \frac{1}{3}\ r^2h}$

Surface area of a right circular cone (with bottom): $\mathbf{SA = \ r^2 + 2\ r\left(\sqrt{r^2+h^2}\right)}$

Surface area of a right circular cone (without bottom): $\mathbf{S = 2\ r\left(\sqrt{r^2+h^2}\right)}$

Volume is a measure of how much will fit in the box and surface area is the amount of material required to make the box.

Exercise 1--Comparison of Volume and Surface Area

Purpose of the Exercise

- To understand volume as a measurement of how much a container will hold

Materials Needed

- Ruler
- Calculator (optional)
- Small cardboard rectangular box (cracker box, cookie box)
- Small cardboard cylinder-shaped (cylindrical) box with lid (oatmeal box)
- Oatmeal (or rice or small beans)--enough to fill the largest container
- Several other small containers of various odd shapes that appear to be about the same size--vases, glasses, and the like.

1. Measure the length, width, and height of the rectangular box and compute its volume and surface area.
   volume = ________________ surface area = ____________________

2. Measure the height and diameter of the cylindrical box and compute its volume and surface area. Be sure to include the lid and the bottom of the box in the surface area.
   volume = ________________ surface area = ____________________

3. Which container will hold more oatmeal? Explain how you made your decision.
   ____

4. Which box requires more materials to make the box? ______________________

5. Fill one of the boxes with oatmeal. Pour the oatmeal from the first box into the other box. Is there enough oatmeal to fill the box or is there too much oatmeal? ________

6. Is this the result you expected by computing the volumes of the boxes? ________

7. Pick up two other different shaped containers. Guess which container has the larger volume. Which container? ________________________

8. Fill one container with oatmeal. Pour the oatmeal from one container to the other to see if you guessed correctly. Was your guess correct? ____________________

9. Estimate the actual volume of each container by pouring the oatmeal from the container to the rectangular box. Using your ruler, measure the height of the oatmeal in the box. You have already measured the length and the width.

   Vol. container 1 = _______________ Vol. container 2 = ____________________

10. Estimate the actual volume of each container by pouring the oatmeal from the container into the cylindrical box. Using your ruler, measure the height of the oatmeal in the box.

    Vol. container 1 = _______________ Vol. container 2 = ____________________

Exercise 2--Which Container Is Bigger?

Purpose of the Exercise

- To practice using the formulas for volume and surface area
- To compare sizes of containers

Materials Needed

- Calculator
- Paper, scissors, tape
- Ruler

In each of the following, compute the volume and surface area for the information given. Write the formula used and make the comparisons as indicated.

1. a. A rectangular box has dimensions: 14.3 in., 12.9 in., and 2.2 in.

   Volume = ______________ Surface area = ______________

   b. A sphere has a diameter of 8.2 inches.

   Volume = ______________ Surface area = ______________

   c. Which holds more oatmeal, the rectangular box or the sphere? ______________

   d. Which takes more materials to make, the rectangular box or the sphere? ________

2. a. A cone (with a bottom) has dimensions: h = 3 inches, r = .74 in.

   Volume = ______________ Surface area = ______________

   b. A sphere has a radius of 2.3 inches.

   Volume = ______________ Surface area = ______________

   c. Which holds more oatmeal, the cone or the sphere? ____________

   d. Which takes more materials to make, the cone or the sphere? _______

3. a. A tin can (with top and bottom) has dimensions: h = 4.25 inches, r = 1.5 in.

   Volume = ______________ Surface area = ______________

   b. A rectangular canister has dimensions: 4 in., 2.7 in., 3.1 in.

   Volume = ______________ Surface area = ______________

   c. Which holds more oatmeal, the tin can or the canister? ________________

   d. Which takes more materials to make, the tin can or the canister? __________

4. Use paper to construct a cylinder and canister with the dimensions given in Problem 3. Compare the conclusions in parts c and d to the cylinder and box you constructed to see if your objects confirm your mathematical conclusions.

Exercise 3--Using All the Formulas

Purpose of the Exercise
- To practice using geometric formulas and units

Materials Needed
- Calculator

In the following problems, A = area, P = perimeter, SA = surface area, V = volume, C = circumference, r = radius, L = length, W = width, and H = height. Be sure to include proper units with your answer.

1. In a trapezoid, h = 45 cm, the two parallel sides are b = 32 cm, d = 53 cm.

   A = ____________ P = ____________

2. In a cube, the side s = 7.9 ft.

   SA = ____________ V = ____________

3. In a sphere, r = 1.2 m.

   SA = ____________ V = ____________

4. In a right circular cone without a bottom, h = 3.4 in., r = 5 in.

   SA = ____________ V = ____________

5. In a circle, r = 12.5 km.

   A = ____________ C = ____________

6. In a triangle, h = 4.9 mi the base, b = 8 mi.

   A = ____________

7. In a square, the length of one side is 92 cm.

   A = ____________ P = ____________

8. In a circle, the diameter is 1.2 miles.

   A = ____________ C = ____________

9. In a right circular cylinder (with top and bottom), h = 3.5 in., r = 1.1 in.

   SA = ____________ V = ____________

10. In a right circular cylinder (with no top and bottom), h = 3.5 in., r = 1.1 in.

    A = ____________ P = ____________

Review of Terms and Concepts

A **linear unit** is a particular length used for measuring length. (page 168)
A **rectangle** is a four-sided closed figure with opposite sides equal in measure having four right angles. (pages 168, 212)
The **dimensions** of rectangles are **length** and **width.** (page 168)
A **square** is a rectangle for which the length equals the width.
Dimensions, perimeter, circumference, and other lengths are measured in **linear units.** (page 168)
Area is the amount of space in a two-dimensional figure and is measured in **square units.** (page 168)

Factors of a natural number are numbers that multiply to give the number. (page 188)
A natural number is a **multiple** of each of its factors. (page 188)
A natural number is **divisible** by each of its factors. (page 188)
Even numbers are natural numbers that are divisible by two (2, 4, 6, 8, ...). (page 188)
Odd numbers are natural numbers that are not even (1, 3, 5, 7, ...). (page 188)
A **prime number** is a natural number that does not factor except for the number itself times 1. (page 188)
A **composite number** is a natural number that is not prime. (page 188)
A **common multiple** of two or more natural numbers is a number that is a multiple of all of the numbers. (page 194)
The **least common multiple (LCM)** of natural numbers is the smallest multiple all of the numbers. (page 194)
A **common factor** of two or more natural numbers is a number that is a factor of all of the numbers. (page 194)
The **greatest common factor (GCF)** of several natural numbers is the largest factor of all of the numbers. (page 194)

A **rectangular box** is a box in which each of the sides are rectangles. (page 197)
The **dimensions** of a rectangular box are **length, width, and height.** (page 197)
A **cube** is a rectangular box in which each of the sides are squares. (Length = width = height). (page 197)
Volume is the amount of space within a three-dimensional structure, measured in **cubic units**. (page 197)
The **surface area** of a rectangular box is the total area of its six sides and is measured in square units. (page 197)

The **diameter** of a circle is the distance across the circle through the center. (page 206)
The **radius** of a circle is the distance from the center to the rim and is half the diameter. (page 206)
The **circumference** of a circle is the distance around the circle. (page 206)

A **right angle** is formed by two perpendicular lines. This is a 90° angle. (page 211)
An **acute angle** is an angle smaller than 90°. (page 211)
An **obtuse angle** is an angle larger than 90°. (page 211)
The **perimeter** of a closed geometric figure is the sum of the lengths of the sides. (page 211)
A **right triangle** is a triangle in which one of the angles is a right angle. (page 211)
The **hypotenuse** of a right triangle is the longest side and is opposite the right angle. The other two sides are called **legs.**

A **line segment** is part of a straight line. (page 212)
A **quadrilateral** is a four-sided closed geometric shape. (page 212)
A corner of a quadrilateral or triangle is called a **vertex.** (page 211, 212)
A **parallelogram** is a quadrilateral in which the opposite sides are parallel. (page 212)
A **rectangle** is a parallelogram in which all of the angles are right angles. (page 212)
A **square** is a rectangle in which all sides have the same length. (page 212)
A **trapezoid** is a quadrilateral in which exactly two of the sides are parallel. (page 212)

Formulas

Area of a rectangle: $\mathbf{A = LW}$. (page 173)
Area of a square: $\mathbf{A = s^2}$. (page 182)
Perimeter of a rectangle: $\mathbf{P = 2L + 2W}$. (page 173)
Perimeter of a square: $\mathbf{P = 4s}$. (page 182)

Volume of a rectangular box: $\mathbf{V = L \cdot W \cdot H}$. (pages 202, 225)
Volume of a cube: $\mathbf{V = s^3}$. (page 225)
Surface area of a rectangular box: $\mathbf{SA = 2L+2W+2H}$.(pages 202, 225)
Surface area of a cube: $\mathbf{SA = 6s^2}$. (page 225)

Circumference of a circle: $\mathbf{C = 2\pi r}$ or $\mathbf{C = \pi d}$, where d is the diameter of the circle. (page 207)

Area of a circle: $\mathbf{A = \pi r^2}$, where **r** is the radius of the circle and π 3.1416. (page 208)

Area of a triangle: $\mathbf{A = \frac{1}{2} b \cdot h}$, where b is the base of the triangle and h is the height. (page 212)

Perimeter of a triangle: $\mathbf{P = a + b + c}$, where a, b, c are lengths of the sides. (page 212)
Area of a parallelogram: $\mathbf{A = b \cdot h}$. (page 212)
Perimeter of a parallelogram: $\mathbf{P = 2a + 2b}$, where a and b are the two different lengths of parallel sides. (page 212)

Area of a trapezoid: $\mathbf{A = \frac{1}{2} h(b + d)}$ where h is the height and b and d are the parallel sides. (page 212)

Perimeter of a trapezoid: $\mathbf{P = a + b + c + d}$, where a, b, c, and d are the lengths of the sides. (page 212)
Pythagorean theorem: $\mathbf{a^2 + b^2 = c^2}$, where c is the hypotenuse of a **right triangle** and a and b are the sides. (page 221)

Volume of a right circular cylinder: $\mathbf{V = \pi r^2 h}$. (page 225)

Surface area of a right circular cylinder (top and bottom included): $\mathbf{SA = 2\pi rh + 2\pi r^2}$. (page 225)

Surface area of a right circular cylinder (without top and bottom): $\mathbf{SA = 2\pi rh}$. (page 225)

Volume of a sphere: $\mathbf{V = \frac{4}{3}\pi r^3}$. (page 225)

Surface area of a sphere: $\mathbf{SA = 4\pi r^2}$. (page 225)

Volume of a right circular cone: $\mathbf{V = \frac{1}{3}\pi r^2 h}$. (page 225)

Surface area of a right circular cone (with bottom): $\mathbf{SA = \pi r^2 + \pi r\left(\sqrt{r^2 + h^2}\right)}$. (page 225)

Surface area of a right circular cone (without bottom): $\mathbf{SA = \pi r\left(\sqrt{r^2 + h^2}\right)}$. (page 225)

Processes

To **prime factor a number,** find two factors that multiply together to get the number. Continue to factor each factor until all of the factors are prime numbers. (page 193)

To find a least common multiple (page 195)

1. Factor each number completely.
2. If a factor appears more than once in one of the numbers, write it in exponential form.
3. The least common multiple is the product of each factor raised to the highest power that appears in any one of the numbers.

To find a greatest common factor (page 195)

1. Factor each number completely.
2. If a factor appears more than once in one of the numbers, write it in exponential form.
3. The greatest common factor is the product of the factors that appear in all numbers. If a factor is raised to a power, use the lowest power that appears in any one number.

Chapter 3 Chapter Test

Part 1: Do not use a calculator.

1. Without using a calculator, prime factor each of the following numbers.

a. 18 = ________________ b. 35 = ____________________

c. 36 = ________________ d. 40 = ____________________

e. 23 = ________________ f. 38 = ____________________

g. 75 = ________________ h. 29 = ____________________

2. Find the least common multiple and greatest common factor of each of the following groups of numbers.

a. 33 and 11

LCM = ____________________

GCF = ____________________

b. 14 and 28

LCM = ____________________

GCF = ____________________

c. 12, 4 and 18

LCM = ____________________

GCF = ____________________

d. 7 and 9

LCM = ____________________

GCF = ____________________

3. In each of the following determine whether the units of measurement are linear, square, or cubic. Circle the correct answer.

(a) The distance from Kansas City to Seattle	linear	square	cubic
(b) The amount of property in Kansas City	linear	square	cubic
(c) The floor space in a building	linear	square	cubic
(d) The amount of space in a storage box	linear	square	cubic
(e) The distance from Earth to the Moon	linear	square	cubic
(f) The amount of space to heat in a room	linear	square	cubic
(g) The area of a rectangle	linear	square	cubic
(h) The length of a side of a triangle	linear	square	cubic
(i) The circumference of a circle	linear	square	cubic
(j) The perimeter of a trapezoid	linear	square	cubic
(k) The area of a circle	linear	square	cubic
(l) The volume of a rectangular box	linear	square	cubic

4. If 1 ft = 12 in.,

1 in. = ______________ ft 1 ft^2 = ______________ in.^2

1 in.^2 = ______________ ft^2 5 ft = ______________ in.

5 in. = ______________ ft 2 ft^2 = ______________ in.^2

5.

The picture on the left is a square that is 1 inch by 1 inch.

The vertical and horizontal lines cut the sides into four equal pieces.

a. The grid cuts the square into how many pieces? ______________

b. What fraction of a square inch is one of the smaller squares? ______________

In parts c-e, include proper units with the answer.

c. What are the dimensions of one of the smaller squares? ______________

d. What is the perimeter of the entire square? ______________

e. What is the area of the entire square? ______________

Part 2: Calculator recommended.

6. In each of the following, compute the area and perimeter of the rectangle and state whether the rectangle is a square. Include appropriate units with your answers.

a. L = 17 ft, W = 27 ft   A = __________   P = __________   Square-- yes no

b. L = 49.6 m, W = 23.5 m   A = __________   P = __________   Square-- yes no

c. L = 19 ft, W = 19 ft   A = __________   P = __________   Square-- yes no

d. L = 0.5 in., W = 0.5 in.   A = __________   P = __________   Square-- yes no

7. The following are dimensions of rectangular boxes. Find the volume and surface area of the box.

(a) L = 18 in., W = 11 in., H = 9 in. V = ________

SA = ________

(b) L = 39 cm, W = 28.3 cm, H = 17.9 cm V = ________

SA = ________

(c) L = 6 ft, W = 1 ft, H = 6 ft V = ________

SA = ________

8. Give the name of each of the following geometric shapes and compute the quantities requested.

a.

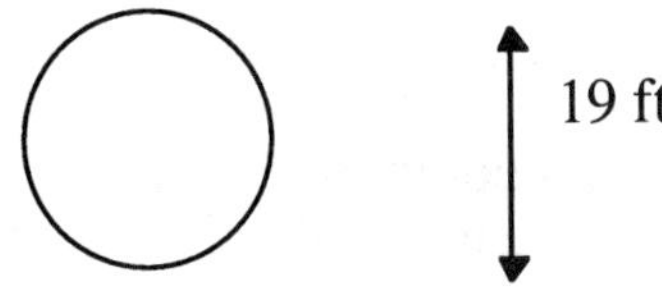

Name of shape ________

Circumference = ________

Area = ________

b.

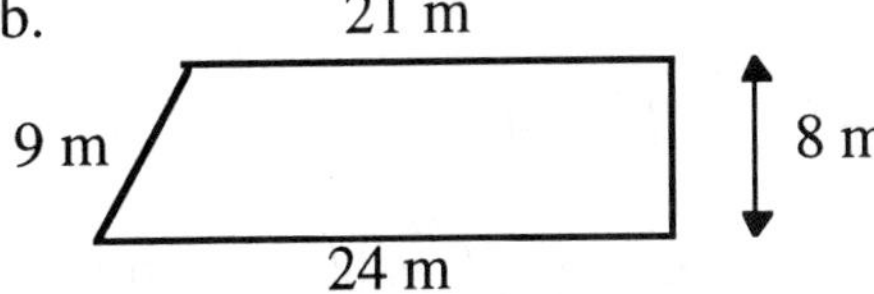

Name of shape ________

Perimeter = ________

Area = ________

c.

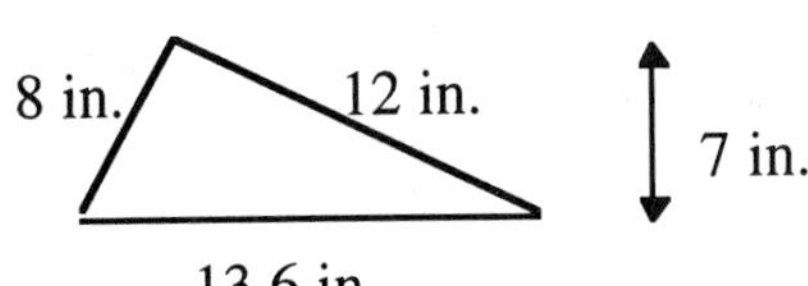

Name of shape ________

Perimeter = ________

Area = ________

d.

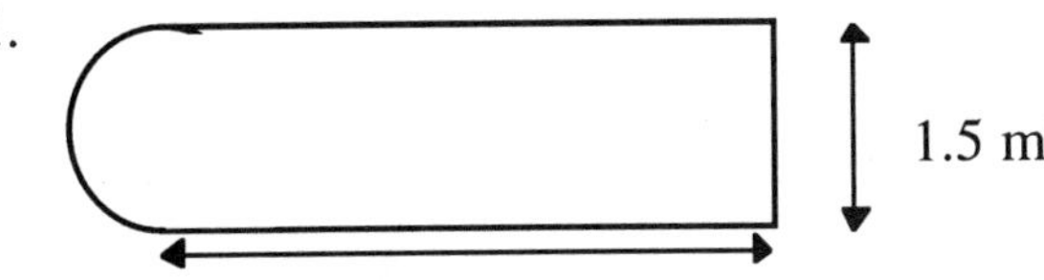

Perimeter = ________

Area = ________

(e)

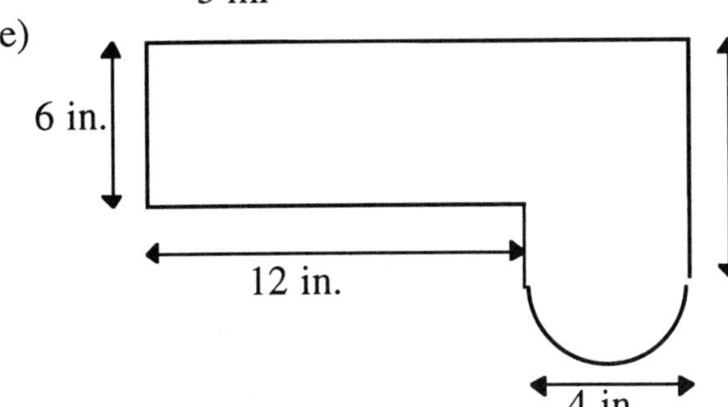

A = ________

P = ________

9. For each of the following, compute the quantities from the information given.

a. A right cylindrical cone where r = 29 in., h = 14 in.

SA = ____________________ V = ____________________

b. A sphere where r = 29.4 in.

SA = ____________________ V = ____________________

c. A cube where s = 43 cm

SA = ____________________ V = ____________________

d. A rectangular box where L = 7m, W = 4 m, H = 2 m.

SA = ____________________ V = ____________________

e. A triangle where a = 12 ft, b = 13 ft, c = 15 ft, h = 11 ft (where b is the base)

SA = ____________________ V = ____________________

10. For each of the following triangles, use the Pythagorean theorem to determine whether or not the triangle is a right triangle with c as the hypotenuse. How did you determine whether or not the triangle was a right triangle?

a. a = 6, b = 8, c = 10. Right triangle? yes no

b. a = 7, b = 12, c = 18. Right triangle? yes no

11. In each of the following events, Donna walked 5 miles and Cheryl walked 3.5 miles. For each event, draw a diagram that shows Donna's and Cheryl's starting positions and their paths and then determine how far apart (along the shortest path) they are after the walk.

a. Both Donna and Cheryl walked on a path that went due North.

b. Donna walked South and Cheryl walked North.

c. Donna walked South and Cheryl walked West.

Chapter 4 Fractions

Key Topics

- Arithmetic with fractions
- Number sense--fractions
- Fractions in real life
- Solving equations with fractions

Materials Needed for the Exercises (one set per group)

- Tracing paper
- Scissors
- Calculator
- 45 beans in a plastic or paper bag
- Cardboard straight-edge
- Ruler or measuring tape (marked in inches)
- Scotch tape or masking tape (optional)
- 20-30 inches of string (optional)
- 3 large sheets of paper--newsprint is a good source
- Snickers bar and knife

Section 4-1 *What Is a Fraction?*

In this chapter we will take a detailed look at a fraction as a part of a linear unit, circle, or rectangle, as well as a part of an object or part of a collection of objects. From an understanding of fractions, we will discover the processes of the arithmetic of fractions.

We also know that decimal parts of numbers also represent parts of a whole, so fractions and decimals are closely related. \$0.50 and \$0.25 are called half-dollars and quarters because 50 cents is one-half of a dollar and 25 cents is one-fourth of a dollar. In Section 4-3 we will learn how to convert fractions to decimal form, but first we will learn about fractions themselves.

Let's now look at an inch and fractions of an inch. When one inch is divided into two equal parts, each of the parts is 1/2 inch.

1 inch   1/2 in   1/2 in

When one inch is divided into three equal parts, each part is 1/3 inch.

1 inch $\frac{1}{3}$in $\frac{1}{3}$in $\frac{1}{3}$in

2/3 inch is two of the one-thirds.

2/3 inch

When continuing this pattern of defining fractions, each of 4 equal parts of an inch is 1/4 inch. Three of those 1/4 inches are 3/4 inches. When the inch is divided into 16 equal parts, each part is 1/16 of an inch. Two of those parts make 2/16 inch, three of those parts make 3/16 inch, and so on. In general, the denominator (number below the fraction bar) of a fraction is the number of equal parts into which the unit is divided. The numerator (number above the fraction bar) indicates the number of the equal parts selected.

When reading a ruler, different size marks distinguish between 1/4, 1/8, and 1/16 of an inch. Let's look at a simplified drawing of an inch as it might appear on your ruler. The shorter marks are the 1/8-in. marks and the larger marks are 1/4-in. marks, which divide the inch into 4 equal parts. If you consider both the long and short marks, they divide the inch into 8 equal parts.

1/2

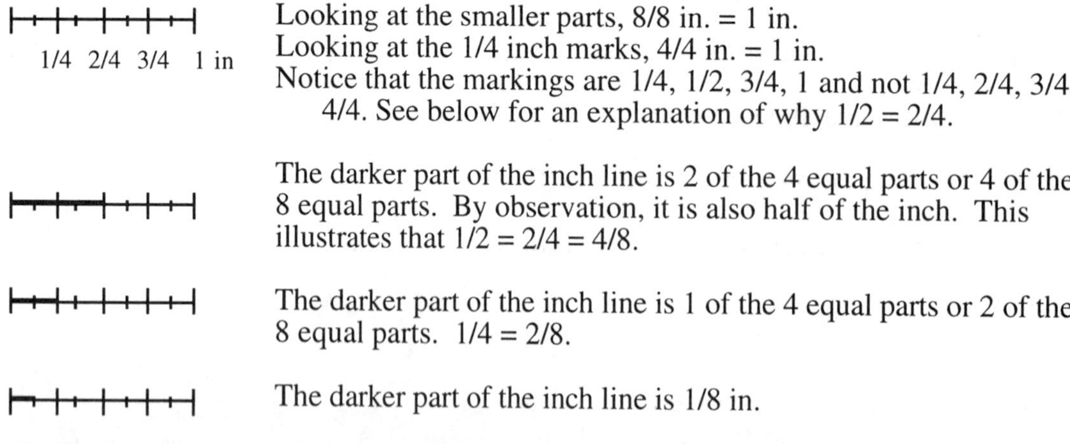

Looking at the smaller parts, 8/8 in. = 1 in.
Looking at the 1/4 inch marks, 4/4 in. = 1 in.
Notice that the markings are 1/4, 1/2, 3/4, 1 and not 1/4, 2/4, 3/4, 4/4. See below for an explanation of why 1/2 = 2/4.

The darker part of the inch line is 2 of the 4 equal parts or 4 of the 8 equal parts. By observation, it is also half of the inch. This illustrates that 1/2 = 2/4 = 4/8.

The darker part of the inch line is 1 of the 4 equal parts or 2 of the 8 equal parts. 1/4 = 2/8.

The darker part of the inch line is 1/8 in.

The darker part of the inch line is 5/8 in. (count the eighths).

We can also look at fractions as parts of almost anything, not just inches. The following exercise is designed to recognize fractions of circles and to compare fractions geometrically.

Exercise 1-- What Is a Fraction?

Purpose of the Exercise

- To see fractions as parts of circles
- To compare sizes of fractions visually
- To practice reading a ruler

Materials Needed

- Tracing paper
- Ruler or measuring tape marked in inches

For each of the following circles, observe how many equal pieces the circle is divided into and what fraction of the whole circle one of these pieces makes. The first one is done as an example.

1.

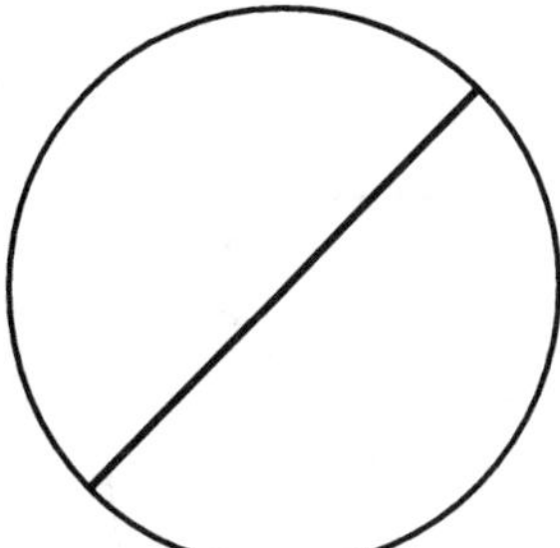

How many equal parts? ___2___

Each piece is ___1/2___ of the circle.

2.

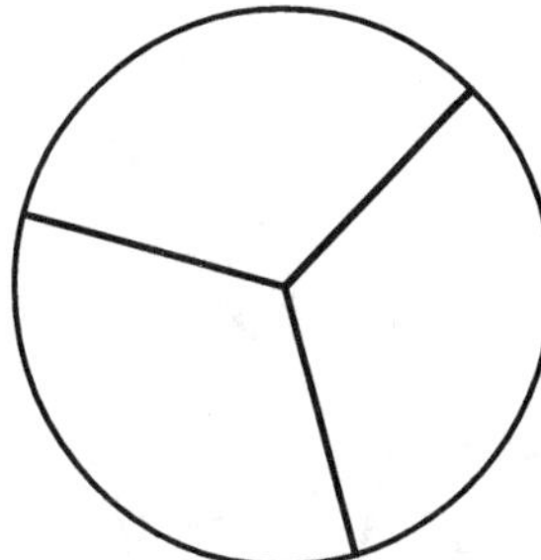

How many equal parts? __________

Each part is __________ of the circle.

3.

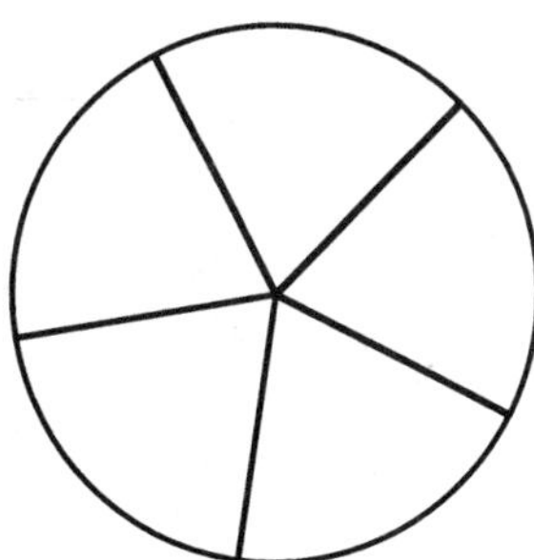

How many equal parts? __________

Each part is __________ of the circle.

4.

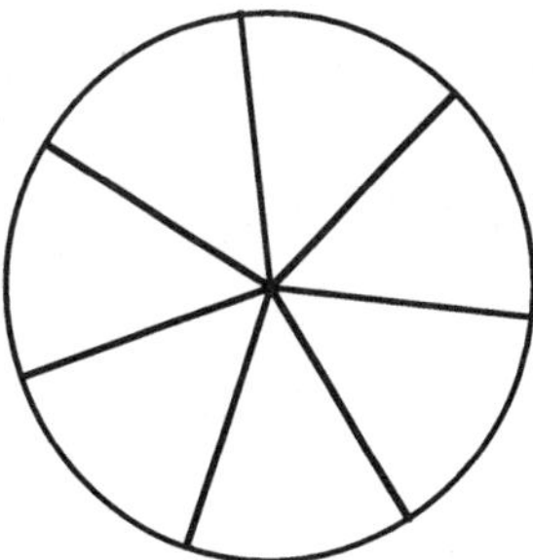

How many equal parts? __________

Each part is __________ of the circle.

5.

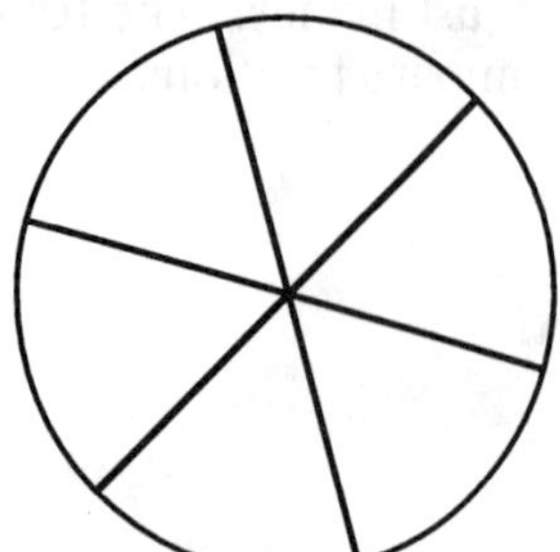

How many equal parts? ______________

Each part is ______________ of the circle.

6.

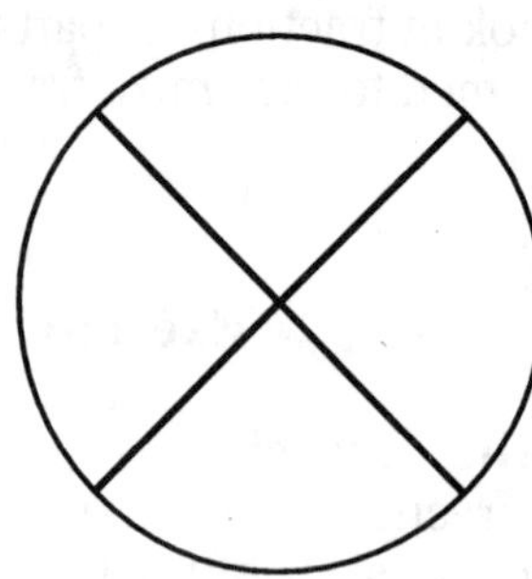

How many equal parts? ______________

Each part is ______________ of the circle.

7. The smallest fraction is the one that represents the smallest piece of the circle. Starting with the smallest fraction, write all of the fractions of the circles in Problems 1 - 6 above in order.

smallest fraction = _________ next smallest = _________ next = _________

next = _________ next = _____________ largest fraction = __________

8.

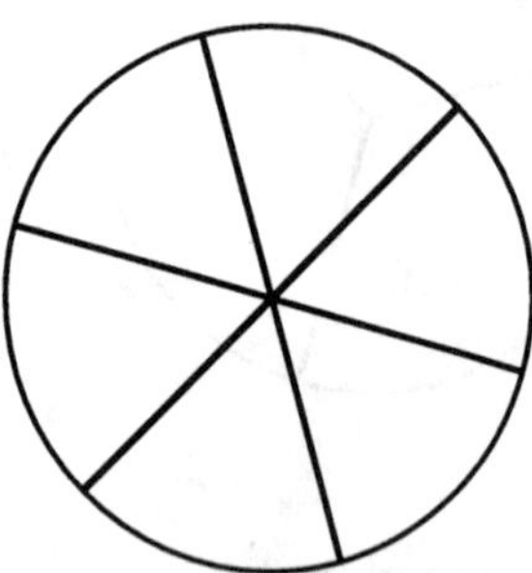

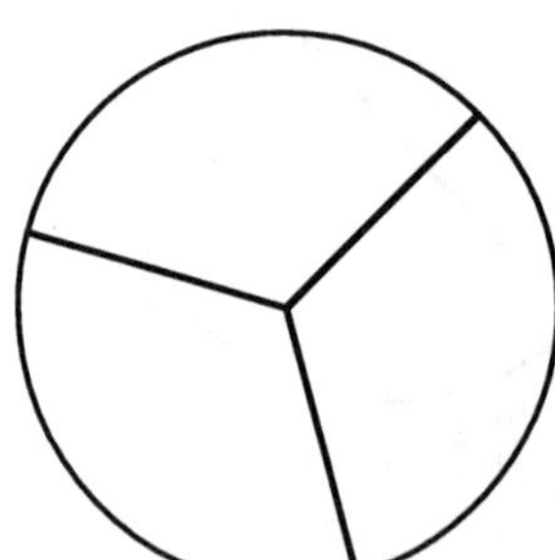

Shade 3/6 of the circle on the left. (This is 3 of the 6 parts or 3 of the one-sixths.)
Shade 2/3 of the circle on the right,. (This is 2 of the 3 parts or 2 of the one-thirds.)

Which is bigger, 3/6 of the circle or 2/3 of the circle? ____________________

9.

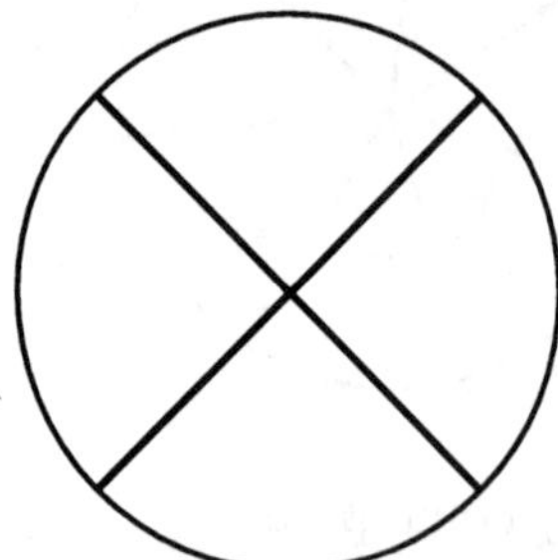

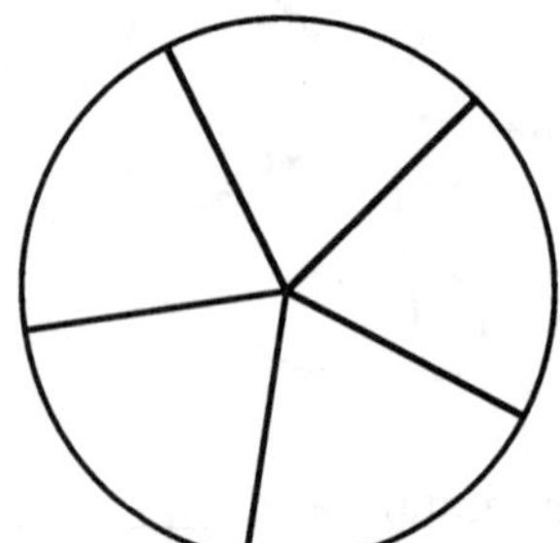

Shade 3/4 of the circle on the left. Shade 3/5 of the circle on the right.

Which is bigger? 3/4 of the circle or 3/5 of the circle? ____________________

10.

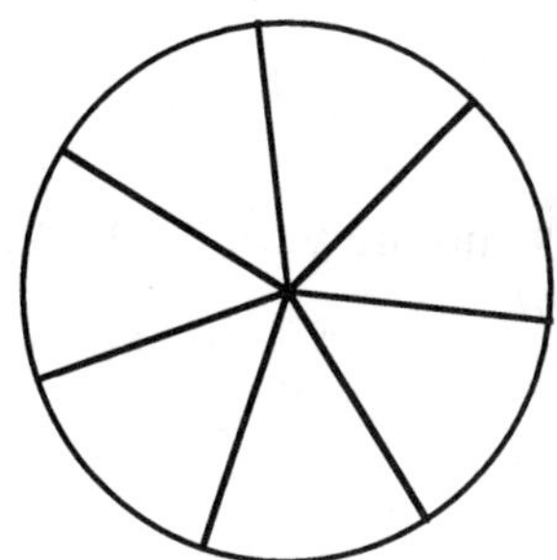

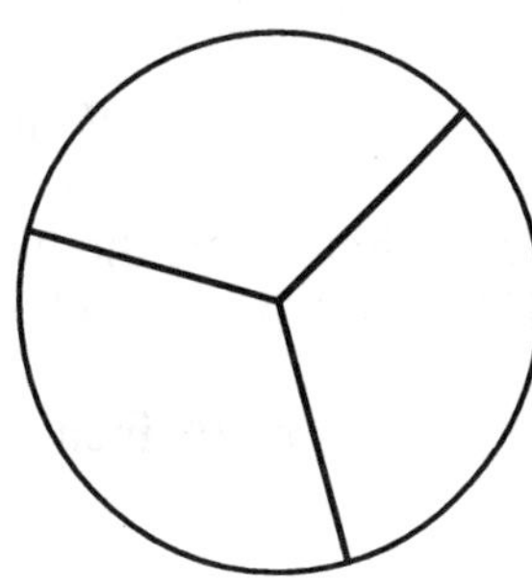

Shade 2/7 of the circle on the left. Shade 1/3 of the circle on the right.

Which is bigger? 2/7 of the circle or 1/3 of the circle? ________________

11. Using the circles in Problems 1 through 6 as an aid, determine which fraction of each of the following pairs is larger. Circle the larger fraction. Use tracing paper to make appropriate illustrations.

a. $\frac{1}{2}$ or $\frac{3}{7}$ b. $\frac{5}{7}$ or $\frac{3}{7}$ c. $\frac{2}{3}$ or $\frac{3}{4}$

d. $\frac{3}{5}$ or $\frac{3}{6}$ e. $\frac{2}{3}$ or $\frac{2}{5}$ f. $\frac{3}{3}$ or $\frac{6}{7}$

g. $\frac{5}{4}$ or $\frac{6}{6}$ h. $\frac{2}{5}$ or $\frac{2}{6}$ i. $\frac{2}{3}$ or $\frac{4}{5}$

12. Using the circles in Problems 1 through 6 as an aid, determine whether the fractions in each of the following pairs are equal or not equal. Use tracing paper to make appropriate illustrations.

a. $\frac{2}{3}$ and $\frac{4}{6}$ Equal Not equal

b. $\frac{5}{5}$ and 1 Equal Not equal

c. $\frac{1}{2}$ and $\frac{2}{4}$ Equal Not equal

d. $\frac{1}{3}$ and $\frac{3}{7}$ Equal Not equal

e. $\frac{2}{6}$ and $\frac{1}{3}$ Equal Not equal

f. $\frac{1}{3}$ and $\frac{3}{1}$ Equal Not equal

Use your ruler to answer the following questions.

13. For 1 inch on your ruler, count the total number of spaces between the marks to determine how the inch is divided.

a. One inch on your ruler is divided into how many parts? __________

b. How long is each part? ______________________________

14. Find the marks that divide the inch into four equal parts and draw 1/4 inch, 2/4 inch, and 3/4 inch below.

1/4 inch   2/4 inch   3/4 inch

15. Find the marks that divide the inch into eight equal parts and draw 1/8 inch, 2/8 inch, 3/8 inch, 4/8 inch, 5/8 inch, 6/8 inch, and 7/8 inch below.

1/8 inch   2/8 inch   3/8 inch   4/8 inch

5/8 inch   6/8 inch   7/8 inch

16. Use your ruler to draw and compare each of the following fractions (as fractions of inches). Are the two fractions of each pair equal? If they are not equal, circle the larger fraction.

a. $\frac{3}{4}$ and $\frac{6}{8}$ Equal   Not Equal   b. $\frac{3}{16}$ and $\frac{3}{8}$ Equal   Not Equal

c. $\frac{3}{4}$ and $\frac{5}{8}$ Equal   Not Equal   d. $\frac{5}{8}$ and $\frac{7}{16}$ Equal   Not Equal

e. $\frac{1}{2}$ and $\frac{2}{4}$ Equal   Not Equal   f. $\frac{5}{8}$ and $\frac{1}{2}$ Equal   Not Equal

g. $\frac{1}{8}$ and $\frac{1}{4}$ Equal   Not Equal   h. $\frac{3}{4}$ and $\frac{7}{8}$ Equal   Not Equal

Now that you have an understanding of what a fraction is, let's learn some terminology regarding fractions.

In the fraction, 1/2, 1 is the **numerator** and 2 is the **denominator**. The inch or some other object is divided into 2 equal pieces. One of those pieces is equal to 1/2.

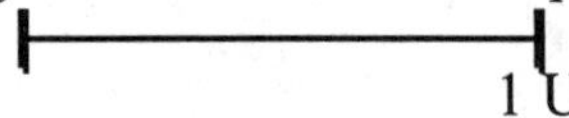

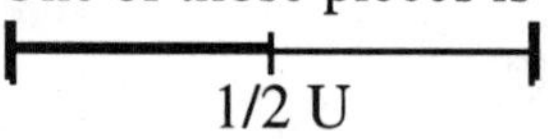

U = unit

Example 1: In the fraction $\frac{7}{8}$, 7 is the numerator and 8 is the denominator. The inch, circle, or other object is divided into 8 equal pieces.

1 U — The marks show the unit divided into 8 parts.

7/8 U is darkened — 7/8 represents 7 of those pieces.

Example 2: In the fraction 15/17, 15 is the numerator and 17 is the denominator. The inch, circle, or other object is divided into 17 equal pieces. 15/17 is 15 of those pieces.

Vocabulary

In more general terms, in the fraction, $\frac{a}{b}$ (or a/b), a is the **numerator** and b is the **denominator**.

The **denominator** is the number of equal parts into which the unit is divided.
The **numerator** is how many of the equal parts are selected.

The terms numerator and denominator, no doubt, came from the words numerate and denomination. To numerate means to count and a denomination is a type of something--as in a type of religion. A denomination can also be something of a particular value or size. For example, the different denominations of bills when talking about money are the \$1-bills, \$5-bills, and so on. In a fraction, the denominator indicates the number of the equal parts in the whole, and the numerator is how many of these parts are specified. For example, 1 of 3 (1/3) or 5 of 7 (5/7).

When a circle or unit is divided into 3 equal parts, each part is $\frac{1}{3}$ of the circle or unit, $\frac{2}{3}$ is two of the parts, and $\frac{3}{3}$ is all three parts so $\frac{3}{3}$ is the entire circle or unit. Stated symbolically, $\frac{3}{3}$ of the circle = 1 entire circle. Stated more abstractly, $\frac{3}{3} = 1$. Similarly, 4/4 = 1, 5/5 = 1 6/6 = 1, and so forth.

It is also possible to represent more than a unit, or one, as a fraction. For example, 4/3 of a circle means four of the three equal size parts. Since three of the parts make up one entire circle, 4/3 is an entire circle plus one more third of another circle of the same size. In other words, $\frac{4}{3} = 1 + \frac{1}{3}$, or $1\frac{1}{3}$.

A fraction such as $\frac{4}{3}$ is called an **improper fraction** because it represents a number more than or equal to one entire unit.

$1\frac{1}{3}$ (read "one and one-third") is called a **mixed number** because it consists of a whole number and a fraction.

We will work more with improper and mixed numbers later, but for now we just need to recognize the meaning of the symbols.

From the money model, we know that 0.10 (or 0.1) is 1/10. We will cover this in more detail later, but for now you need to know that $0.1 = \frac{1}{10}$, $0.2 = \frac{2}{10}$, $1.3 = 1\frac{3}{10}$, and so on.

Exercise 2--Making a Ruler--Fractions, Decimals, and Linear Units (Group)

Purpose of the Exercise
- To help build an understanding of linear units
- To see fractions and decimals as measurements of lengths

Materials Needed
- Cardboard straight-edge with no markings (recommended size 2" by 17")
- 1 to 3 large sheets of paper--a large pad of newsprint is a good source
- Tape--to tape sheets of paper together if needed (optional)
- A string 20 to 30 inches long (optional)

Have one person in the group put his/her hand palm down on a piece of paper and mark the width across the widest part of the hand on the paper. This measurement, called a hand (abbreviated **H**), will be used as your standard unit. Do not convert your length to inches or centimeters. (Note that you could just as well have used your foot or the length of your arm or some other width or length as a standard.)

1. Make a ruler on your cardboard straight edge. Mark as many **hands** as will fit.
2. Use your ruler and draw 1 H on a piece of paper and divide the unit into four equal parts. Label 1/4 H, 2/4 H, and 3/4 H. See the picture below or ask the instructor for help if you do not understand what this means.
3. Draw 1 H on a piece of paper and divide the unit into 2 equal parts. Label the 1/2 **H** mark.
4. Draw 1 H on a piece of paper and divide the unit into 8 equal parts. Label the 1/8 **H**, 2/8 **H**, and so on.
5. Draw 1 H on a piece of paper and divide the unit into 10 equal parts. Label the 0.1 or 1/10 **H**, 0.2 or 2/10 **H**, and so on.
6. Draw 1 H on a piece of paper and divide the unit into 3 equal parts. Label the 1/3 **H**, 2/3 **H**, and so on.

   Helpful Hint: Take the time needed on Questions 1 through 6 to get each of the parts as equal as possible. You may want to fold the unit in halves, thirds, and so on, to get the equal parts or you might try to guess how long a part of a unit (say 1/10 H) would be, and then copy 10 of the 1/10 H's (side by side) to determine if your guess is too small or too large and adjust your guess accordingly.

   Your marked units should look something like the following picture, but on a larger scale.

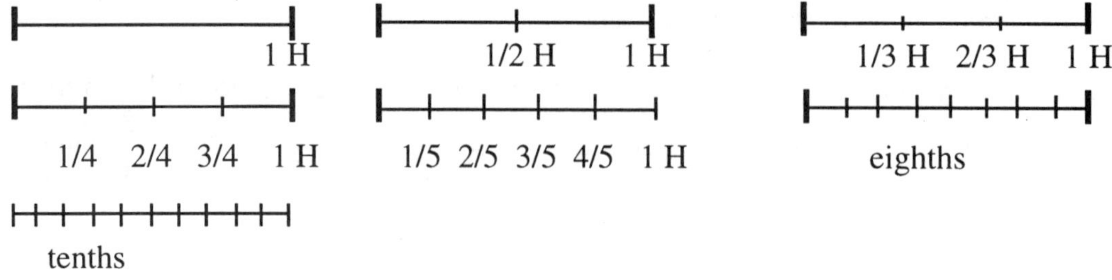

7. On a separate piece of paper, draw and label the length of each of the following. You might choose one of the columns; do the other columns only if you need more practice.

$\frac{1}{3}$ H	$\frac{1}{5}$H	$\frac{1}{2}$ H	$\frac{1}{4}$ H
$\frac{3}{4}$ H	$\frac{2}{3}$H	$\frac{2}{5}$H	$\frac{3}{5}$ H
$\frac{5}{4}$ H	$\frac{6}{5}$H	$\frac{7}{4}$ H	$\frac{8}{7}$H
$1\frac{1}{4}$ H	$1\frac{1}{5}$ H	$1\frac{1}{8}$H	$1\frac{1}{7}$ H
1.4 H	1.5 H	1.8 H	1.7 H
$\frac{4}{10}$ H	$\frac{7}{10}$ H	$\frac{5}{10}$ H	$\frac{6}{10}$ H
.4 H	.7 H	.5 H	.6 H
.8 H	.9 H	.3 H	.2 H
1.5 H	1.6 H	1.4 H	1.9 H
2.5 H	2.6 H	2.4 H	2.9 H

Now you can use your ruler to compare the size of fractions.

Exercise 3--Which is Bigger, $\frac{1}{3}$ or $\frac{1}{4}$? (Group)

Purpose of the Exercise

- To compare visually the sizes of common fractions and numbers with decimal points

Materials Needed

- A blank straight-edge or tracing paper

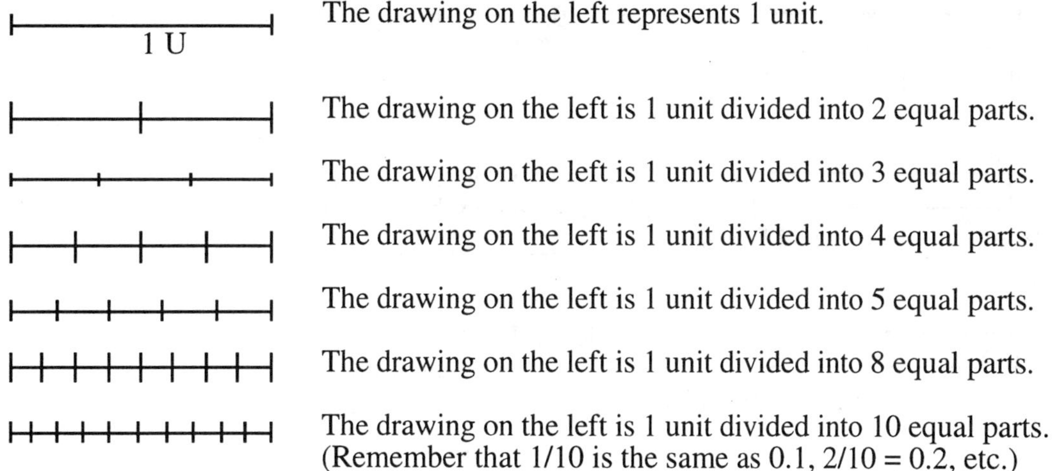

The drawing on the left represents 1 unit.

The drawing on the left is 1 unit divided into 2 equal parts.

The drawing on the left is 1 unit divided into 3 equal parts.

The drawing on the left is 1 unit divided into 4 equal parts.

The drawing on the left is 1 unit divided into 5 equal parts.

The drawing on the left is 1 unit divided into 8 equal parts.

The drawing on the left is 1 unit divided into 10 equal parts. (Remember that 1/10 is the same as 0.1, 2/10 = 0.2, etc.)

Use tracing paper or a straight-edge to transfer the measurements above to the drawing below. The measurements shown represent 1 unit. Based on the picture, determine which number is bigger. The first one is done as an example.

1. Shade 1/4 U on the segment.
Shade 1/3 U on the segment.
Which is bigger, 1/4 U or 1/3 U? 1/3 U is bigger

2. Shade 2/3 U on the segment.
Shade 3/4 U on the segment.
Which is bigger, 2/3 U or 3/4 U? ______________

3. Shade 3/8 U on the segment.
Shade 3/4 U on the segment.
Which is bigger, 3/8 U or 3/4 U? ______________

4. Shade 3/10 U on the segment.
Shade 2/5 U on the segment.
Which is bigger, 3/10 U or 2/5 U? ______________

5. Shade 5/8 U on the segment.
Shade 1/2 U on the segment.
Which is bigger, 5/8 U or 1/2 U? ______________

6. Shade 7/8 U on the segment.
Shade 2/3 U on the segment.
Which is bigger, 7/8 U or 2/3 U? ______________

7. Shade 3/5 U on the segment.
Shade 3/4 U on the segment.
Which is bigger, 3/5 U or 3/4 U? ______________

8. Shade 0.7 U on the segment.
Shade 1/3 U on the segment.
Which is bigger, 0.7 U or 1/3 U? ______________

9. Shade 0.5 U on the segment.
Shade 2/5 U on the segment.
Which is bigger, 0.5 U or 2/5 U? ______________

10. Shade 0.2 U on the segment.
Shade 1/2 U on the segment.
Which is bigger, 0.2 U or 1/2 U? ______________

Comparing two fractions with the same denominator is simple because the pieces are the same size. In this case, the numerator determines which fraction is larger because the numerator is the count of the number of equal pieces.

Example 3: 3/5 is bigger than 2/5 because the denominators are the same but the first numerator, 3, is larger than the second, 2.

7/10 is bigger than 4/10 because the denominators are the same but 7 is bigger than 4.

When comparing two fractions with the same numerator, we look to the denominator to determine which fraction is bigger. Since the denominator determines the size of each equal piece, the larger the denominator, the smaller the piece. For example, suppose you have two identical pies. If one is divided into 8 slices, and the other is divided into 5 slices, which will have the smaller slices? We know that the one divided into 8 pieces has the smaller pieces. In other words, $\frac{1}{8} < \frac{1}{5}$.

Example 4: 3/5 is smaller than 3/4 because each fifth is smaller than each of the three fourths.

1/8 is larger than 1/10 since the denominator, 8, divides the unit into 8 equal parts while the denominator, 10, divides the unit into 10 equal parts.

When the numerators and denominators are both different, or when we compare a fraction to a decimal, the comparison becomes more difficult. At this point, we make a visual estimate to determine which number is bigger. Throughout the chapter, we will find other ways to compare numbers.

In this chapter, we will work with fractions in several settings. Geometric pictures help us understand the concept of fractions, but fractions are used in many other areas of mathematics and life. For example, money or debt may be divided among family members or business partners. Each person gets a certain fraction of the total amount of money or has to pay a certain fraction of the debt.

Example 5: If a dollar is divided into four equal parts, each part is 25¢, so 1/4 of a dollar is 25¢. (This is why a 25-cent coin is called "a quarter.") Similarly, a half-dollar is a 50-cent coin since 50 cents is 1/2 of a dollar.

If a million dollars is divided into four equal parts, each part is $250,000. A house that costs "a quarter of a million dollars" cost $250,000.

In the following sections, we will learn how to reduce, build up, add, subtract, multiply, and divide fractions and to convert between mixed numbers and improper fractions. We will also use fractions in problems with real data.

Section 4-2 *Equivalent Fractions and Improper Fractions*

As we have already observed, sometimes two different fractions represent the same length or the same part of a circle.

For example, 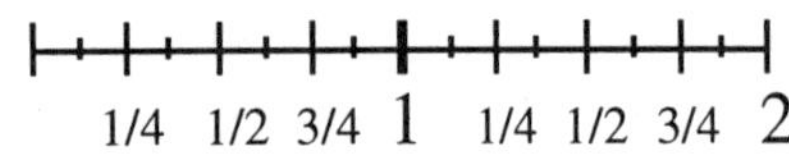

1/2 = 2/4 3/4 = 6/8

We will now learn a more formal process to recognize equivalent fractions and to "reduce" certain fractions to a simpler form. We will also work with fractions that represent more than the whole object or length. The fraction, 3/2 inch means three of the 1/2 inch segments. Since 2/2 in. = 1 in., it stands to reason that 3/2 inches is longer than 1 inch.

1/4 1/2 3/4 1 1/4 1/2 3/4 2

1/2 in. 1 in. 3/2 in. Three 1/2-in. segments is 1/2 inch more than 1 inch.

Vocabulary

A fraction is in **reduced (simplest) form** when it is written with the smallest possible numerator and denominator. For example: $\frac{5}{10}$ in reduced form is $\frac{1}{2}$.

When the numerator of a fraction is larger than or equal to the denominator, the fraction is greater than or equal to 1. Such a fraction is called an **improper fraction.**

Equivalent fractions are fractions that represent the same amount.

$\frac{5}{10}$ and $\frac{1}{2}$ are equivalent fractions since $\frac{5}{10} = \frac{1}{2}$. (If a pie is divided into 10 slices and you have 5 slices, you have 5 out of 10 slices or 1/2 the pie. There are 5 slices left, which is the other half of the pie.

In the following exercises, you will divide line segments into parts to recognize equivalent fractions. You could also use circles or rectangles for illustrations.

Exercise 1--Equivalent Fractions and Reducing Fractions

Purpose of the Exercise

- To recognize equivalent fractions
- To discover a process for reducing fractions

1.

a. Divide the line segment into 4 equal pieces and shade 2 of the pieces.

b. How much of the entire line segment is shaded? ______________________________

c. $\frac{2}{4}$ = __________ Write a simpler fraction that represents the entire shaded part.

2. ├──────────────┤

a. Divide the line segment into 6 equal pieces and shade 3 of the pieces.

b. How much of the entire segment is shaded? ________________

c. $\frac{3}{6}$ = ____________ (in simpler form)

3. ├──────────────┤

a. Divide the line segment into 6 equal pieces and shade 2 of the pieces.

b. How much of the entire rectangle is shaded? ________________

c. $\frac{2}{6}$ = ____________ (in simpler form)

4. ├──────────────┤

a. Divide the line segment into 6 equal pieces and shade 5 of the pieces.

b. How much of the entire rectangle is shaded? ________________

c. Can $\frac{5}{6}$ be written as a simpler fraction? **Explain.** ________________

5. ├──────────────┤

a. Divide the line segment into 8 equal pieces and shade 4 of the pieces.

c. How much of the entire rectangle is shaded? ________________

d. $\frac{4}{8}$ = ____________ (in simpler form)

6. ├──────────────┤

a. Divide the line segment into 8 equal pieces and shade 2 of the pieces.

b. How much of the entire rectangle is shaded? ________________

c. $\frac{2}{8}$ = ____________ (in simpler form)

7. ├──────────────┤

a. Divide the line segment into 8 equal pieces and shade 7 of the pieces.

b. How much of the entire rectangle is shaded? ________________

c. Can $\frac{7}{8}$ be written as a simpler fraction? **Explain.** ________________

8. Does $\frac{7}{14} = \frac{1}{2}$? **Illustrate** your answer.

9. Does $\frac{6}{9} = \frac{12}{18}$? **Illustrate** your answer.

10. ├──────────────┤

a. Shade $\frac{1}{2}$ of the line segment.

b. If you divided the line segment into 8 parts, how many parts would you have to shade to represent $\frac{1}{2}$ of the line segment? ________ $\frac{1}{2} = \frac{?}{8}$ Write the fraction ___________

c. If you divided the line segment into 16 parts, how many parts would you have to shade to represent $\frac{1}{2}$ of the line segment? ________ $\frac{1}{2} = \frac{?}{16}$ Write the fraction ___________

d. If you divided the line segment into 40 parts, how many parts would you have to shade to represent $\frac{1}{2}$ of the line segment? ________ $\frac{1}{2} = \frac{?}{40}$ Write the fraction ___________

e. If you divided the line segment into 40 parts, how many parts would you have to shade to represent the <u>entire line segment</u>? ______ Write the fraction _________________

f. If you divided the line segment into 2156 parts, how many parts would you have to shade to represent the <u>entire line segment</u>? ______ Write the fraction _________________

g. If you divided the line segment into 7 pieces, how many pieces would it take to be the same amount as <u>2 entire line segments</u>? ______ Write the fraction _________________

h. If you divided the line segment into 8 pieces, how many pieces would it take to be the same amount as <u>2 entire line segments</u>? ______ Write the fraction _________________

11. In each of the following, replace the question mark with the correct number. Draw a sketch as needed to illustrate.

a. $\frac{2}{3} = \frac{?}{6}$ b. $\frac{1}{2} = \frac{?}{12}$ c. $\frac{1}{2} = \frac{?}{130}$

d. $\frac{3}{13} = \frac{?}{26}$ e. $\frac{1}{2} = \frac{?}{22}$ f. $\frac{1}{2} = \frac{?}{14}$

g. $\frac{1}{3} = \frac{?}{6}$ h. $\frac{2}{3} = \frac{?}{9}$ i. $\frac{2}{3} = \frac{?}{12}$

In each case above, you started with a simplified fraction and you "built the fraction up" to an equivalent fraction with a larger numerator an denominator.

12. Describe what process you used to build up fractions in Problem 11.

____

____

____

13. Reduce the following fractions by replacing the question mark with the correct number. Draw a sketch as needed to illustrate.

a. $\frac{6}{12} = \frac{?}{2}$ b. $\frac{2}{10} = \frac{?}{5}$ c. $\frac{6}{8} = \frac{?}{4}$

14. Describe the process you used to reduce the fractions in Problem 13.

____

____

____

15. Reduce each of the following fractions. Use an illustration if necessary.

a. $\frac{8}{14}$ = ________ b. $\frac{3}{12}$ = ________ c. $\frac{2}{6}$ = ________

d. $\frac{9}{27}$ = ________ e. $\frac{22}{44}$ = ________ f. $\frac{22}{46}$ = ________

g. $\frac{5}{15}$ = ________ h. $\frac{3}{9}$ = ________ i. $\frac{12}{18}$ = ________

j. $\frac{6}{18}$ = ________ k. $\frac{8}{18}$ = ________ l. $\frac{2}{20}$ = ________

m. $\frac{20}{30}$ = ________ n. $\frac{70}{90}$ = ________ o. $\frac{10}{50}$ = ________

16. What whole numbers do each of the following fractions represent?

a. $\frac{3}{3}$ = ________ b. $\frac{6}{3}$ = ________ c. $\frac{9}{3}$ = ________

d. $\frac{14}{2}$ = ________ e. $\frac{5}{5}$ = ________ f. $\frac{2738}{2738}$ = ________

g. $\frac{10}{10}$ = ________ h. $\frac{20}{10}$ = ________ i. $\frac{99}{99}$ = ________

j. $\frac{18}{18}$ = ________ k. $\frac{18}{9}$ = ________ l. $\frac{18}{6}$ = ________

Summary Observations

- Fractions can represent a part of an object, an entire object, or more than one object.
- If the numerator and the denominator are the same number but not zero, the fraction represents one entire object. Each of the following fractions is equal to 1.

  $\frac{42}{42}$ $\quad {}^{20987}/_{20987}$ $\quad \frac{2001}{2001}$
- If the numerator < denominator, the fraction is less than one whole object. Each of the following fractions is less than 1. $\frac{127}{900}$ $\quad {}^{377}/_{378}$ $\quad {}^{1000}/_{1001}$ $\quad$ 99/100
- If the numerator > the denominator, the fraction is greater than one. Each of the following fractions is greater than 1: ${}^{99}/_{98}$ $\quad {}^{20987}/_{876}$ $\quad \frac{9}{5}$ $\quad \frac{2002}{2001}$
- If the numerator is twice the denominator, the fraction equals 2. Each of the following fractions is equal to 2: ${}^{90}/_{45}$ $\quad \frac{300}{150}$ $\quad \frac{202}{101}$ $\quad \frac{4000}{2000}$

- If the numerator > twice the denominator, the fraction is greater than two. If the numerator < twice the denominator, the fraction is less than two.

  Each of the following fractions is greater than 2: $^{115}/_{45}$ $\frac{321}{150}$ $^{219}/_{101}$ $\frac{5021}{2000}$

  Each of the following fractions is less than 2 (but greater than 1): 5/4 $\frac{200}{153}$ $\frac{5007}{4087}$

Generalizing these findings, a fraction is equal to three if the numerator is three times the denominator. A fraction is less than three is the numerator is less than three times the denominator, and so on.

Exercise 2--Estimating the Size of Fractions

Purpose of the Exercise
- To practice estimating a "ballpark size" of a fraction

Materials Needed
- Calculator

In each of the following, determine whether the numerator is larger than, less than, or equal to the denominator. Then **without a calculator**, estimate the size of each of the fractions and decide in which of the categories below the fraction fits.

(a) less than 1 (b) equal to 1 (c) between 1 and 2
(d) equal to 2 (e) between 2 and 3 (f) equal to 3
(g) greater than 3

1. _____ 7/9	2. _____ 22/9	3. _____ 2/17	4. _____ 28/15
5. _____ 30/15	6. _____ 31/15	7. _____ 3046/2579	8. _____ 999/100
9. _____ 999/1000	10. _____ 999/999	11. _____ 60/10	12. _____ 60/30
13. _____ 60/35	14. _____ 60/20	15. _____ 60/25	16. _____ 299/40
17. _____ 402/299	18. _____ 789/789	19. _____ 856/789	20. _____ 789/856
21. _____ $\frac{45}{15}$	22. _____ $^{42}/_{15}$	23. _____ $\frac{50}{15}$	24. _____ $^{17}/_{145}$
25. _____ $\frac{145}{17}$	26. _____ 18/5	27. _____ 5/18	28. _____ 11/9
29. _____ 9/11	30. _____ 17/210	31. _____ 210/17	32. _____ 53/20
33. _____ 99/67	34. _____ 67/22	35. _____ 505/201	36. _____ 7/987
37. _____ 987/7	38. _____ 36/12	39. _____ 8/4	40. _____ 4/8

41. Using a calculator, convert each of the fractions above to decimal form to check your answers. To convert 9/11 to decimal form, divide 9 by 11. On your calculator, key the following strokes: [9] [÷] [11] [=] 9/11 0.82.

Now that we have visually reduced and built up fractions to get equivalent fractions, let's look at a more formal process for finding equivalent fractions.

Process: Reducing a Fraction

Step 1: Prime factor the numerator and denominator of the fraction.
Step 2: Find the greatest common factor (GCF) of the numerator and denominator.
Step 3: Divide the numerator and the denominator by the GCF.
Step 4: Multiply the remaining factors in the numerator.
Step 5: Multiply the remaining factors in the denominator.

Note: If preferred, steps 2 and 3 can be replaced with the step, "Divide all factors that are common to both the numerator and denominator." Sometimes this is easier than finding the greatest common factor.

Example 1: Reduce 2/4 to lowest terms.

$\frac{2}{4} = \frac{(2)(1)}{(2)(2)}$ Step 1. Prime factor the numerator and the denominator.

$= \frac{\boxed{(2)}(1)}{\boxed{(2)}(2)}$ Step 2. The GCF is 2.

$= (1)\left(\frac{1}{2}\right)$ Step 3. Divide numerator and denominator by the GCF.

$= \frac{1}{2}$ There are no remaining factors to multiply.

Example 2: Reduce 180/420 to lowest terms.

In this case, we will factor the numerator and denominator to find the greatest common factor, and then rewrite the fraction with numerator and denominator in factored form with the GCF as one of the factors. This is a slight variation of the above process.

$180 = 2^2(3^2)(5)$ $\quad 420 = (2^2)(3)(5)(7)$ $\quad \text{GCF} = (2^2)(3)(5) = 60$

$180 = (60)(3)$ $\quad 420 = (60)(7)$

$\frac{180}{420} = \frac{(60)(3)}{(60)(7)}$ Factor the GCF out of the numerator and out of the denominator.

$= \frac{3}{7}$ Divide the GCF out of the numerator and denominator.

Notice: Not all fractions can be reduced.

If the numerator and denominator are prime numbers, the fraction cannot be reduced any further. For example, both the numerator and denominator of $\frac{3}{11}$ are prime numbers. $\frac{3}{11}$ is not reducible.

If there are no common factors (other than 1) in the numerator and denominator, the fraction does not reduce. For example, $\frac{20}{21} = \frac{(2)(2)(5)}{(3)(7)}$. There are no factors common to both the numerator and denominator so $\frac{20}{21}$ is already reduced (cannot be reduced.)

Process to Build up a Fraction

Multiply both the numerator and the denominator by the same number. The resulting fraction is equivalent to the original fraction because you are multiplying the fraction by 1. $\frac{a}{a} = 1, a \ 0$ so $(1)\left(\frac{x}{y}\right) = \left(\frac{a}{a}\right)\left(\frac{x}{y}\right)$

By multiplying both sides we get $\frac{x}{y} = \frac{ax}{ay}$

Example 3: Building up fractions.

$\frac{5}{7} = \frac{5(4)}{7(4)} = \frac{20}{28}$ $\quad$ $8/11 = 8(5)/11(5) = 40/55$ $\quad$ $\frac{7}{9} = \frac{7(10)}{9(10)} = \frac{70}{90}$ $\quad$ $\frac{1}{4} = \frac{1(100)}{4(100)} = \frac{100}{400}$

Exercise 3--Reducing Fractions

Purpose of the Exercise

- To practice reducing fractions

Materials Needed

- Calculator

Without a calculator, factor the numerators and denominators to reduce the following fractions. In Problems 1 through 9, list the GCF of the numerator and denominator. Check the answers by putting both the reduced and unreduced fractions in decimal form and comparing the numbers. If a fraction is already reduced, state that. The first one is done as an example.

1. $\frac{4}{10} = \frac{\cancel{(2)}(2)}{\cancel{(2)}(5)} = \frac{2}{5}$ GCF 2
2. $\frac{25}{15} =$ ________ GCF ________
3. $\frac{15}{25} =$ ________ GCF ________
4. $\frac{18}{12} =$ ________ GCF ________
5. $\frac{7}{17} =$ ________ GCF ________
6. $\frac{22}{11} =$ ________ GCF ________
7. $\frac{4}{12} =$ ________ GCF ________
8. $\frac{12}{4} =$ ________ GCF ________
9. $\frac{37}{11} =$ ________ GCF ________
10. $\frac{15}{30} =$ ________
11. $\frac{27}{9} =$ ________
12. $\frac{27}{6} =$ ________
13. $\frac{6}{21} =$ ________
14. $\frac{25}{75} =$ ________
15. $\frac{15}{75} =$ ________
16. $\frac{20}{75} =$ ________
17. $\frac{75}{20} =$ ________
18. $\frac{30}{48} =$ ________
19. $\frac{21}{12} =$ ________
20. $\frac{44}{55} =$ ________
21. $\frac{44}{88} =$ ________
22. $\frac{7}{63} =$ ________
23. $\frac{11}{47} =$ ________
24. $\frac{10}{21} =$ ________
25. $\frac{7}{10} =$ ________
26. $\frac{20}{25} =$ ________
27. $\frac{21}{25} =$ ________
28. $\frac{44}{35} =$ ________
29. $\frac{35}{44} =$ ________
30. $\frac{28}{35} =$ ________

Section 4-3 *Decimals, Fractions, and Rational and Irrational Numbers*

Fractions can be written as decimal numbers and many times it is convenient to convert fractions to decimal form for ease in computation. In Exercise 2 from the previous section, you checked to see if two fractions were equivalent by putting both fractions in decimal form to compare answers. In that set of exercises, you converted a fraction to decimal form by dividing the numerator by the denominator. Although that is the process, let's review the model of decimal numbers as money to get a more intuitive feel for converting fractions to decimals and decimals to fractions.

Example 1: We know from our experience with money that
1/2 dollar is 50 cents (or $0.50), which means that 1/2 = 0.50. (We even use the names, 50 cents and a half-dollar interchangeably.)
A quarter , or 1/4 dollar, is 0.25, so 1/4 = 0.25.
A dime is 1/10 of a dollar and is expressed as 0.10, so 1/10 = 0.10.
A penny is 1/100 of a dollar and is expressed as 0.01, so 1/100 = 0.01.

Example 2: Since the symbols, $\frac{1}{2}$, $1 \div 2$, and $2\overline{)1}$ all mean "1 divided by 2," **we do the indicated division to convert a fraction to a decimal.**

$$\frac{1}{2} = 0.5 \text{ because } \begin{array}{r} 0.5 \\ 2\overline{)1.0} \end{array}$$

Process for Converting Fractions to Decimals
Divide the numerator of the fraction by the denominator. (This can be done by long division or on a calculator.)
The quotient (answer) is the fraction expressed as a decimal.
The denominator of the fraction is the **divisor** of the related division problem.

When converting a fraction (with integer numerator and denominator) to a decimal, the answer may come out exactly after a certain number of decimal places or a pattern may repeat infinitely.

Vocabulary
When dividing 3 into 1 to get 1/3, notice that the remainder never comes out to be 0. The remainder always comes out to be 1. The process of division could continue indefinitely always getting a remainder of 1 and the next digit in the quotient being 3. When this situation happens, we have what is called a **repeating decimal**. We indicate a repeating decimal by putting a line segment over the repeating pattern the last time we write the pattern. For example, 1/3 = 0.333333... so we would write, $1/3 = 0.\overline{3}$. 1/11 = 0.09090909..., so we would write, $1/11 = 0.\overline{09}$.
When converting a fraction to a decimal, if we eventually get a 0 remainder and the division process ends, the decimal is called a **terminating decimal.** For example, 2/5 = 0.4 and 7/16 = 0.4375. 2/5 and 7/16 are both **terminating decimals.**

When working with fractions as decimals, we generally work with decimal approximations of repeating decimals and also of terminating decimals that have a lot of decimal places before terminating. To use a calculator to compute the decimal equivalent of a fraction, input the numerator, the divided by sign (÷ or /), the denominator, and then press the equal (enter) key. This is the same process for division as in Chapter 1--the numerator is the dividend and the denominator is the divisor.

Exercise 1--Division and Fractions as Decimals

Purpose of the Exercise
- To practice writing fractions as decimals

Materials Needed
- Calculator

1. **Without a calculator**, do the following divisions. Write each quotient and fraction in decimal form. Check whether your answer is reasonable by using the money model. The first one is done as an example.

	decimal form		decimal form
a. $4\overline{)1}$ =	0.25	$\frac{1}{4}$ =	0.25

Using money as a model, 1/4 dollar is 25 cents or \$0.25, so the answer is reasonable.

b. $10\overline{)1}$ =	________	$\frac{1}{10}$ =	________
c. $100\overline{)1}$ =	________	$\frac{1}{100}$ =	________
d. $5\overline{)2}$ =	________	$\frac{2}{5}$ =	________
e. $4\overline{)3}$ =	________	$\frac{3}{4}$ =	________

2. In each of the following, divide to convert the fraction to a decimal. Keep dividing until the remainder is zero or until you have a repeating pattern. If the fraction comes out as a repeating decimal, draw a line over the pattern. **Show your work and do not use a calculator.**

a. $\frac{1}{8}$ = ________ b. $\frac{1}{5}$ = ________

c. $\frac{4}{3}$ = ________ d. $\frac{2}{3}$ = ________

e. $\frac{14}{11}$ = ________ f. $\frac{2}{11}$ = ________

g. $\frac{3}{10}$ = ________ h. $\frac{33}{100}$ = ________

3. **Use a calculator** to convert each of the following fractions to decimals. If there are more than 5 decimal places in the quotient, round off the answer to 5 decimal places. Circle the word "exact" if the answer is exact. Otherwise indicate that the answer is rounded. The first one is done as an example.

a. 12/7 =	1.71429	exact	(rounded)
b. 22/9 =	________	exact	rounded
c. 2/19 =	________	exact	rounded
d. 81/8 =	________	exact	rounded
e. 12/7 =	________	exact	rounded

f. 22/9 = ______________ exact rounded

g. 8/397 = ______________ exact rounded

h. 397/8 = ______________ exact rounded

i. 66/11 = ______________ exact rounded

j. 11/66 = ______________ exact rounded

k. 79/19 = ______________ exact rounded

l. 15/12 = ______________ exact rounded

m. 1/15 = ______________ exact rounded

n. 4/100 = ______________ exact rounded

o. 298/100 = ______________ exact rounded

p. 298/99 = ______________ exact rounded

q. 55/75 = ______________ exact rounded

r. 129/10 = ______________ exact rounded

s. 44/29 = ______________ exact rounded

t. 13/3 = ______________ exact rounded

u. 4/5 = ______________ exact rounded

Rational and Irrational Numbers

Vocabulary

A **rational number** is a number that can be represented by a repeating decimal or a terminating decimal. Any rational number can also be represented as a fraction where the numerator and denominator are integers.

An **irrational number** is a number that cannot be represented as a repeating or a terminating decimal.

When writing an irrational number in decimal form, we generally round it to a certain decimal place. In this way, we approximate an irrational number as a rational number. The number of decimal places is decided by the accuracy needed in the situation.

Most of the numbers we have worked with so far are rational numbers. The following numbers are examples of rational numbers. Some have terminating decimals; some have repeating decimals.

Example 4: The following are examples of rational numbers written as fractions and in decimal form. In each case, the rational number is categorized as having a repeating or terminating decimal.

Fraction form	Decimal form	Terminating? Repeating?
3/5	0.6	terminating decimal
7/11	$0.\overline{63}$	repeating decimal
63/100	0.63	terminating decimal
15/7	$2.\overline{142857}$	repeating decimal
$8\frac{122}{125}$	8.976	terminating decimal
84/2	42 (or 42.0)	terminating decimal

Example 5: The following are examples of irrational numbers rounded to 4 decimal places.

The Number	Approximation
π	3.1416
$\sqrt{7}$	2.6458
$\sqrt{245}$	15.6525
2π	6.2832
$\pi/2$	1.5708
$\sqrt{2}$	1.4142
$4\sqrt{87}$	37.3095

Example 6: To get rational approximations for π, first use the π button on your **calculator to find π.** If you have difficulty finding the button, consult your calculator manual or ask your instructor.

$\pi \approx 3.14159$ (to 5 decimal places)

$\pi \approx 3.14$ (to 2 decimal places)

$\pi \approx 3.141592654$ (to 9 decimal places)

$\pi \approx 3.1$ (to 1 decimal places)

To use the calculator to approximate 2 : Enter [2] [x] [π] [=]

$2\pi \approx 6.28319$ (to 5 decimal places)

$2\pi \approx 6.2832$ (to 2 decimal places)

$2\pi \approx 6.283185307$ (to 9 decimal places)

$2\pi \approx 6.3$ (to 1 decimal places)

When computing square roots on a scientific calculator that displays one number at a time, remember you put in the number first, then the square root sign. To approximate $\sqrt{5}$ using the calculator, do the following key strokes:

[5] [$\sqrt{\ }$] [=] This should give 2.236067978. (Some calculators will have fewer digits; some will have more.)

When computing square roots on a calculator that displays the entire calculation on the screen at once, remember you put in the square root sign first, then the number. To approximate $\sqrt{5}$ using the calculator, do the following key strokes:

[$\sqrt{\ }$] [5] [=] This should give 2.236067978. (Some calculators will have fewer digits; some will have more.)

Example 7: Approximations for $\sqrt{5}$.

$\sqrt{5} \approx 2.24$ rounded to 2 decimal places.

$\sqrt{5} \approx 2.2361$ rounded to 4 decimal places.

$\sqrt{5} \approx 2.23607$ rounded to 5 decimal places.

$\sqrt{5} \approx 2.2360680$ rounded to 6 decimal places.

Exercise 2--Which Is Bigger, 22/7 or π?

Purpose of the Exercise

- To approximate irrational numbers as rational numbers
- To compare the sizes or two numbers (rational or irrational)

Materials Needed

- Calculator

1. Use your calculator to convert each of the following irrational numbers to decimal form. Write each number rounded off to 2 decimal places; then to 4 decimal places.

The Number	Rounded to 2 Decimal Places	Rounded to 4 Decimal Places
a. $\sqrt{73}$	________________	________________
b. $3\sqrt{73}$	________________	________________
c. $\pi/4$	________________	________________
d. 7π	________________	________________
e. $\sqrt{978}$	________________	________________

2. Use the calculator to approximate each of the following pairs of numbers. Compare the two numbers in their decimal form and **circle the larger number**. Record the number of decimal places you had to use to determine which number was larger. The first one is done as an example.

a. 17.3 and (104/6) — # of decimal places 2

$104/6 \approx 17.3333$ — 17.3 and 17.3333 are the same until the second decimal place.

b. 22/7 and π — # of decimal places ______

c. π and 3.14 — # of decimal places ______

d. 2π and 6.28 — # of decimal places ______

e. $\frac{14142}{10001}$ and $\sqrt{2}$ — # of decimal places ______

f. $\frac{91}{89}$ and $\frac{89}{91}$ — # of decimal places ______

g. $\sqrt{5}$ and 2.36 — # of decimal places ______

h. 3/4 and 151/200 — # of decimal places ______

i. 150/17 and $\sqrt{78}$ — # of decimal places ______

Section 4-4 *Multiplying Fractions*

In arithmetic, the word "of" can indicate multiplication. For example 3 groups of 7 squares is 21 squares. We use the word "of" frequently when talking about fractions. As we already observed, one-quarter of a million dollars is $250,000. In this section we will find how to multiply $\frac{1}{4}$ and 1,000,000 to get 250,000.

Looking at a ruler, we observe that $\frac{1}{2}$ of $\frac{1}{2}$ inch is $\frac{1}{4}$ inch--that is if 1/2 inch is divided into two equal parts, each part is 1/4 inch. Also note that $\frac{1}{2}$ of $\frac{1}{4}$ inch is $\frac{1}{8}$ inch--that is if 1/4 inch is divided into 2 equal parts, each part is 1/8 inch. (Check these statements out with a ruler or the illustration below.) These observations would indicate that

$$\left(\frac{1}{2}\right)\left(\frac{1}{2}\right) = \frac{1}{4} \quad \text{and} \quad \left(\frac{1}{2}\right)\left(\frac{1}{4}\right) = \frac{1}{8}$$

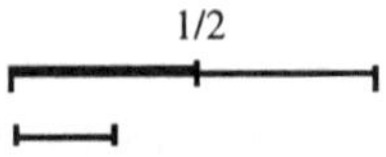

1/2

$\frac{1}{2}$ of $\frac{1}{2}$ -in = $\frac{1}{4}$ in.

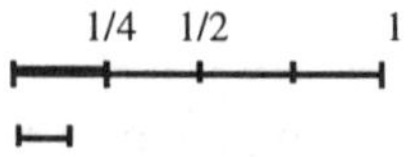

1/4 1/2 1

$\frac{1}{2}$ of $\frac{1}{4}$ -in = $\frac{1}{8}$ in.

The following exercise is designed to use line segments and shadings on line segments to discover rules for multiplying fractions. The exercise uses linear units for illustrations, but rectangles or circles could illustrate the concept just as well.

Exercise 1--Taking Parts of Parts of a Line Segment (Group)

Purpose of the Exercise

- To discover a process for multiplying fractions
- To present information to the class (if Problem 6 is assigned)

Materials Needed

- One Snickers bar for each group (if Problem 6 is assigned)
- One knife for each group (if Problem 6 is assigned)

1. 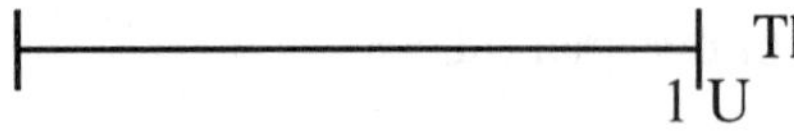 This line segment represents 1 unit.

   1 U

   a. Divide the line segment into 2 equal parts and shade $\frac{1}{2}$ U.

   b. Divide the shaded area into 3 equal parts and darkly shade $\frac{1}{3}$ of $\frac{1}{2}$ U.

   c. What fraction of the entire line segment is $\frac{1}{3}$ of $\frac{1}{2}$ U? ____________________

   d. $\left(\frac{1}{3}\right)\left(\frac{1}{2}\right) =$ ____________________

2. ⊢――――――⊣ The line segment represents 1 unit.
1 U

a. Divide the line segment into 3 equal parts and shade $\frac{1}{3}$ U.

b. Divide the shaded area into 3 equal parts and darkly shade $\frac{1}{3}$ of $\frac{1}{3}$ U.

c. What fraction of the entire line segment is $\frac{1}{3}$ of $\frac{1}{3}$ U? ____________

d. $\left(\frac{1}{3}\right)\left(\frac{1}{3}\right) =$ ____________

3. ⊢――――――⊣ The line segment represents 1 unit.
1 U

a. Divide the line segment into 5 equal parts and shade $\frac{2}{5}$ U.

b. Divide the shaded area into 2 equal parts and darkly shade $\frac{1}{2}$ of $\frac{2}{5}$ U.

c. What fraction of the entire line segment is $\frac{1}{2}$ of $\frac{2}{5}$ U? ____________

d. $\frac{1}{2} \cdot \frac{2}{5} =$ ____________

4. ⊢――――――⊣ The line segment represents 1 unit.
1 U

a. Divide the line segment into 5 equal parts and shade $\frac{2}{5}$ U.

b. Divide the shaded area into 3 equal parts and darkly shade $\frac{1}{3}$ of $\frac{2}{5}$ U.

c. What fraction of the entire line segment is $\frac{1}{3}$ of $\frac{2}{5}$ U? ____________

d. $\frac{1}{3} \cdot \frac{2}{5} =$ ____________

5. ⊢――――――⊣ The line segment represents 1 unit.
1 U

a. Divide the line segment into 4 equal parts and shade $\frac{1}{4}$ U.

b. Divide the shaded area into 2 equal parts and darkly shade $\frac{1}{2}$ of $\frac{1}{4}$ U.

c. What fraction of the entire line segment is $\frac{1}{2}$ of $\frac{1}{4}$ U? ____________

d. $\left(\frac{1}{2}\right)\left(\frac{1}{4}\right) =$ ____________

6. Each group take one Snickers bar. The instructor will assign you one of the Problems 1-5 above. Do parts a and b by making appropriate cuts on the Snickers bar; then answer the c and d parts based on your observations. Choose a spokesperson to present the problem to the class.

Process for Multiplying Fractions

Step 1: Factor each numerator and each denominator.
Step 2: Divide factors that are common to the numerators and the denominators (of either fraction or you can divide across fractions.)
Step 3: Multiply the remaining factors in the numerators.
Step 4: Multiply the remaining factors in the denominators.

Note: An alternative for steps 1 and 2 is to start with step 2 and divide a common factor out of the numerator and denominator. After the division, see if there is another common factor. If so, divide that common factor from the numerator and denominator. Continue this process until there are no common factors left; then proceed with steps 3 and 4.

Example 1: Compute 8/9 of 15/14.

$\left(\frac{8}{9}\right)\left(\frac{15}{14}\right)=\left(\frac{2\cdot 2\cdot 2}{3\cdot 3}\right)\left(\frac{3\cdot 5}{2\cdot 7}\right)$ Step 1: Factor each numerator and each denominator.

$=\left(\frac{2}{2}\right)\left(\frac{3}{3}\right)\left(\frac{2\cdot 2\cdot 5}{3\cdot 7}\right)$ Step 2: Rearrange the fractions so that common factors are together.

$=(1)(1)\left(\frac{2\cdot 2\cdot 5}{3\cdot 7}\right)$ Step 3: Divide common factors.

$=\frac{20}{21}$ Steps 4 and 5: Multiply the remaining factors in the numerator and the denominator.

Example 2: Compute 2/9 of 4/3.

$\left(\frac{2}{9}\right)\left(\frac{4}{3}\right)=\left(\frac{2}{3\cdot 3}\right)\left(\frac{2\cdot 2}{3}\right)$ Step 1: Factor each numerators and each denominator.

$=\frac{2\cdot 2\cdot 2}{3\cdot 3\cdot 3}$ Step 2: There are no common factors to rearrange. Write the 2 fractions as one fraction.

$=\frac{8}{27}$ Steps 3 and 4: Multiply the factors in the numerator and the denominator.

Example 3: Compute the product of 5 and 7/15.

$(5)\left(\frac{7}{15}\right)=\left(\frac{5}{1}\right)\left(\frac{7}{15}\right)$ Before multiplying a whole number by a fraction, rewrite the whole number as a fraction; 5 = 5/1.

$=\left(\frac{5}{5}\right)\left(\frac{7}{3}\right)$ Steps 1 and 2 : Factor each numerator and rearrange the factors, to divide out the common factors.

$=(1)\left(\frac{7}{3}\right)=\frac{7}{3}$ Steps 3 and 4: Multiply the remaining fractions

In all mathematical computations, it is important to show the setup and calculations. When translating word problems to mathematical symbols, it is particularly important to also use clearly stated sentences and labels not only to help clarify the thinking in the problem, but also to communicate how the problem was analyzed and set up. Although the preceding problems just involved calculations, the calculations must be written and shown in an orderly fashion. Notice that the problem was stated; then as the multiplication and simplification were done, each new step was shown on a different line with the "=" signs lined up.

In the following examples, we not only show the work in an orderly fashion but also paraphrase the question and information to make the arithmetic setup easier. In word problems or real life, answers do not always come out exact so we round off answers. We should always check the answers to determine whether they are correct and reasonable. When answers are rounded off, the check may be slightly off. Sometimes round-off errors are of no significance; other times, we may adjust some of the answers up or down so that the numbers add up.

Example 4: Five people shared in the cost of buying lottery tickets and won $755. If the winnings were divided equally among them, how much was each share?

Paraphrase: Since there were five people, each person's share was 1/5 of $755.

$$\left(\frac{1}{5}\right)(755) = \left(\frac{1}{5}\right)\left(\frac{755}{1}\right)$$
$$= \frac{755}{5}$$
$$= 151$$

Each person's share of the winnings was $151.

Check: Five people each won $151. (5)($151) = $755, the total amount won, so the answer checks.

Example 5: A 48-ounce cake is to be cut into 16 equal pieces. How much will each piece weigh?

Paraphrase: Each piece would be 1/16 of the cake so would weigh 1/16 of 48 ounces.

$$\left(\frac{1}{16}\right)(48) = \left(\frac{1}{16}\right)\left(\frac{48}{1}\right)$$
$$= \frac{(3)(16)}{16}$$
$$= 3$$

Each piece would weigh 3 ounces.

Check: There are 16 pieces at 3 ounces each. (16)(3 oz) = 48 ounces total.

Example 6: The 12 employees at the *We-Customize* office decided to order pizza to celebrate the end of a very busy season. The pizza was delivered and the total bill came to $87.27 including tax and tip. Since the season had been so productive, the boss agreed to pay for half of the cost of the pizza and let the 12 employees split the rest of the cost. How much money did the boss chip in and how much did each employee chip in for the pizza?

Paraphrase: The boss chipped in 1/2 of $87.27.
Each employee paid 1/12 of the amount left over after the boss paid his share.

Boss's share: $\left(\frac{1}{2}\right)(87.27) = \$43.64$ — The exact amount was $43.645, but since we do not have the "mill" coin anymore, round the answer to the nearest penny.

The money left over: $87.27 - $43.64 = $43.63

Each employee's share: $\left(\frac{1}{12}\right)(43.63)$ $3.64 each

Check: The total price, $87.27 should be the boss's share plus the employees' shares.
43.64 + 12(3.64) = $87.32

This amount is 5 cents too much, so either the delivery woman gets the 5 cents or the boss or some of the employees pay slightly less.

Example 7: Another way to analyze Example 6.

Paraphrase: The boss chipped in 1/2 of $87.27.

After the boss paid his half, each employee paid 1/12 of the other 1/2 of $87.27.

Boss's share: $\left(\frac{1}{2}\right)(87.27) = \$43.64$

Employee's share: $\left(\frac{1}{12}\right)$ of $\left(\frac{1}{2}\right)$ of 87.27 is

$$\left(\frac{1}{12}\right)\left(\frac{1}{2}\right)(87.27) = \left(\frac{1}{24}\right)(87.27)$$

$3.64 each

Exercise 2--Drill on Multiplication of Fractions

Purpose of the Exercise

- To practice skills in multiplying and reducing fractions
- To translate problems stated in words as products of fractions and to solve the problems

Materials Needed

- Calculator

Without a calculator, do the following multiplication exercises and reduce answers where possible. Do your work on a separate sheet of paper. Use this paper to record answers.

1. $\frac{1}{2} \times \frac{1}{2}$ = ____________
2. $\frac{1}{2}\left(\frac{1}{3}\right)$ = ____________
3. $\frac{1}{2} \times \frac{2}{5}$ = ____________
4. $\frac{2}{5} \times \frac{1}{2}$ = ____________
5. $\frac{2}{15} \cdot \frac{4}{3}$ = ____________
6. $\frac{1}{10} \cdot \frac{7}{9}$ = ____________
7. $3 \times \frac{1}{2}$ = ____________
8. $3 \times \frac{5}{7}$ = ____________
9. $\frac{7}{15} \times \frac{3}{14}$ = ____________
10. $\frac{5}{9} \cdot \frac{2}{7}$ = ____________
11. $\frac{4}{5} \cdot \frac{2}{3}$ = ____________
12. $\frac{4}{5} \times \frac{15}{2}$ = ____________
13. $\frac{7}{8} \times 4$ = ____________
14. $\frac{44}{85} \cdot \frac{2}{3}$ = ____________
15. $\left(\frac{4}{3}\right)(7)$ = ____________
16. $18 \cdot \frac{2}{3}$ = ____________
17. $22 \cdot \frac{2}{11}$ = ____________
18. $\left(\frac{9}{10}\right)\left(\frac{1}{2}\right)$ = ____________
19. $\left(\frac{7}{10}\right)\left(\frac{5}{21}\right)$ = ____________
20. $\left(\frac{19}{2}\right)(6)$ = ____________
21. $\frac{5}{7} \cdot \frac{5}{7}$ = ____________
22. $\frac{5}{7} \times \frac{7}{5}$ = ____________
23. $\frac{217}{99} \times \frac{99}{217}$ = ____________
24. $\frac{1}{2} \times \frac{100}{17}$ = ____________
25. $\frac{3}{8} \cdot \frac{2}{7}$ = ____________
26. $15 \cdot \frac{2}{3}$ = ____________
27. $\left(\frac{4}{6}\right)\left(\frac{15}{2}\right)$ = ____________
28. $\frac{3}{7} \cdot \frac{8}{9}$ = ____________

Translate each of the following word problems to a product(s) involving fractions and compute the answer(s). Use a calculator as needed to put answers in decimal form. Then round off the answers. On a separate sheet of paper, show how you paraphrased the problem. Clearly show and label the work and the answers. Check your answers.

29. Eight vendors would like to display their products at a conference. Approximately 300 square feet of space is available for setting up the displays. If the vendors each get the same amount of space, how many square feet of area can each vendor use?

30. Andrea went on a business trip and wanted to bring souvenirs to her three children. She has $52 to spend. Approximately how much can she spend on each child?

31. Harry bought a box of 140 daffodil bulbs and plans to plant about one-third of the bulbs on his lot. The rest, he will divide evenly among his three children, Audrey, Malcolm, and Elizabeth. Approximately how many bulbs does each person get to plant?

32. When Ms. Trailor died, her estate was worth $78,345 after the probate charges were paid. According to her will, 1/6 of the money was to be left to establish a scholarship fund. The remaining money was to be divided evenly among her three children. How much money was set up to establish the scholarship fund? How much money did each of her children inherit?

33. Henry, George, Carolyn, Terry, Suellen, and Claire pool their money to buy stock household items in bulk and then split the cost and the products evenly among them. An order arrived with 100 pounds of flour, 80 pounds of sugar, 50 pounds of oatmeal, 10 pounds of cheddar cheese, and 10 pounds of provolone cheese. How many pounds of each item does each person get?

34. A recipe for coffee cake is as follows.

1/2	cup shortening	1	cup sugar
2	eggs	1	teaspoon vanilla
1/4	teaspoon lemon juice	1	teaspoon baking powder
2	cups flour	1	teaspoon soda
1	cup sour cream		

Mix together and pour into a greased and floured tube pan. Sprinkle with the following:

1/2	cup chopped nuts	1/4	cup brown sugar
1	tablespoon sugar	2	teaspoons cinnamon.

Bake at 350° for 40 minutes or until toothpick comes out clean.

a. Suppose you want to make two coffee cakes but only mix the ingredients once. Write the recipe for doubling all of the ingredients.

b. Write the recipe for the coffee cake if you want to make only half as much cake

35. (Project) Check on directions for making oatmeal or pancakes from a mix to which you only add water. What is the smallest size serving it gives directions for? Decide how to prepare the mix to make 1 serving, 2 servings, and 3 servings. When using kitchen measuring utensils, your measuring cups only come in certain sizes, such as 1/3 cup, 1/4 cup, and so on. If you take 1/2 of 3/4, for example, multiplication gives you 3/8. Since you do not have a 3/8 cup size, determine which of the sizes is closest to 3/8 size.

When doing arithmetic with fractions, we may also have to deal with a combination of negative and positive numbers. The rules of signs for fractions are the same as the rules we already know. Let's review the rules of signs for multiplication and division.

Rules of Signs for Multiplication

A **positive** number times a **positive** number = a **positive** number.	(+)(+) = +
A **positive** number times a **negative** number = a **negative** number.	(+)(−) = −
A **negative** number times a **positive** number = a **negative** number.	(−)(+) = −
A **negative** number times a **negative** number = a **positive** number.	(−)(−) = −

Rules of Signs for Division

A **positive** number divided by a **positive** number = a **positive** number.	(+)/(+) = +
A **positive** number divided by a **negative** number = a **negative** number.	(+)/(−) = −
A **negative** number divided by a **positive** number = a **negative** number.	(−)/(+) = −
A **negative** number divided by a **negative** number = a **positive** number.	(−)/(−) = +

Example 8: Determine whether each of the following are negative or positive.

$\frac{-5}{16}$ (−)/(+) The answer is negative.

$\frac{-98}{-87}$ (−)/(−) The answer is positive.

$\left(\frac{-5}{16}\right)\left(\frac{-98}{-87}\right)$ (−)(+) The answer is negative.

The first fraction is negative and the second fraction is positive, so the answer is negative.

Example 9: Compute the product $4\left(-\frac{3}{16}\right)\left(\frac{-2}{-9}\right)$.

There are 3 factors: the first is positive, the second is negative, and the third is positive (+)(−)(+). This makes the answer negative, so we can now work with the fractions without worrying about the signs. Rewriting 4 as 4/1 and working the problem without the signs, we get

$$\left(\frac{4}{1}\right)\left(\frac{3}{16}\right)\left(\frac{2}{9}\right)=\left(\frac{3}{4}\right)\left(\frac{2}{9}\right)=\frac{1}{6}$$

so that

$$4\left(-\frac{3}{16}\right)\left(\frac{-2}{-9}\right) = -\frac{1}{6}$$

Now it is time to practice what we have learned.

Exercise 3--Multiplication of Positive and Negative Fractions

Purpose of Exercise

- To apply the rules of signs to multiplication of fractions

In each of the following multiplications, determine if the product is positive or negative. Circle the correct answer.

1. $\left(\frac{137}{5}\right)\left(\frac{-4}{5}\right)$ positive negative
2. $-\left(\frac{7}{5}\right)\left(\frac{4}{5}\right)$ positive negative
3. $\left(\frac{-57}{52}\right)\left(\frac{41}{59}\right)$ positive negative
4. $\left(\frac{-57}{52}\right)\left(\frac{-41}{59}\right)$ positive negative
5. $\left(\frac{-17}{5}\right)\left(\frac{4}{5}\right)$ positive negative
6. $-\left(\frac{-17}{5}\right)\left(\frac{4}{5}\right)$ positive negative
7. $-(27)\left(\frac{-41}{59}\right)$ positive negative
8. $(-27)\left(\frac{-41}{59}\right)$ positive negative
9. $-\left(\frac{-17}{5}\right)\left(-\frac{4}{5}\right)$ positive negative
10. $\frac{18}{-9}$ positive negative
11. $\frac{-21}{-33}$ positive negative
12. $-\frac{-21}{-33}$ positive negative

Compute the following products. Reduce fractions where appropriate. (Hint: Determine the sign of the answer first; then multiply the fractions.)

13. $\left(\frac{7}{5}\right)\left(\frac{-4}{5}\right) =$ ______
14. $\left(-\frac{15}{5}\right)\left(\frac{-4}{5}\right) =$ ______
15. $(-100)\cdot\left(-\frac{2}{25}\right) =$ ______
16. $(-5)\cdot\left(\frac{2}{5}\right) =$ ______
17. $-\left(\frac{7}{5}\right)\left(\frac{4}{5}\right) =$ ______
18. $\left(\frac{14}{3}\right)\cdot\left(\frac{2}{7}\right) =$ ______
19. $\left(\frac{-100}{85}\right)(5) =$ ______
20. $-\left(\frac{300}{-400}\right)(-1) =$ ______
21. $\left(\frac{-4}{7}\right)\left(-\frac{5}{6}\right) =$ ______
22. $\left(\frac{137}{5}\right)\left(\frac{-4}{5}\right) =$ ______
23. $-\left(\frac{-9}{5}\right)\left(-\frac{25}{3}\right) =$ ______
24. $-\left(\frac{-17}{5}\right)\left(-\frac{4}{5}\right) =$ ______
25. $15\left(\frac{3}{5}\right) =$ ______
26. $-\left(\frac{1}{5}\right)\left(-\frac{7}{3}\right) =$ ______

Section 4-5 *Adding and Subtracting Fractions with the Same Denominator*

When working with signed numbers, we used number lines to add numbers. The sum $1 + 2 = 3$ can be illustrated as below. Starting at 0, a length of 1 unit added to a length of 2 units ends on the unit 3. By using the letter U, we are just indicating that each length is of a particular unit that we generically abbreviate as "U." We could work instead with standard units such as inches, centimeters, and so forth.

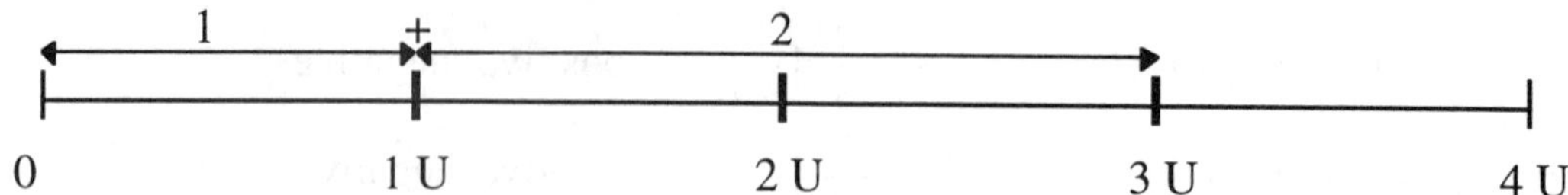

We can also use a number line to illustrate subtraction, such as $3 - 2 = 1$. We start with the length 3 units, then "back up" 2 units. We end up on the 1 unit mark.

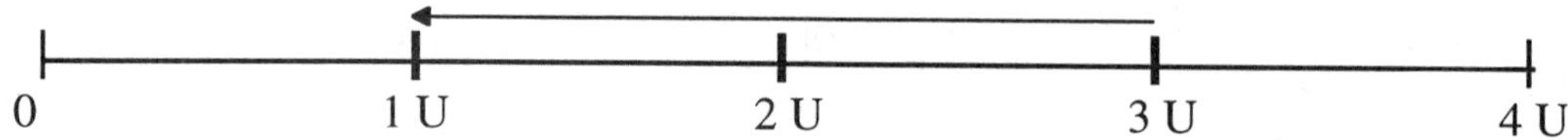

As with whole numbers, we can use the number line or ruler to compute or estimate sums or differences of fractions.

Example 1: To illustrate 1/3 + 1/3, start at the 0 mark and draw two 1/3 units next to each other.

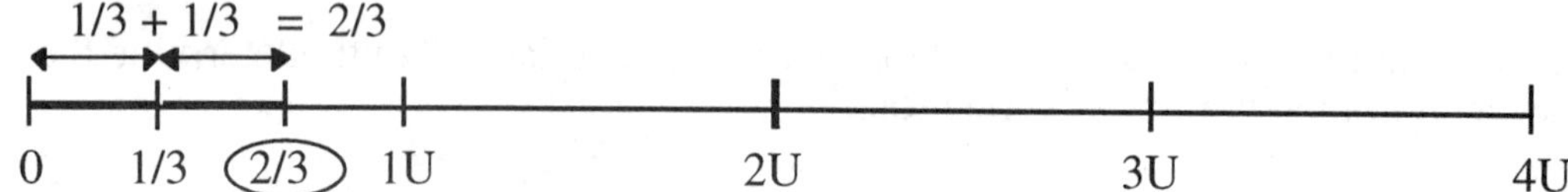

Example 2: To illustrate 5/8 - 3/8, start at the 5/8 mark, and then back up (go left) 3/8 units.

5/8 - 3/8 = 2/8 or 1/4

0 2/8 5/8 1 U 2 U 3 U 4 U

Example 3: To illustrate $\frac{7}{8} + \frac{3}{8}$, start at the $\frac{7}{8}$ mark and go forward (right) $\frac{3}{8}$units.

$\frac{7}{8} + \frac{3}{8} = \frac{10}{8}$

We can also see visually that $\frac{7}{8} + \frac{3}{8} = 1 + \frac{2}{8}$ since the arrow ends up two spaces past 1 unit.

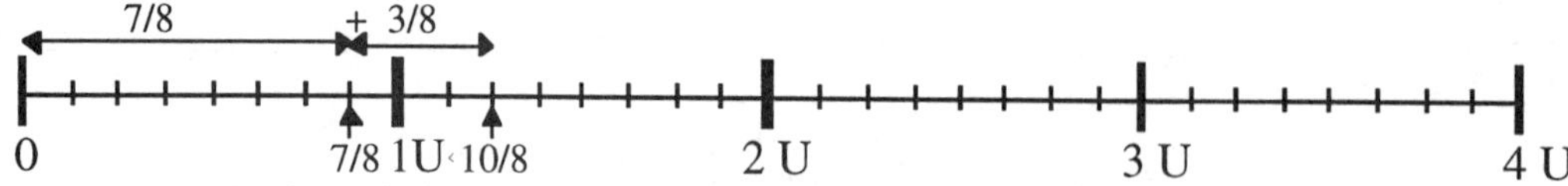

$1 + \frac{2}{8}$ is more concisely written as $1\frac{2}{8}$ and $1 + \frac{1}{4}$ is more concisely written as $1\frac{1}{4}$. Since 2/8 reduces to 1/4 and 10/8 reduces to 5/4, we can make the following observations from the ruler above:

$$1\frac{2}{8} = \frac{10}{8} \quad \text{or in reduced form} \quad 1\frac{1}{4} = \frac{5}{4}$$

Vocabulary

A number in the form, $1\frac{2}{8}$ or $1\frac{1}{4}$ or $3\frac{4}{5}$, is called a **mixed number**.

These numbers are read as "one and two-eighths," "one and one-fourth," and "three and four-fifths," respectively.

$1\frac{2}{8}$ means $1 + \frac{2}{8}$ $\quad$ $1\frac{1}{4}$ means $1 + \frac{1}{4}$ $\quad$ $3\frac{4}{5}$ means $3 + \frac{4}{5}$

As we observed from Example 3, mixed numbers can also be written as improper fractions. Examples 4 and 5 show how to use a ruler to convert between mixed numbers and improper fractions.

Example 4: Convert $\frac{7}{4}$ to a mixed number. Mark 7/4 on the ruler.

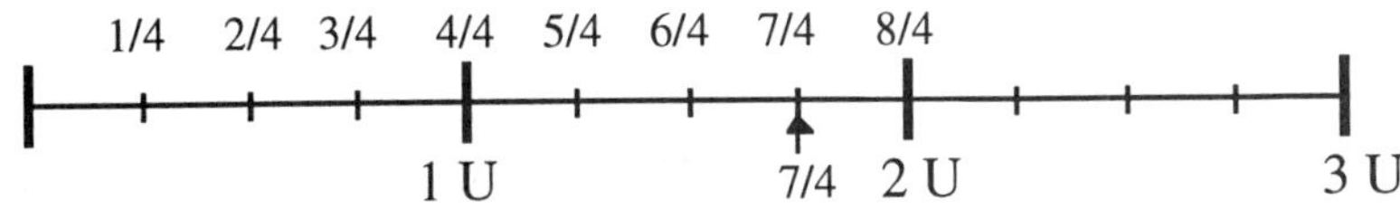

We can see that $\frac{7}{4} = 1 + \frac{3}{4}$

Equivalently, $\frac{7}{4} = \frac{4}{4} + \frac{3}{4}$ $\quad$ (since $\frac{4}{4}$U = 1 U)

$= 1\frac{3}{4}$.

Example 5: Convert the mixed number $2\frac{1}{5}$ to an improper fraction.

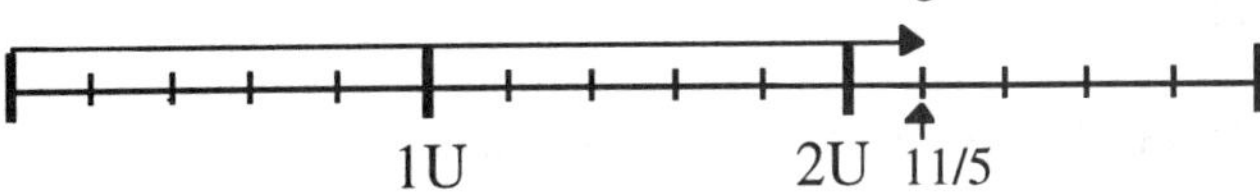

This illustrates that $2\frac{1}{5} = \frac{11}{5}$. $\quad$ (Count the 5ths to see this.)

We could also approach this with arithmetic.

$$2\frac{1}{5} = 1 + 1 + \frac{1}{5}$$

$$= \frac{5}{5} + \frac{5}{5} + \frac{1}{5}$$

$$= \frac{11}{5}$$

In the following set of exercises we will use a number line to discover how to add and subtract fractions with the same denominator, to convert between mixed numbers and improper fractions and to add whole numbers to fractions.

Exercise 1--Adding and Subtracting Fractions with the Same Denominator

Purpose of Exercise

- To understand adding and subtracting fractions on the number line
- To learn an arithmetic process for adding fractions with the same denominator
- To use a number line to convert between improper fractions and mixed numbers

1. ├──────────┤ a. Divide the line segment into 4 equal parts.

b. How much is each part? ______________

c. Shade two of the pieces. How much of the line is shaded? ______________

d. $\frac{1}{4} + \frac{1}{4} =$ ______________ (Write the answer as a fraction in 2 ways.)

2. ├──────────┤ a. Divide the line segment into 5 equal pieces.

b. Shade three of the pieces. How much is shaded? ______________

c. $\frac{1}{5} + \frac{1}{5} + \frac{1}{5} =$ ______________ $\frac{2}{5} + \frac{1}{5} =$ ______________

3. ├──────────┤ a. Divide the line segment into 4 equal pieces.

b. Shade three of the pieces. How much of the total is shaded? ______________

c. It you cut 2 of the shaded pieces from the line, how much of the original line segment is represented by the remaining shaded area? ______________

d. $\frac{3}{4} - \frac{2}{4} =$ ______________

4. ├──────────┤ a. Divide the line segment into 5 equal pieces.

b. Shade three of the pieces. How much is shaded? ______________

c. It you cut two of the shaded pieces from the line, how much of the original line segment is represented by the remaining shaded area? ______________

d. $\frac{3}{5} - \frac{2}{5} =$ ______________

5. a. Divide each unit of the line segment below into 5 equal pieces.

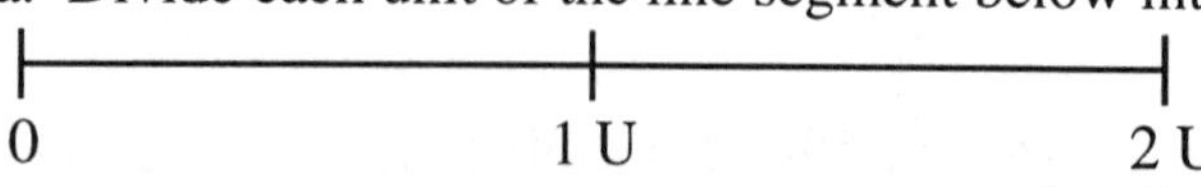

b. Use the illustration to explain why $1 + \frac{3}{5} = \frac{8}{5}$ ______________

____

c. Use the illustration to explain why $1\frac{3}{5}$ and $\frac{8}{5}$ are the same number. ______________

____

6. a. Divide each unit of the line segment below into 4 equal pieces.

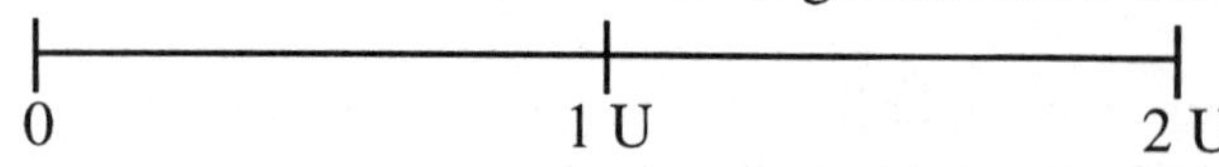

b. Use the illustration to explain why $1 + \frac{3}{4} = \frac{7}{4}$ ______________

____

c. Use the illustration to explain why $1\frac{3}{4}$ and $\frac{7}{4}$ are the same number. ______________

____

Illustrate each of the following sums and differences of fractions on the ruler shown and visually compute the answer. Reduce answers if possible.

7. 3/4 + 1/4 = ________________

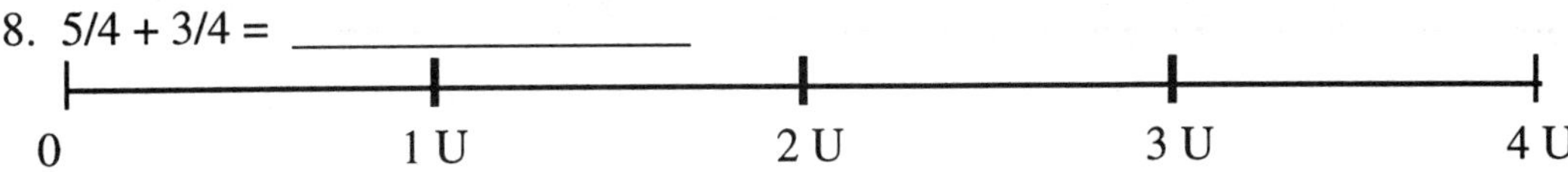

8. 5/4 + 3/4 = ________________

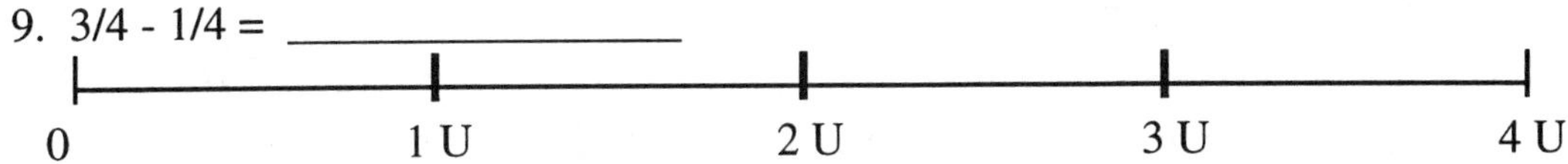

9. 3/4 - 1/4 = ________________

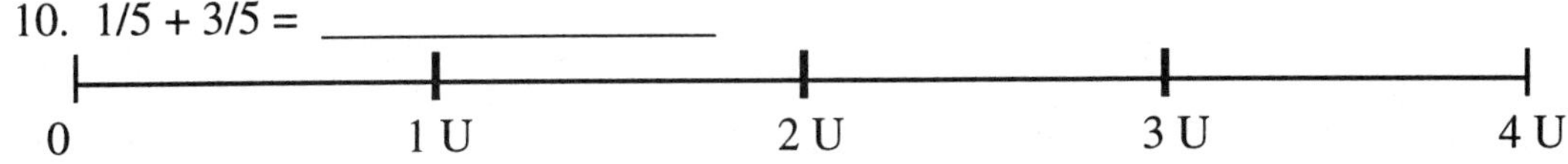

10. 1/5 + 3/5 = ________________

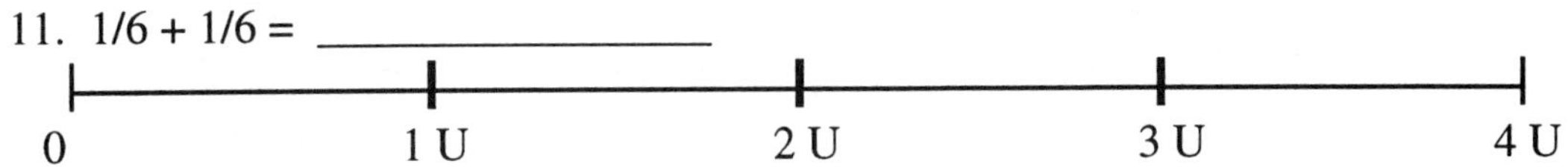

11. 1/6 + 1/6 = ________________

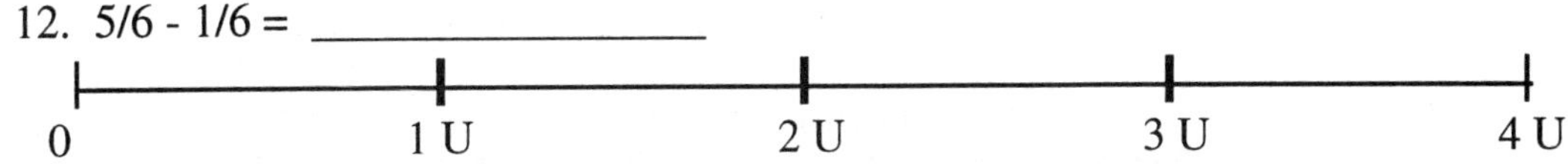

12. 5/6 - 1/6 = ________________

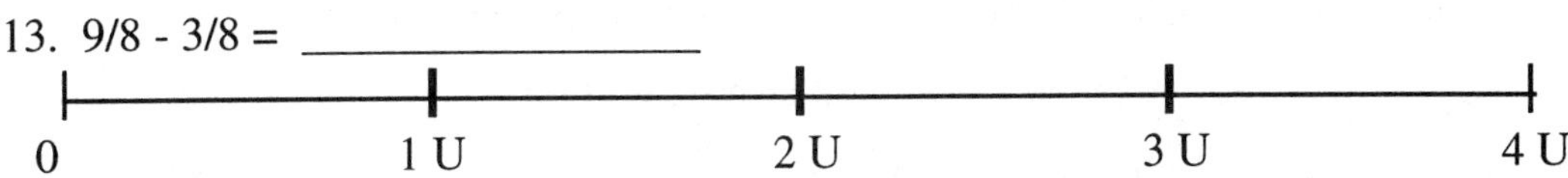

13. 9/8 - 3/8 = ________________

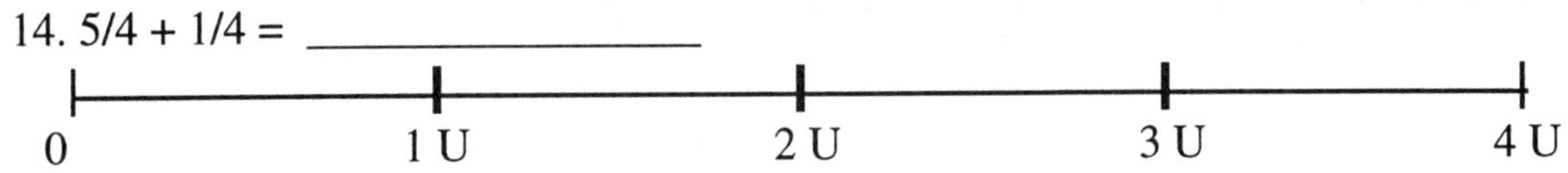

14. 5/4 + 1/4 = ________________

0 1 U 2 U 3 U 4 U

15. Summarize your findings from Problems 1-14 above. Describe how to add and subtract two fractions that have the same denominator.

____

____

____

In each of the following, illustrate the fraction on the number line; then use the ruler to convert the mixed number to an improper fraction or the improper fraction to a mixed number.

16. Convert 11/8 to a mixed number.

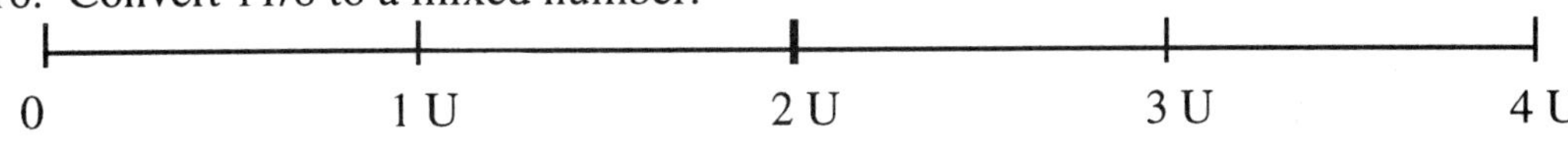

17. Convert $1\frac{2}{3}$ to an improper fraction.

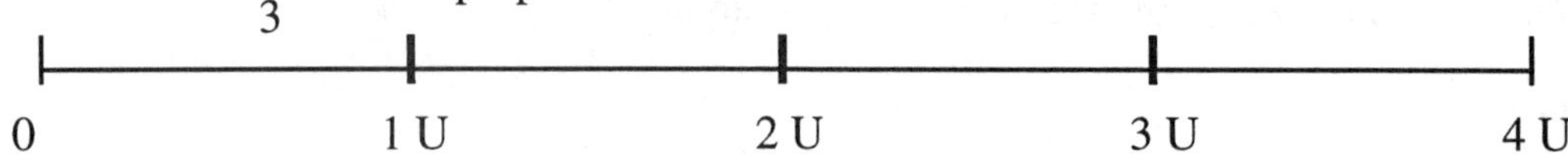

18. Convert $3\frac{5}{6}$ to an improper fraction.

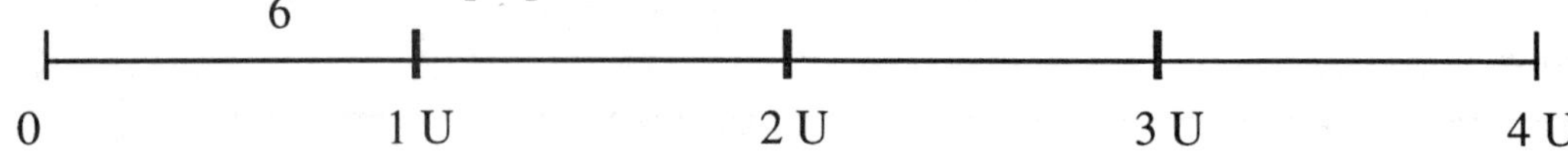

19. Convert 11/4 to a mixed number.

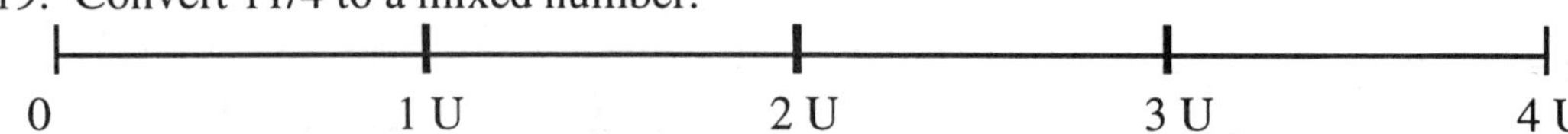

20. Describe how to convert a mixed number to an improper fraction.

____

____

21. Describe how to convert an improper fraction to a mixed number.

____

____

Now let's summarize and generalize the findings from the above exercises. First we will record the process for adding and subtracting fractions with the same denominator.

> **Process for Adding or Subtracting Fractions with the Same Denominator**
>
> To **add fractions** with the same denominator, just **add the numerators** and keep the denominators the same. Reduce the answer if appropriate.
>
> $\frac{3}{5}+\frac{4}{5}=\frac{7}{5}$ The answer could also be written as $1\frac{2}{5}$.
>
> To **subtract fractions** with the same denominator, just **subtract the numerators** and keep the denominators the same. Reduce the answer if appropriate.
>
> $\frac{7}{9}-\frac{1}{9}=\frac{6}{9}$ which reduces to $\frac{2}{3}$
>
> **Process for Adding a Fraction to a Whole Number or Subtracting a Fraction from a Whole Number**
>
> First, change the whole number to fraction form, selecting the denominator to be the same as that of the other fraction; then add the fractions with the same denominator. Reduce the answer if possible. (As you recall, there are many ways to write any whole number as a fraction. $1=\frac{11}{11}$ or $\frac{6}{6}$, and $2=\frac{24}{12}$ or $\frac{30}{15}$, for example.)

Example 6: Subtract: 3 - 4/5.

$3-\frac{4}{5}=\frac{15}{5}-\frac{4}{5}$ 3 written as a fraction with denominator 5 is 15/5.

$=\frac{11}{5}$ Subtract the numerators. The answer cannot be reduced.

Example 7: Add 2 to 7/8 and write the sum as an improper fraction.

$2 + \frac{7}{8} = \frac{16}{8} + \frac{7}{8}$ 2 written as a fraction with denominator 8 is 16/8.

$= \frac{23}{8}$ Add the numerators. The answer cannot be reduced.

Notice, another way to state this is $2\frac{7}{8} = \frac{23}{8}$.

To Convert a Mixed Number to an Improper Fraction
Rewrite the mixed number as a whole number plus a fraction and follow the process above. Roughly check on a number line to see if your answer is reasonable.

Example 8: Convert $2\frac{7}{8}$ to an improper fraction. The work the same as in Example 7.

$$2\frac{7}{8} = 2 + \frac{7}{8}$$
$$= \frac{16}{8} + \frac{7}{8}$$
$$= \frac{23}{8}$$

To Convert an Improper Fraction to a Mixed Number
Use a numeric process to divide the denominator into the numerator.
The quotient is the whole number part of the mixed number and the remainder is the numerator of the fractional part.
Roughly check on a number line, with circles or rectangles to see if your answer is reasonable.

Example 9: Convert 23/7 to a mixed number.

$$\begin{array}{r} 3 \\ 7\overline{)23} \\ \underline{21} \\ 2 \end{array}$$

The quotient is 3 and remainder is 2.

$\frac{23}{7} = 3\,\frac{2}{7}$

Example 9: Convert 23/7 to a mixed number using a calculator.
First, do the division, 23/7 3.29.
The whole number part of the calculator answer is 3 so 3 is the whole number in the mixed number. $\frac{23}{7} = 3 +$ some fraction.
To get the fractional part, express the whole number part as a fraction with denominator 7. $3 = \frac{21}{7}$

$\frac{23}{7}$ is only $\frac{2}{7}$ more than $\frac{21}{7}$. ($\frac{2}{7}$ above came from the remainder.)

$$\frac{23}{7} = \frac{21}{7} + \frac{2}{7}$$
$$= 3 + \frac{2}{7}$$
$$= 3\frac{2}{7}$$

Exercise 2--Practice with Improper Fractions, Mixed Numbers, and Adding Fractions

Purpose of the Exercise

- To practice converting between improper fractions and mixed numbers
- To practice adding and subtracting fractions with the same denominator
- To practice adding and subtracting fractions and whole numbers

Materials Needed

- Calculator

Use the above processes to convert the improper fractions to mixed numbers and the mixed numbers to improper fractions. On a separate sheet of paper, illustrate with a number line as needed. Reduce answers as needed.

1 a. $1\frac{2}{3} =$ ________ b. $\frac{14}{8} =$ ________ c. $\frac{8}{3} =$ ________

d. $2\frac{1}{5} =$ ________ e. $3\frac{1}{2} =$ ________ f. $\frac{9}{4} =$ ________

g. $1\frac{5}{7} =$ ________ h. $\frac{7}{3} =$ ________ i. $\frac{5}{2} =$ ________

2. Convert the following mixed numbers to improper fractions and improper fractions to mixed numbers. Use a calculator as needed.

a. $\frac{45}{16} =$ ________________ b. $7\frac{15}{29} =$ ________________

c. $13\frac{9}{11} =$ ________________ d. $\frac{92}{13} =$ ________________

e. $\frac{59}{17} =$ ________________ f. $\frac{5792}{100} =$ ________________

g. $7\frac{3}{20} =$ ________________ h. $\frac{289}{4} =$ ________________

i. $17\frac{3}{5} =$ ________________ j. $22\frac{2}{13} =$ ________________

k. $\frac{215}{7} =$ ________________ l. $\frac{7903}{879} =$ ________________

m. $\frac{7903}{18} =$ ________________ n. $100\frac{18}{25} =$ ________________

o. $17\frac{1}{2} =$ ________________ p. $\frac{20071}{2} =$ ________________

q. $98\frac{11}{37} =$ ________________ r. $\frac{92}{7} =$ ________________

3. Convert the mixed numbers and improper fractions above to decimal form to check your work.

Without a calculator, compute the following sums and differences. Reduce the answers when appropriate.

4. $\frac{1}{6}+\frac{1}{6}$ = ______________________

5. $\frac{1}{6}+\frac{1}{6}+\frac{1}{6}$ = ______________________

6. $\frac{2}{6}+\frac{1}{6}$ = ______________________

7. $\frac{3}{6}-\frac{1}{6}$ = ______________________

8. $7-\frac{5}{9}$ = ______________________

9. $\frac{4}{5}+\frac{1}{5}$ = ______________________

10. $\frac{4}{5}-\frac{2}{5}$ = ______________________

11. $\frac{2}{7}+\frac{3}{7}$= ______________________

12. $\frac{3}{17}+\frac{8}{17}$ = ______________________

13. $\frac{8}{17}+\frac{12}{17}$ = ______________________

14. $\frac{29}{100}-\frac{13}{100}$ = ______________________

15. $\frac{5798}{3526}-\frac{4786}{3526}$ = ______________________

16. $\frac{876}{279}-\frac{500}{279}$ = ______________________

17. $\frac{78}{59}+\frac{10}{59}$ = ______________________

18. $\frac{24}{24}-\frac{7}{24}$ = ______________________

19. $1-\frac{7}{24}$ = ______________________

20. $\frac{19}{19}+\frac{3}{19}$ = ______________________

21. $1+\frac{3}{19}$ = ______________________

22. $\frac{79}{279}+\frac{279}{279}$ = ______________________

23. $\frac{79}{279}+1$ = ______________________

24. $\frac{12}{4}-\frac{1}{4}$ = ______________________

25. $3-\frac{1}{4}$ = ______________________

26. $\frac{14}{7}+\frac{3}{7}$ = ______________________

27. $2+\frac{3}{7}$ = ______________________

28. $2-\frac{3}{7}$ = ______________________

29. $\frac{12}{3}-\frac{2}{3}$ = ______________________

30. $4-\frac{2}{3}$ = ______________________

31. $1-\frac{2}{87}$ = ______________________

The rules of signs for adding and subtracting are the same for fractions as the rules of signs we have already covered for decimal numbers (including whole numbers).

Rules of Signs for Addition of Fractions

When adding **two positive fractions**, just add the fractions. The **sum is positive.**

$$\frac{2}{4}+\frac{1}{4}=\frac{3}{4} \qquad \frac{3}{5}+\frac{1}{5}=\frac{4}{5} \qquad \frac{2}{5}+\frac{7}{5}=\frac{9}{5}$$

When adding **two negative fractions**, add the two fractions. The **sum is negative.**

$$\left(-\frac{2}{4}\right)+\left(-\frac{1}{4}\right)=-\frac{3}{4} \qquad \frac{-3}{5}+\frac{-1}{5}=\frac{-4}{5}=-\frac{4}{5} \qquad \frac{-2}{5}+\frac{-7}{5}=\frac{-9}{5}=-\frac{9}{5}$$

When adding a **positive fraction and a negative fraction**, subtract the absolute value of the two fractions and keep the sign of the fraction with the larger absolute value.

$$\frac{2}{4}+\left(-\frac{1}{4}\right)=\frac{1}{4} \qquad -\frac{2}{4}+\frac{1}{4}=-\frac{1}{4} \qquad \frac{2}{7}+\left(\frac{-5}{7}\right)=\frac{-3}{7}$$

Rules of Signs for Subtraction of Fractions

When subtracting a **positive fraction from a negative fraction,** add the absolute value of the two fractions. The **answer is negative.**

$$-\frac{2}{7}-\frac{4}{7}=-\frac{6}{7} \qquad \frac{-5}{8}-\frac{2}{8}=-\frac{7}{8} \qquad -\frac{4}{9}-\frac{7}{9}=\frac{-11}{9}$$

When subtracting a **negative fraction from a positive or negative fraction**, change the subtraction to addition of the opposite--instead of subtracting a negative fraction, add the positive fraction. **Then use the rules of adding fractions as outlined above.**

$$-\frac{4}{9}-\frac{-7}{9}=-\frac{4}{9}+\frac{7}{9}=\frac{3}{9}=\frac{1}{3} \qquad \frac{2}{3}-\left(\frac{-2}{3}\right)=\frac{2}{3}+\frac{2}{3}=\frac{4}{3}$$

In the following table are examples of arithmetic involving fractions and signed numbers.

Problem	Positive or Negative Answer?	Subtraction Changes to Addition? Yes/No	Answer and Work
1. $\frac{2}{5}-\frac{3}{5}$	negative	no	$\frac{2-3}{5}=-\frac{1}{5}$
2. $\frac{7}{5}-\frac{4}{5}$	positive	no	$\frac{7-4}{5}=\frac{3}{5}$
3. $-\frac{4}{3}-\frac{9}{3}$	negative	no	$\frac{-4-9}{3}=\frac{-13}{3}$
4. $\frac{-4}{21}-\left(\frac{-15}{21}\right)$	positive	yes	$\frac{-4}{21}+\left(\frac{+15}{21}\right)=\frac{11}{21}$
5. $-\frac{255}{391}+\frac{259}{391}$	positive	addition	$\frac{-255+259}{391}=\frac{4}{391}$
6. $\frac{-41}{19}+\frac{23}{19}$	negative	addition	$\frac{-41+23}{19}=\frac{-18}{19}$
7. $\frac{22}{35}-\left(-\frac{15}{35}\right)$	positive	yes	$\frac{22}{35}+\left(+\frac{15}{35}\right)=\frac{37}{35}$
8. $\frac{-7}{11}-\frac{-4}{11}$	negative	yes	$\frac{-7}{11}+\frac{+4}{11}=\frac{-3}{11}$
9. $-\frac{7}{11}-\frac{4}{11}$	negative	no	$\frac{-7-4}{11}=\frac{-11}{11}=-1$
10. $\frac{-7}{11}+\frac{-4}{11}$	negative	addition	$\frac{-7+(-4)}{11}=\frac{-11}{11}=-1$

Exercise 3--Add/Subtract Fractions and Rules of Signs

Purpose of the Exercise

- To reinforce rules of signs for addition and subtraction of fractions.

In each of the following, fill in the columns as indicated. Refer to the examples in the table above for a guide.

Problem	Positive or Negative Answer?	Subtraction Changes to Addition? Yes/No	Answer and Work
1. $\frac{5}{7}-\frac{2}{7}$			
2. $\frac{7}{5}-\frac{3}{5}$			
3. $\frac{255}{391}-\frac{259}{391}$			
4. $-\frac{255}{391}-\left(\frac{259}{391}\right)$			
5. $\frac{14}{25}-\frac{-4}{25}$			
6. $-\frac{5}{11}+\left(-\frac{3}{11}\right)$			
7. $\frac{5}{11}+\left(-\frac{3}{11}\right)$			
8. $-\frac{2}{15}-\left(-\frac{3}{15}\right)$			
9. $\frac{100}{89}-\frac{200}{89}$			
10. $\frac{84}{25}-\frac{17}{25}$			
11. $\frac{17}{25}-\frac{84}{25}$			
12. $-\frac{129}{32}-\left(\frac{-21}{32}\right)$			
13. $-\frac{129}{32}-\frac{21}{32}$			
14. $-\frac{129}{32}+\frac{21}{32}$			
15. $-\frac{84}{105}+\frac{2}{105}$			
16. $\frac{84}{105}+\frac{-2}{105}$			
17. $\frac{84}{105}-\frac{-2}{105}$			
18. $\frac{4}{5}+\frac{-4}{5}$			
19. $\frac{1}{10}-\frac{7}{10}$			
20. $-\frac{100}{89}-\frac{200}{89}$			

Section 4-6 *Adding and Subtracting Fractions with Different Denominators*

Using a Ruler to Estimate Sums and Differences of Fractions and Decimals

We have already learned to represent sums and differences of numbers on a ruler or number line. Before jumping into the arithmetic involved in adding and subtracting fractions with different denominators, we will first take a visual look at such additions and subtractions. As always, it is important to know what a reasonable answer is before doing the calculations. Whether we add and subtract fractions using pencil-and-paper methods or a calculator, it is easy to make a mistake. A seemingly small mistake can cause a large error, so it is helpful to see what a reasonable answer is before we begin.

To represent addition of lengths on a ruler, draw the two lengths side by side so they make up one line segment. The sum is the length of the combined line segment. To represent subtraction on a ruler, start with the first number and back up (go left) the length subtracted.

Example 1: Add: $\frac{1}{2} + \frac{1}{4}$ on the number line.

1/2 U 1 U 1/4 U 1 U

3/4 U

$$\frac{1}{2} + \frac{1}{4} = \frac{3}{4}$$

1/2 + 1/4 ended up on one of the marks when we divided the unit into 4 equal pieces. Not all sums or differences are that easy to determine exactly visually; however, the number line is valuable in estimating harder to find answers.

Example 2: Use the number line to estimate $\frac{3}{4} + \frac{2}{3}$.

The diagram on the left below is one unit divided into fourths and on the right is one unit divided into thirds. From the diagrams, we will mark 3/4 U and 2/3 U and then transfer the lengths to the number line below, starting with 3/4 U and from the 3/4 spot, go right another 2/3 U.

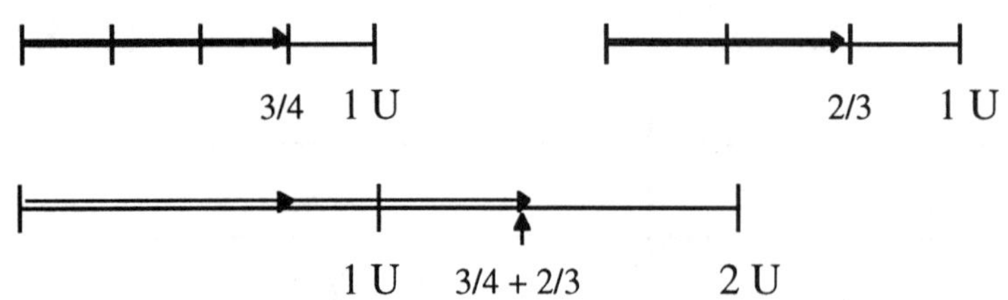

From the number line, it is easy to see that $\frac{3}{4} + \frac{2}{3}$ is greater than 1 and less than $1\frac{1}{2}$.

A **common mistake** in adding $\frac{3}{4} + \frac{2}{3}$ is to add the numerators and add the denominators to get $\frac{5}{7}$ which is less than 1. The ruler shows us that $\frac{5}{7}$ is obviously a wrong answer.

Example 3: Use a number line to estimate $1.6 - \frac{3}{4}$.

Remember that 0.6 is the same as 6/10 so, for 1.6, our length will be 1 + 6/10. We will measure 6/10 from the unit below divided into tenths and measure 1.6 on our ruler to get the starting point.

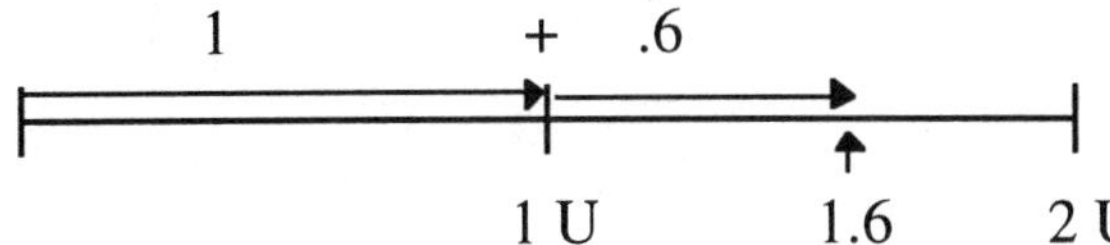

Now we will measure 3/4 from the diagram below and back up 3/4 units from the 1.6 mark.

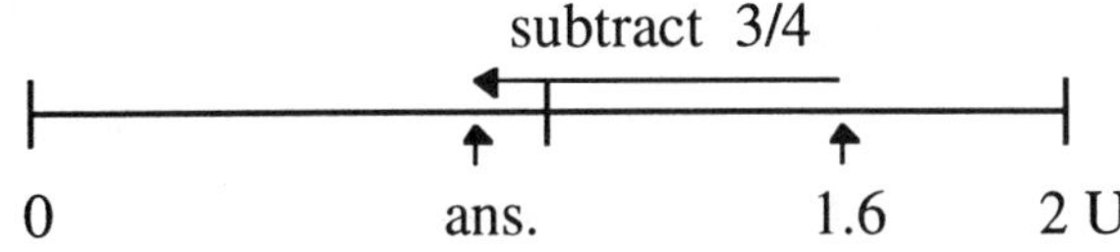

Observing from the ruler, $1.6 - \frac{3}{4}$ is slightly less than 1. To get a fairly accurate answer, mark the length from the 0 point to the answer; then compare that length to the units below. It appears that 9/10 is a close approximation.

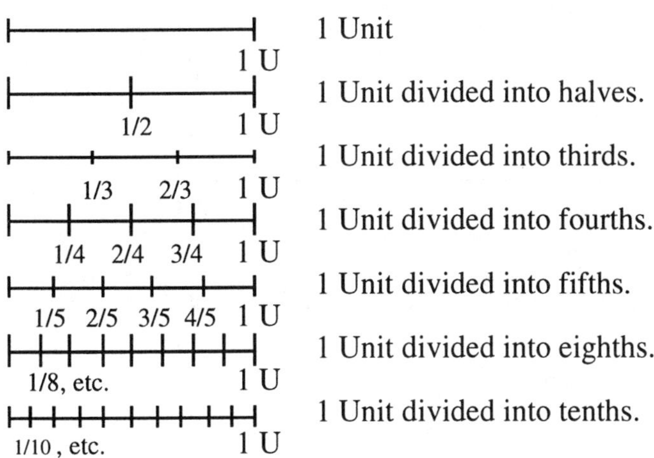

1 Unit

1 Unit divided into halves.

1 Unit divided into thirds.

1 Unit divided into fourths.

1 Unit divided into fifths.

1 Unit divided into eighths.

1 Unit divided into tenths.

Exercise 1--Estimating Sums and Differences of Fractions and Decimals

Purpose of the Exercise

- To visually estimate sums and differences of fractions and decimal numbers

In each of the following, mark on the ruler shown to represent the sums or differences. For accuracy, transfer lengths from the above diagram. From the list of possible answers given, choose and circle the best answer. The first one is done as an example.

Possible Answers

1. 1/3 + 1/3 — 2/6 — 1/9 — (2/3)

1/3 + 1/3 — 1 U — 2 U — 3 U — 4 U

2. 3/4 + 2/3 — 1.3 — 5/7 — 1 5/12

1 U — 2 U — 3 U — 4 U

3. 3/8 + 3/4 — 6/12 — 9/8 — 9/4

1 U — 2 U — 3 U — 4 U

4. 1/2 + 5/10 — 1 — 6/12 — 6/10

1 U — 2 U — 3 U — 4 U

5. 5/4 + 5/4 — 10/8 — 2 — 2 1/2

1 U — 2 U — 3 U — 4 U

6. 11/8 + 7/4 — 3 1/8 — 9/4 — 18/12

1 U — 2 U — 3 U — 4 U

7. 2 - 5/4 — 3/4 — 7/4 — 3 1/4

1 U — 2 U — 3 U — 4 U

8. 2 - 9/8 — 25/8 — -7/8 — 7/8

1 U — 2 U — 3 U — 4 U

9. 0.2 + 0.5 — 1 — 7 — 0.7

1 U — 2 U — 3 U — 4 U

10. 1.5 - 0.2 — 1.7 — 1.3 — 0.9

1 U — 2 U — 3 U — 4 U

11. 1.5 + 1/2 — 1.6 — 2.3 — 2

1 U — 2 U — 3 U — 4 U

12. 3.1 + 9/4 — 5 — 4 — 6

1 U — 2 U — 3 U — 4 U

13. 3 3/4 + 2 1/2 — 5 — 6 — 7

1 U — 2 U — 3 U — 4 U

Now that we know how to estimate an answer to a sum or difference of fractions, it is time to learn the process for adding and subtracting fractions with different denominators. As you discovered from the above exercises, the process is not just simple addition or subtraction of numerators and denominators.

When adding and subtracting fractions with the same denominator, we were adding and subtracting pieces of the same size. That makes the process simple--just add or subtract the number of pieces (i.e., add or subtract the numerators.) In fractions with different denominators, however, we are working with different size pieces of the whole. For that reason, the process of addition is more complicated. We have to break the pieces down further until we are adding the same size pieces. But this is moving ahead too fast. Before we cover the process of finding the common denominator (that breaks both fractions into the same size pieces), let's work with dividing rectangles.

To illustrate how to add $\frac{1}{2}$ and $\frac{3}{5}$, first divide the left rectangle below into 2 equal parts using horizontal lines. Shade one of the parts to represent the fraction $\frac{1}{2}$. Divide the second rectangle into 5 equal parts using horizontal lines as shown below. Shade 3 of the parts to represent $\frac{3}{5}$.

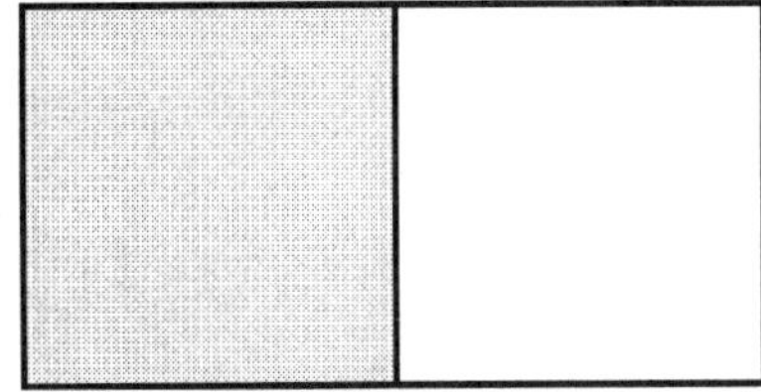

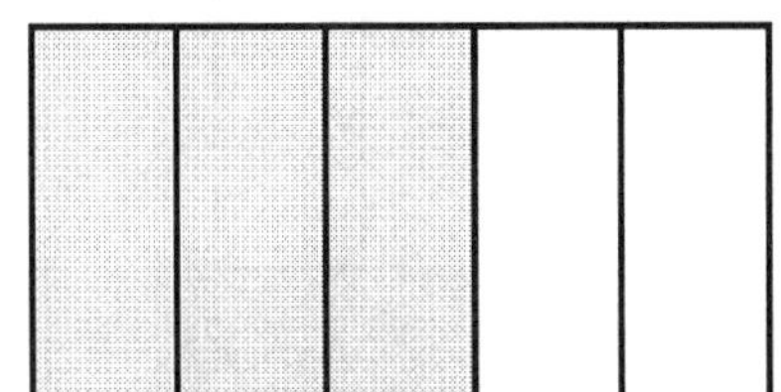

To add $\frac{1}{2}$ and $\frac{3}{5}$, we need the same size pieces, not halves or fifths. To get the same size pieces, we have to divide both the $\frac{1}{2}$ and the $\frac{3}{5}$ into smaller uniform size pieces as shown below.

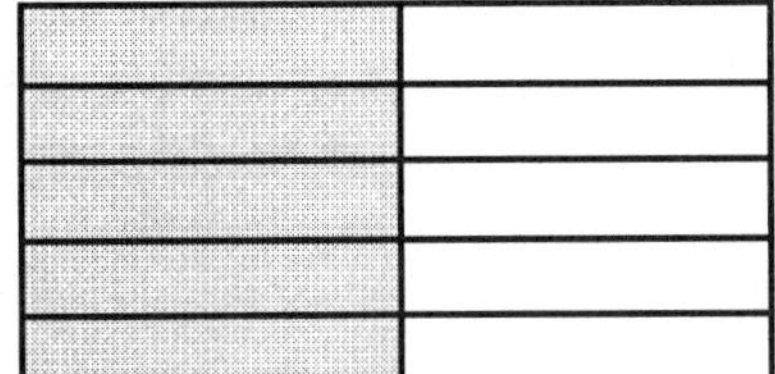

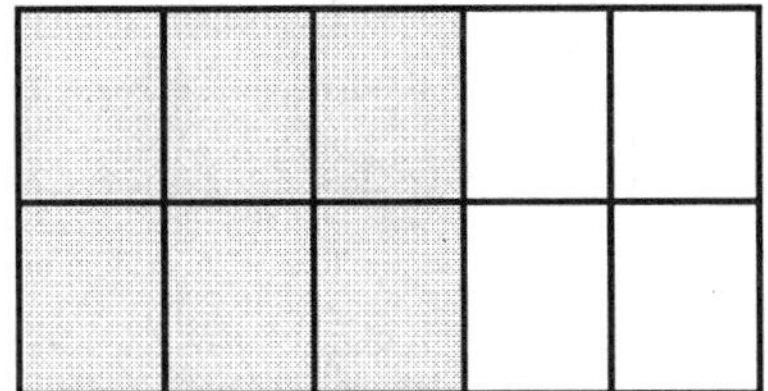

Notice that both of the two rectangles above are now divided into 10 pieces. Although the small pieces are different shapes in the two rectangles, they represent the same area since each small piece is $\frac{1}{10}$ of the same size rectangle.

We already know from the section on equivalent fractions that

$\frac{1}{2}$ is the same as $\frac{5}{10}$ and $\frac{3}{5}$ is the same as $\frac{6}{10}$.

Confirm this with the diagrams above.

Therefore, to add $\frac{1}{2}$ and $\frac{3}{5}$ is the same as adding $\frac{5}{10}$ and $\frac{6}{10}$.

$\frac{1}{2}+\frac{3}{5} = \frac{5}{10}+\frac{6}{10}= \frac{11}{10}$ or $1\frac{1}{10}$

Exercise 2--Different Shapes Representing the Same Size

Purpose of the Exercise
- To understand that fractions may be represented in different shapes

Materials Needed
- Tracing paper and scissors

1. 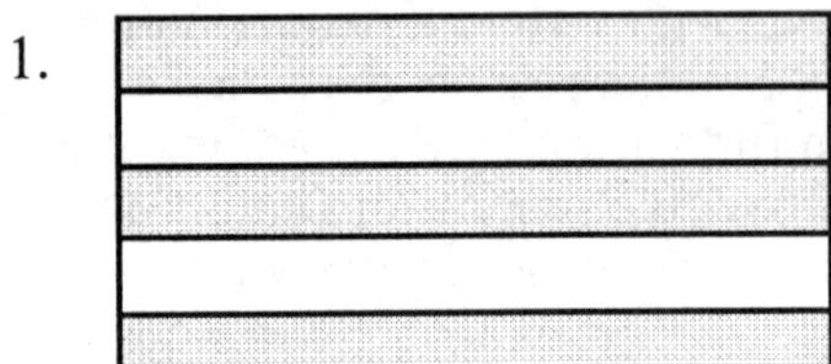 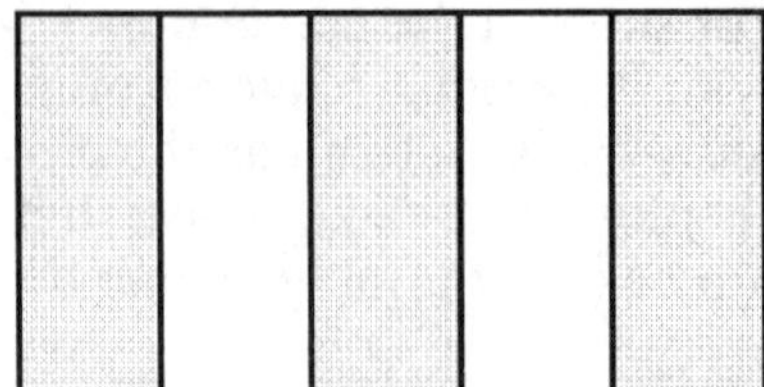

The above drawings represent the same size rectangles divided into 5 equal parts. Although the parts are different shapes, they are the same size. Trace the two different shaded shapes that represent 1/5 of the rectangle and confirm that they are the same size by cutting one shape to fit into the other.

Exercise 3--Adding Fractions with Different Denominators (Group)

Purpose of the Exercise
- To visually understand adding fractions with different denominators.
- To discover the process of adding fractions with different denominators.

1. 

a. Using vertical lines, divide the left rectangle into 2 equal parts and divide the right rectangle into 3 equal parts. Shade $\frac{1}{2}$ of the left rectangle and $\frac{1}{3}$ of the right rectangle. Determine from the shadings whether $\frac{1}{2} + \frac{1}{3}$ is larger than 1, smaller than 1, or about the same as 1. ______________________

b. Using horizontal lines, divide the left rectangle into 3 equal parts and divide the right rectangle into 2 equal parts. Into how many total parts does the grid divide each rectangle? ______________

c. Rewrite $\frac{1}{2} + \frac{1}{3}$ as the sum of fractions with the same denominator. __________

d. $\frac{1}{2} + \frac{1}{3} =$ ______________________.

2. 

a. Using vertical lines, divide the left rectangle into 4 equal parts and divide the right rectangle into 5 equal parts. Shade $\frac{3}{4}$ of the left rectangle and $\frac{2}{5}$ of the right rectangle.

Is $\frac{3}{4} + \frac{2}{5}$ larger than 1, smaller than 1, or about the same as 1? __________

b. Using horizontal lines, divide the left rectangle into 5 equal parts and divide the right rectangle into 4 equal parts. Into how many total parts does the grid divide each rectangle? ____________________

c. Rewrite $\frac{3}{4} + \frac{2}{5}$ as a sum of fractions with the same denominator. ______________

d. $\frac{3}{4} + \frac{2}{5} =$ ____________________

3. 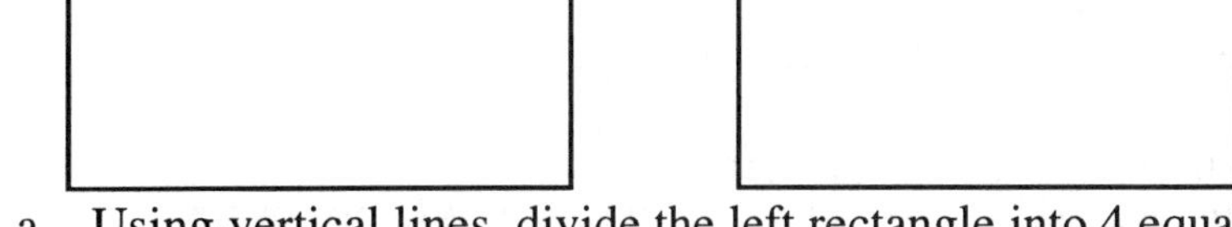

a. Using vertical lines, divide the left rectangle into 4 equal parts and the right rectangle into 3 equal parts. Shade $\frac{3}{4}$ on the left rectangle and $\frac{1}{3}$ on the right rectangle. Is $\frac{3}{4} + \frac{1}{3}$ larger than 1, smaller than 1, or about the same as 1? _____________

b. Using horizontal lines, divide the left rectangle into 3 equal parts and the right rectangle into 4 equal parts.
Into how many total parts does the grid divide each rectangle? __________________

c. Rewrite $\frac{3}{4} + \frac{1}{3}$ as the sum of fractions with the same denominator. ________________

d. $\frac{3}{4} + \frac{1}{3} =$ ____________________

4. 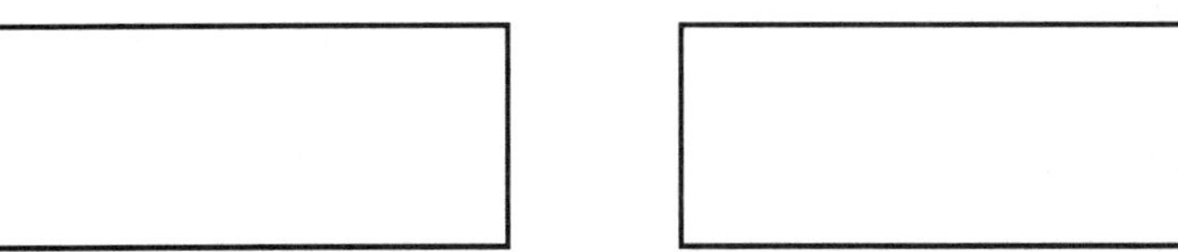

a. Using vertical lines, divide the left rectangle into 5 equal parts and the right rectangle into 4 equal parts. Shade $\frac{3}{5}$ of the left rectangle and $\frac{3}{4}$ of the right rectangle. Is $\frac{3}{5} + \frac{3}{4}$ larger than 1, smaller than 1, or about the same as 1? _____________

b. Using horizontal lines, divide the left rectangle into 4 equal parts and divide the right rectangle into 5 equal parts.
Into how many total parts does the grid divide each rectangle? __________________

c. Rewrite $\frac{3}{5} + \frac{3}{4}$ as the sum of fractions with the same denominator. ______________

d. $\frac{3}{5} + \frac{3}{4} =$ ____________________

Vocabulary

A **common denominator** is a number that all of the denominators will divide evenly into. Each denominator is a factor of a common denominator.

A **least common denominator** (abbreviated LCD) is the smallest number that all denominators will divide into. Note: The least common denominator is the least common multiple of all of the denominators.

Finding the Least Common Denominator

Step 1: Prime factor each denominator.

Step 2: If a factor appears more than once in any one denominator, write that factor in exponential notation.

Step 3: The least common denominator is the product of the factors that appear in all of the denominators. If a factor has an exponent in at least one of the denominators, include that factor to the highest power that appears in any one denominator.

(If the denominators are small numbers, you may be able to find the LCD by guessing.)

Example 4: Find the least common denominator of the fractions, 2/9, 3/5 and 5/6.

The three denominators are 9, 5, and 6.

$9 = 3^2$ 5 is prime $6 = 2 \cdot 3$

The least common denominator = $(3^2)(5)(2) = 90$

The factor, 3, appears as factors of 9 as well as in 6. 3 is a factor twice of the number, 9, and only once of the number 6. In the LCD, it must be included twice because that is the most number of times it appears in any one denominator.

Process for Adding or Subtracting Fractions with Different Denominators

To add or subtract fractions with different denominators,

Step 1: Find the least common denominator.

Step 2: Find an equivalent fraction for each fraction, so that all fractions have the same denominator.

Step 3: Follow the process and rules of signs for adding and subtracting fractions with the same denominators.

Step 4: Reduce the answer if possible.

Example 5: Add and subtract as indicated, $\frac{4}{9} - \frac{7}{5} + \frac{5}{6}$.

9 = (3)(3), 6 = (3)(2), and 5 is prime. LCD = (3)(3)(2)(5) = 90.

To build up a fraction, for example, $\frac{7}{5}$, divide the denominator into 90 to find the factor for building up the fraction. $\frac{7}{5} = \frac{(7)(18)}{(5)(18)} = \frac{126}{90}$. Similarly, we build up 4/9 and 5/6.

$$\frac{4}{9} - \frac{7}{5} + \frac{5}{6} = \frac{40}{90} - \frac{126}{90} + \frac{75}{90}$$

$$= \frac{-86}{90} + \frac{75}{90}$$

$$= \frac{-11}{90}$$

This answer cannot be reduced.

Adding and Subtracting Mixed numbers

To add $4\frac{2}{3}$ and $2\frac{1}{2}$, lets picture the mixed numbers with rectangles. $4\frac{2}{3}$ is 4 whole rectangles plus $\frac{2}{3}$ of a 5th rectangle. $2\frac{1}{2}$ represents 2 whole rectangles plus $\frac{1}{2}$ of another rectangle.

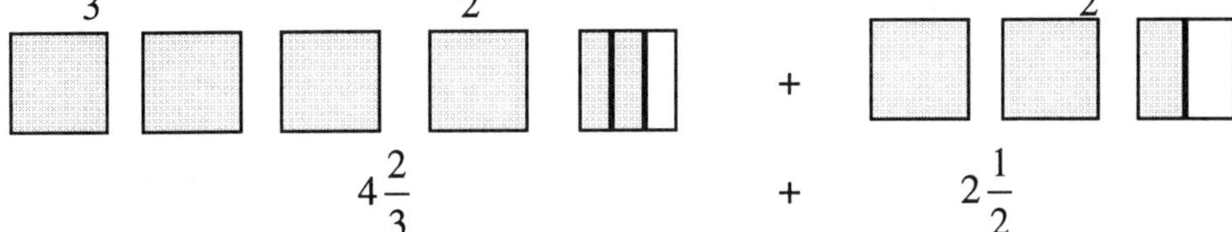

$4\frac{2}{3} \quad + \quad 2\frac{1}{2}$

First, we will get a ballpark estimate by observing the diagrams. The answer is obviously larger than 6 because there are 6 rectangles in the sum plus some fractional parts. The fractional parts add up to more than one rectangle, so the answer is larger than 7 but smaller than 8.

We will learn two processes for adding mixed numbers.

To Add Mixed Numbers (method one)

1. Add the fractional parts. If the answer is an improper fraction, convert the answer to a mixed number.
2. Add the whole number parts of the two original numbers to the whole number part of the sum of the fractions from Step 1.
3. The sum is the whole number parts + the fractional part.

To add $4\frac{2}{3} + 2\frac{1}{2}$, the whole number parts are 4 and the fractional parts are $\frac{2}{3}$ and $\frac{1}{2}$.

Step 1: Add the fractional parts. $\frac{2}{3} + \frac{1}{2} = \frac{4}{6} + \frac{3}{6} = \frac{7}{6} = 1\frac{1}{6}$

Step 2: Add the whole number parts: $4 + 2 + 1 = 7$ (The 1 came from step 1.)

Step 3: The sum is: $7 + \frac{1}{6}$ or $7\frac{1}{6}$

To Add Mixed Numbers (method two)

Step 1: Change each mixed number to an improper fraction.
Step 2: Add the improper fractions.
Step 3: Convert the answer to a mixed number (if that is the desired form.)

To add $4\frac{2}{3} + 2\frac{1}{2}$,

Step 1: Convert each mixed number to an improper fraction: $4\frac{2}{3} = \frac{14}{3}$

$2\frac{1}{2} = \frac{5}{2}$

Steps 2 and 3: $\frac{14}{3} + \frac{5}{2} = \frac{28}{6} + \frac{15}{6} = \frac{43}{6} = 7\frac{1}{6}$

To subtract improper fractions, either method above can be used. When the whole number part is a one- or two-digit number, there is probably less confusion in general with the second method. As always, the rules of signs apply.

Exercise 4--Addition and Subtraction of Fractions

Purpose of the Exercise

- To sharpen skills on getting ballpark answers when adding and subtracting fractions that are positive, negative, mixed, improper, and proper
- To practice adding and subtracting fractions--mixed, with same or different denominators, and positive and negative

Materials Needed

- Calculator

In each of the following, circle the best answer from the choices. **Do not use a calculator.**

1.	$\frac{1}{9}-\frac{11}{12}=$	$-\frac{10}{3}$	$-\frac{29}{36}$	$\frac{29}{36}$
2.	$\frac{1}{10}-\frac{1}{22}=$	$\frac{12}{220}$	$-\frac{12}{220}$	0
3.	$2\frac{17}{19}+9\frac{1}{5}=$	$11\frac{18}{24}$	$12\frac{9}{95}$	$11\frac{89}{95}$
4.	$4\frac{21}{32}-5\frac{2}{5}=$	$9\frac{41}{160}$	$-1\frac{41}{160}$	$-1\frac{19}{27}$
5.	$4\frac{21}{32}+5\frac{2}{5}=$	$9\frac{41}{160}$	$10\frac{9}{160}$	$9\frac{23}{37}$
6.	$-3\frac{4}{5}-\frac{1}{3}=$	$-2\frac{2}{15}$	$-3\frac{3}{2}$	$-4\frac{2}{15}$
7.	$6\frac{4}{5}-8\frac{1}{3}=$	$-2\frac{3}{15}$	$-2\frac{7}{15}$	$15\frac{2}{15}$
8.	$\frac{5}{21}-7=$	$-\frac{2}{21}$	$-\frac{12}{21}$	$-6\frac{16}{21}$
9.	$-\frac{5}{21}-7=$	$-7\frac{5}{21}$	$-6\frac{16}{21}$	$-\frac{12}{21}$
10.	$-17+\frac{4}{5}=$	$-\frac{13}{5}$	$-17\frac{4}{5}$	$-16\frac{1}{5}$
11.	$-17-\frac{4}{5}=$	$-\frac{13}{5}$	$-17\frac{4}{5}$	$-16\frac{1}{5}$

12. After estimating the answers, check your answers with a calculator by representing each number in decimal form and computing the answer. Be careful with signs.

Without a calculator, add or subtract each of the following fractions as indicated. Be careful with signs. Show the work involved and reduce answers as needed.

13. $\frac{7}{10}-\frac{2}{5}=$ ______________________ 14. $\frac{2}{3}-\frac{4}{5}=$ ______________________

15. $2\frac{1}{3}+4\frac{2}{5}=$ ______________________ 16. $2\frac{1}{3}-4\frac{2}{5}=$ ______________________

17. $4-\frac{8}{9}=$ ______________________ 18. $\frac{8}{9}-4=$ ______________________

19. $-\frac{10}{13}-\frac{3}{7}=$ ______________________ 20. $\frac{5}{6}+\frac{3}{8}=$ ______________________

21. $\frac{5}{6}-\frac{3}{8}=$ ______________________ 22. $7+\frac{4}{7}=$ ______________________

The Distributive Law

All of the laws of arithmetic previously covered hold true for fractions. The commutative and associative laws for addition and multiplication allow us to change the order and groupings of numbers added or multiplied. The distributive law also holds true for fractions. The rules for arithmetic sometimes look more confusing when they involve fractions so let's review them in the context of fractions.

The following examples show computations involving fractions and the distributive law.

Example 6: We will compute $4\left(\frac{5}{2}-\frac{1}{4}\right)$ two different ways.

1. Do the computations inside the parentheses first; then multiply the answer by 4.

$$4\left(\frac{5\cdot 2}{2\cdot 2}-\frac{1}{4}\right)=4\left(\frac{10}{4}-\frac{1}{4}\right)$$
$$=4\left(\frac{9}{4}\right)$$
$$=\left(\frac{4\cdot 9}{4}\right)$$
$$=9$$

2. Apply the distributive law first and then add or subtract.

$$4\left(\frac{5}{2}\right)-4\left(\frac{1}{4}\right)=\frac{4}{1}\left(\frac{5}{2}\right)-\frac{4}{1}\left(\frac{1}{4}\right)$$
$$=\frac{(2)(2)(5)}{2}-\frac{4}{1}\left(\frac{1}{4}\right)$$
$$=\frac{(2)(5)}{1}-1$$
$$=10-1$$
$$=9$$

Example 7: Use the distributive law to multiply.

$$7\left(\frac{3x}{7}-5\right) = 7\left(\frac{3x}{7}\right)-7(5)$$
$$= \frac{(7)(3x)}{7}-35$$
$$= 3x - 35$$

Example 8: Use the distributive law to multiply.

$$\frac{1}{4}(3x+8) = \frac{1}{4}(3x)+\frac{1}{4}(8)$$
$$= \frac{3x}{4}+2$$

Exercise 5--The Distributive Law in Problems with Fractions

Purpose of the Exercise

- To practice using the distributive law when there are fractions in the expression
- To observe that in some cases, the fractions disappear after using the distributive law and reducing the resulting fractions

In each of the following, use the distributive law to multiply; then reduce each resulting fraction if possible. Use your own paper and clearly show each step.

1. $8\left(\frac{15}{8}-\frac{1}{2}\right)$
2. $8\left(\frac{15}{8}-2\right)$
3. $8\left(\frac{15}{8}x-2\right)$
4. $-8\left(\frac{15}{8}-2t\right)$
5. $-6\left(\frac{6}{5}-\frac{5m}{6}\right)$
6. $\frac{1}{3}(-3y-15)$
7. $\frac{1}{3}\left(\frac{3}{7}+\frac{1}{3}r\right)$
8. $-15\left(\frac{3}{5}-\frac{2m}{3}\right)$
9. $-15\left(\frac{3y}{5}-\frac{2m}{3}+\frac{1}{15}\right)$
10. $\frac{1}{11}(-11-11w)$
11. $\frac{1}{11}(-11-w-22t)$
12. $\frac{-4}{3}\left(\frac{1}{2}-3x\right)$
13. $-\frac{4}{3}\left(\frac{1}{2}-3x\right)$
14. $-\frac{3}{4}\left(\frac{4}{3}-\frac{4}{3}x\right)$
15. In which of the problems above did the fractions disappear?

 ____

16. In each of the following, find a number to multiply by that will clear out the fractions. Use the distributive law to multiply. Reduce the resulting fractions to see if you made a good choice.

 a. $\_\_\_\_\left(\frac{4}{3}-\frac{1}{3}\right)$ b. $\_\_\_\_\left(\frac{4}{3}-\frac{1}{6}\right)$

c. ______ $\left(\frac{4}{3}-\frac{1}{2}\right)$

d. ______ $\left(\frac{4}{3}x-1\right)$

e. ______ $\left(\frac{4}{7}-\frac{2}{14}a\right)$

f. ______ $\left(-\frac{4}{7}+\frac{2w}{3}\right)$

17. In each of the following, determine whether you would use the distributive law or the associative law to perform the arithmetic. Do the arithmetic. The first two are done as examples. Use a separate sheet of paper and show your work.

a. $(4)\left(\frac{2}{5}\right)\left(\frac{-3}{4}\right)$ — The three numbers are multiplied, so apply the associative law.

$(4)\left(\frac{2}{5}\right)\left(\frac{-3}{4}\right) = \left(\frac{4}{1}\right)\left[\left(\frac{2}{5}\right)\left(\frac{-3}{4}\right)\right]$ — Group with brackets. In the second two fractions (as grouped), divide the factor 2 in the numerator and 4 in the denominator by the common factor 2.

$= \left(\frac{4}{1}\right)\left[\left(\frac{1}{5}\right)\left(\frac{-3}{2}\right)\right]$ — Multiply the second two fractions (as grouped).

$= \left(\frac{4}{1}\right)\left(\frac{-3}{10}\right)$ — Divide the factor 4 in the numerator and the factor 10 in the denominator by the common factor 2.

$= \left(\frac{2}{1}\right)\left(\frac{-3}{5}\right)$ — Multiply the fractions.

$= \frac{-6}{5}$

b. $(4)\left(\frac{2}{5}-\frac{3}{4}\right)$ — Multiplication and subtraction mixed--use the distributive law.

$(4)\left(\frac{2}{5}-\frac{3}{4}\right) = (4)\left(\frac{2}{5}\right)-(4)\left(\frac{3}{4}\right)$ — Multiply the fractions

$= \frac{8}{5}-3$ — Write 3 as 15/5 and subtract the fractions.

$= \frac{8}{5}-\frac{15}{5}$

$= \frac{-7}{5}$

c. $\left(\frac{1}{2}\right)\left(\frac{4}{3}\right)\left(\frac{-5}{6}\right)$

d. $\left(\frac{1}{2}\right)\left(\frac{4}{3}-\frac{5}{6}\right)$

e. $7\left(\frac{4}{5}-\frac{3}{7}\right)$

f. $7\left(\frac{4}{5}+\frac{3}{7}\right)$

g. $8\left(\frac{x}{8}+\frac{3}{4}\right)$

h. $8\left(\frac{x}{8}-\frac{3}{4}\right)$

i. $8\left(\frac{x}{8}\right)\left(-\frac{3}{4}\right)$

j. $\frac{2}{3}\left(\frac{x}{8}\right)\left(-\frac{3}{4}\right)$

Section 4-7 *Division of Fractions*

Division involving fractions comes in two basic types--(1) a number divided by a fraction and (2) a fraction divided by a whole number. As we will learn, the process for both types is basically the same. In this section, we will first get a sense of what division involving fractions means.

In the first exercise set, we will build on our previous knowledge of fractions and division to develop an understanding of division involving fractions and discover a procedure for dividing fractions.

Exercise 1--Division of Fractions; Understanding the Concept (Group)

Purpose of the Exercise

- To get a hands-on sense of dividing a fraction of something by a whole number
- To discover a process for dividing fractions

Materials Needed

- A 12-inch ruler

Do Problems 1-4 on the unit shown. Problem 1 is done as an example.

1. a. Divide the unit into two halves.

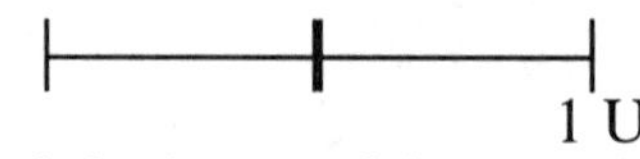

b. Divide both of the halves into 3 equal pieces and shade one of the resulting pieces.

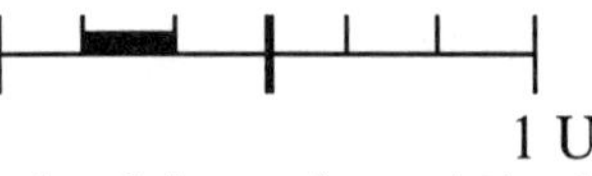

c. How much of the entire unit is shaded? ______1/6______

d. Write a product of fractions that gets the same result as $\frac{1}{2} \div 3 = \frac{1}{2} \cdot \frac{1}{3} = \frac{1}{6}$

2. a. Divide the unit into three thirds.

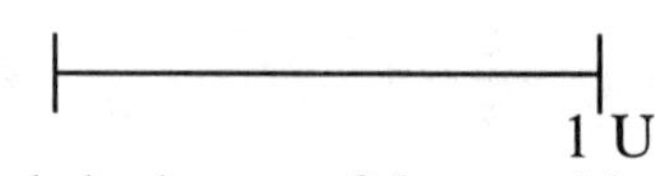

b. Divide each of the thirds into 2 equal pieces and shade one of the resulting pieces.

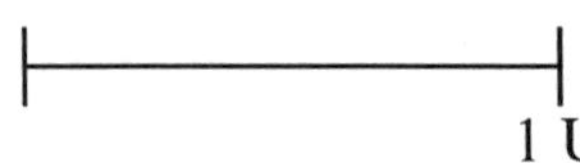

c. How much of the entire unit is shaded? ________________

d. Write a multiplication of fractions problem that gets the same result as $\frac{1}{3} \div 2 =$ ______

3. a. Divide the unit into fourths.

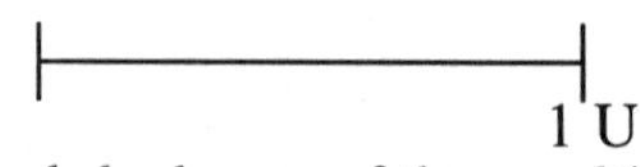

b. Divide each of the fourths into 3 equal pieces and shade one of the resulting pieces.

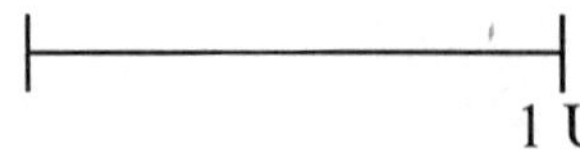

c. How much of the entire unit is shaded? ________________

d. Write a multiplication of fractions problem that gets the same result as $\frac{1}{4} \div 3 =$ ______

4. a. Divide the unit into fifths.

1 U

b. Divide each of the fifths into 2 equal pieces and shade one of the resulting pieces.

1 U

c. How much of the entire unit is shaded? ______________________

d. Write a multiplication of fractions problem that gets the same result as $\frac{1}{5} \div 2 =$ ____

5. To summarize observations from above, write each of the following division problems as a product of two fractions; then compute the product, reducing the answer when appropriate.

	product	reduced answer		product	reduced answer
a. $\frac{1}{2} \div 2 =$	____________	= ______	b. $\frac{1}{3} \div 2 =$	____________	= ______
c. $\frac{1}{3} \div 5 =$	____________	= ______	d. $\frac{1}{3} \div 6 =$	____________	= ______
e. $\frac{1}{4} \div 5 =$	____________	= ______	f. $\frac{1}{5} \div 7 =$	____________	= ______

In the following problems you are to make observations about the ruler. The following examples will help in answering the questions.

Example 1: If a 12-inch ruler were cut into 2-inch pieces, you would have six pieces. Confirm this on your ruler. Observe that 12 inches ÷ 2 inches gives the 6 pieces, so the division that gives the number of pieces is $\frac{12}{2} = 6$.

Example 2: If a 12-inch ruler were cut into 3-inch pieces, there would be four pieces. Confirm this on your ruler. Observe that 12 inches ÷ 3 inches yields 4 pieces so the division that gives the number of pieces is 12 ÷ 3 = 4.

6. Count how many 4-inch pieces are in the 12-inch ruler. ______________ pieces

   Write the division problem that gives you the answer. ______________

7. Count how many 6-inch pieces are in the 12-inch ruler. ______________ pieces

   Write the division problem that gives you the answer. ______________

8. Count how many 1/2-inch pieces are in the 12-inch ruler. ______________ pieces

   Write the division problem that gives you the answer. ______________

   Write a related product that gives the same answer. ______________

9. Count how many 1/4-inch pieces are in the 12-inch ruler. ______________ pieces

   Write the division problem that gives you the answer. ______________

   Write a related product that gives the same answer. ______________

10. Count how many 3/4-inch pieces are in the 12-inch ruler. _______________ pieces
Write the division problem that gives you the answer. _______________________
Write a related product that gives the same answer. _______________________

11. Count how many $1\frac{1}{2}$-inch pieces are in the 12-inch ruler. _______________ pieces
Write the division problem that gives you the answer. (Express $1\frac{1}{2}$ as a mixed number in the problem.) _______________________
Write a related product that gives the same answer. _______________________

12. Count how many 1/2 inch pieces are in 3/2 (or $1\frac{1}{2}$) inches. _____________ pieces
(Write 3/2 as an improper fraction instead of a mixed number.)
Write the division problem that gives you the answer. _______________________
Write a related product that gives the same answer. _______________________

13. Count how many 1/4 inch pieces are in 5/2 (or $2\frac{1}{2}$) inches. _____________ pieces
(Write 5/2 as an improper fraction instead of a mixed number.)
Write the division problem that gives you the answer. _______________________
Write a related product that gives the same answer. _______________________

14. How many 1/3 inch pieces are in 7/3 (or $2\frac{1}{3}$) inches. _____________ pieces
(Write 7/3 as an improper fraction instead of a mixed number.)
Write the division problem that gives you the answer. _______________________
Write a related product that gives the same answer. _______________________

15. How many 1/16 inch pieces are in 5/2 (or $2\frac{1}{2}$) inches. _____________ pieces
(Write 5/2 as an improper fraction instead of a mixed number.)
Write the division problem that gives you the answer. _______________________
Write a related product that gives the same answer. _______________________

16. How many 3/16 inch pieces are in 9/2 (or $4\frac{1}{2}$) inches. _____________ pieces
(Write 9/2 as an improper fraction instead of a mixed number.)
Write the division problem that gives you the answer. _______________________
Write a related product that gives the same answer. _______________________

17. Summarize your observations about division involving fractions. Rewrite each of the following divisions as products and compute the answers.

a. $12 \div \frac{1}{2} =$ _____________ b. $12 \div \frac{1}{3} =$ _____________

c. $12 \div \frac{1}{4} =$ _____________ d. $12 \div \frac{1}{8} =$ _____________

e. $12 \div \frac{3}{4} =$ ____________　　f. $\frac{3}{2} \div \frac{1}{2} =$ ____________

g. $\frac{5}{2} \div \frac{1}{4} =$ ____________　　h. $\frac{9}{2} \div \frac{1}{8} =$ ____________

18. Describe in words how to divide a number by a fraction. ____________________

____

Challenging Problems

19. Suppose you have a stick of wood that is 1 inch by 1 inch by 4 feet long. You want to cut the stick to make as many 1 in. x 1 in. x 1 in. cubes as possible. (Remember that a cube is a three-dimensional object in which all dimensions are the same.)

a. If you could make the cuts without losing any wood to sawdust, how many cubes can you make from the 4 foot stick? Show your computations.

____

b. If the saw made a 1/16-inch blade-cut, how many actual inches of the 4 foot length would be required to make one cube? Show your computations.

____

c. How many 1-inch cubes can you cut from the 4 foot stick of wood (as described above) when the blade cut is 1/6 inch? Discard any part of a cube that might be left over. Show the mathematical computations.

____

20. a. Suppose you want 800 cubes that are 1 in. x 1 in. x 1 in. If your lumber yard has only 1 in. x 1 in. wood in 8 foot lengths, and a saw blade makes a 1/16-in. cut, how many pieces of wood would you have to buy? Show computations.

____

b. How much wood would be left over? ____________________

21. a. Suppose you want 800 cubes that are 1/2 in. x 1/2 in. x 1/2 in. If your lumber yard has only 1/2 in. x 1/2 in. wood in 8 foot lengths, and a saw blade makes a 1/16-in. cut, how many pieces of wood would you have to buy? Show computations.

____

b. How much wood would be left over? ____________________

Division involving fractions can be restated as multiplication as in the following examples. To compute the answer, follow the process for multiplication of fractions.

Example 3:

$$\frac{3}{4} \div 2 = \frac{3}{4} \cdot \frac{1}{2} = \frac{3}{8}$$

$$\frac{4}{5} \div 4 = \frac{4}{5} \cdot \frac{1}{4} = \frac{4}{4} \cdot \frac{1}{5} = (1)\left(\frac{1}{5}\right) = \frac{1}{5}$$

$$10 \div \frac{2}{3} = 10 \cdot \frac{3}{2} = \frac{10}{1} \cdot \frac{3}{2} = 15$$

$$\frac{3}{5} \div 7 = \frac{3}{5} \cdot \frac{1}{7} = \frac{3}{35}$$

$$\frac{7}{9} \div 9 = \frac{7}{9} \cdot \frac{1}{9} = \frac{7}{81} \qquad 9 \div \frac{7}{9} = 9 \cdot \frac{9}{7} = \frac{9}{1} \cdot \frac{9}{7} = \frac{81}{7}$$

$$\frac{6}{11} \div 11 = \frac{6}{11} \cdot \frac{1}{11} = \frac{6}{121} \qquad \frac{6}{11} \div 6 = \frac{6}{11} \cdot \frac{1}{6} = \frac{6}{6} \cdot \frac{1}{11} = \frac{1}{11}$$

As we observed in the previous exercise, division by a whole number can be rewritten as multiplication by 1/(the number) and division by a fraction can be restated as multiplication by the fraction with the numerator and denominator reversed. Describing the procedure in such a way is cumbersome, so let's learn some terminology to help us in describing the process.

> **Definition**
> The **reciprocal** of a fraction is another fraction with the numerator and denominator reversed.

Reciprocals come in pairs. For example the reciprocal of $\frac{2}{3}$ is $\frac{3}{2}$ and the reciprocal of $\frac{3}{2}$ is $\frac{2}{3}$. In other words, $\frac{2}{3}$ and $\frac{3}{2}$ are reciprocals of each other. Remember that $\frac{3}{1}$ and 3 are same number, so 3 and $\frac{1}{3}$ are reciprocals of each other.

Exercise 2--Reciprocals

Purpose of the Exercise

- To practice finding the reciprocal of a number

Write the reciprocal of each of the following fractions.

1. The reciprocal of $\frac{2}{3}$ is __________
2. The reciprocal of $\frac{17}{5}$ is __________
3. The reciprocal of $\frac{145}{87}$ is __________
4. The reciprocal of $\frac{871}{45}$ is __________
5. The reciprocal of $\frac{2759}{2}$ is __________
6. The reciprocal of $\frac{801}{35}$ is __________
7. The reciprocal of $\frac{1}{6}$ is __________
8. The reciprocal of $\frac{1}{17}$ is __________
9. The reciprocal of $\frac{6}{1}$ is __________
10. The reciprocal of $\frac{17}{1}$ is __________
11. The reciprocal of 29 is __________
12. The reciprocal of 1/29 is __________

Compute the following divisions.

13. 2/1 = __________
14. 76/1 = __________
15. 298/1 = __________
16. 906/1 = __________
17. Any whole number can be written as a fraction. Explain how this is done.

____

Now that we know what a reciprocal is, we can record the process for division when the divisor or the dividend is a fraction.

Process for Dividing Fractions

Find the reciprocal of the divisor; then restate the problem as a multiplication of fractions as shown symbolically below and follow the procedure for multiplying fractions.

Symbolically, $\frac{a}{b} \div \frac{c}{d} = \frac{a}{b} \cdot \frac{d}{c}$.

Example 4:

$45 \div \frac{3}{5} = 45\left(\frac{5}{3}\right)$	The divisor is $\frac{3}{5}$, so multiply 45 by the reciprocal of $\frac{3}{5}$.
$= (15)(3)\left(\frac{5}{3}\right)$	Factor 45 to divide out common factors.
$= (15)\left(\frac{5}{1}\right)$	Divide the numerator and denominator by the common factor 3.
$= 75$	Multiply 15 by 5. Recall that 5/1 = 5.

Example 5:

$12 \div \frac{5}{3} = 12\left(\frac{3}{5}\right)$	The divisor is $\frac{5}{3}$ so multiply 45 by the reciprocal of $\frac{5}{3}$.
$= \frac{36}{5}$	

Example 6:

$$\frac{3}{4} \div \frac{5}{7} = \frac{3}{4}\left(\frac{7}{5}\right) = \frac{21}{20} \qquad \frac{3}{4} \div \left(\frac{1}{2}\right) = \frac{3}{4}\left(\frac{2}{1}\right) = \frac{3}{2 \cdot 2} \cdot \frac{2}{1} = \frac{3}{2}$$

The rules of signs for division are the same as division of decimal numbers, including whole numbers. Here is a review of those rules.

Rules of Signs for Division

The rules of signs for fractions are the same as the rules of signs developed from the bean arithmetic.

A **positive** number divided by a **positive** number = a **positive** number.
A **positive** number divided by a **negative** number = a **negative** number.
A **negative** number divided by a **positive** number = a **negative** number.
A **negative** number divided by a **negative** number = a **positive** number.

Example 7:

$$45 \div \frac{-3}{5} = 45\left(-\frac{5}{3}\right) = 15 \cdot 3 \cdot \left(-\frac{5}{3}\right) = -75$$

$$-12 \div \frac{5}{3} = -12\left(\frac{3}{5}\right) = -\frac{36}{5}$$

$$\left(-\frac{3}{4}\right) \div \left(-\frac{5}{7}\right) = \left(-\frac{3}{4}\right) \div \left(-\frac{7}{5}\right) = \frac{21}{20}$$

$$-\frac{3}{4} \div \left(-\frac{1}{2}\right) = \left(-\frac{3}{4}\right) \cdot (-2) = \frac{3}{2}$$

Exercise 3--Division of Fractions

Purpose of the Exercise

- Practice division of fractions and rules of signs regarding division

Without a calculator, restate the following division problems as division; then compute the answer.

Restated as Multiplication — Answer

1. $\frac{2}{3} \div \frac{3}{8} =$ ____________ = ____________
2. $\frac{6}{7} \div \frac{7}{12} =$ ____________ = ____________
3. $35 \div \frac{5}{4} =$ ____________ = ____________
4. $\frac{17}{5} \div 2 =$ ____________ = ____________
5. $\frac{4}{5} \div 5 =$ ____________ = ____________
6. $\frac{3}{7} \div \frac{1}{3} =$ ____________ = ____________
7. $\frac{2}{9} \div 2 =$ ____________ = ____________
8. $\frac{1}{3} \div \frac{1}{3} =$ ____________ = ____________
9. $\frac{7}{8} \div \frac{7}{8} =$ ____________ = ____________
10. $\frac{3}{4} \div \frac{4}{3} =$ ____________ = ____________
11. $\frac{7}{100} \div 50 =$ ____________ = ____________
12. $\frac{11}{19} \div \frac{1}{11} =$ ____________ = ____________
13. $\frac{7}{100} \div \frac{1}{50} =$ ____________ = ____________
14. $\frac{2}{13} \div 13 =$ ____________ = ____________

For each of the following divisions, determine whether the quotient is positive or negative.

15. $\left(\frac{137}{5}\right) \div \left(\frac{-4}{5}\right)$ positive negative
16. $-\left(\frac{7}{5}\right) \div \left(\frac{4}{5}\right)$ positive negative
17. $\left(\frac{-57}{52}\right) \div \left(\frac{41}{59}\right)$ positive negative
18. $\left(\frac{-57}{52}\right) \div \left(\frac{-41}{59}\right)$ positive negative
19. $\left(\frac{-17}{5}\right) \div \left(\frac{4}{5}\right)$ positive negative
20. $-\left(\frac{-17}{5}\right) \div \left(\frac{4}{5}\right)$ positive negative
21. $-(27) \div \left(\frac{-41}{59}\right)$ positive negative
22. $(-27) \div \left(\frac{-41}{59}\right)$ positive negative
23. $-\left(\frac{-17}{5}\right) \div \left(-\frac{4}{5}\right)$ positive negative
24. $\frac{18}{-9}$ positive negative
25. $\frac{-21}{-33}$ positive negative
26. $-\frac{-21}{-33}$ positive negative

For each of the following, first determine the sign of the quotient; then divide or reduce the fractions as indicated.

27. $\left(\frac{7}{5}\right)\div\left(\frac{-4}{5}\right)=$ ____________

28. $\left(-\frac{15}{5}\right)\div\left(\frac{-4}{5}\right)=$ ____________

29. $(-100)\div\left(-\frac{2}{25}\right)=$ ____________

30. $(-5)\div\left(\frac{2}{5}\right)=$ ____________

31. $-\left(\frac{7}{5}\right)\div 70=$ ____________

32. $\left(\frac{14}{3}\right)\div(-6)=$ ____________

33. $\frac{-14}{-7}=$ ____________

34. $\frac{27}{-15}=$ ____________

35. $\frac{27}{15}=$ ____________

36. $\frac{-100}{-85}=$ ____________

37. Without a calculator, add, subtract, multiply, or divide the following fractions as indicated.

a. $\frac{4}{5}-\frac{2}{3}=$ ____________

b. $\frac{27}{53}-\frac{32}{53}=$ ____________

c. $\left(\frac{25}{3}\right)\left(\frac{12}{15}\right)=$ ____________

d. $1-\frac{15}{22}=$ ____________

e. $(3)\left(\frac{4}{5}\right)=$ ____________

f. $3+\frac{12}{5}=$ ____________

g. $\left(\frac{5}{9}\right)\div(3)=$ ____________

h. $\frac{3}{4}-2=$ ____________

38. In each of the following, determine if the answer is negative or positive.

a. $\left(\frac{12}{5}\right)\left(\frac{-99}{287}\right)$ positive negative

b. $-\left(\frac{-78}{99}\right)\left(\frac{12}{11}\right)$ positive negative

c. $\left(\frac{12}{5}\right)-\left(\frac{-99}{287}\right)$ positive negative

d. $\left(\frac{-78}{99}\right)+\left(-\frac{12}{11}\right)$ positive negative

e. $\left(\frac{-14}{99}\right)\div\left(-\frac{22}{87}\right)$ positive negative

f. $\frac{-10}{17}-\left(-\frac{7}{9}\right)$ positive negative

39. In each of the following, without doing the computations, determine whether the answer is

(i) greater than 0 but less than or equal to 1
(ii) greater than 1 but less than or equal to 2
(iii) greater than 2 but less than or equal to 3
(iv) none of the above

a. ____________ $\frac{4}{5}-\frac{2}{3}$

b. ____________ $3-\frac{17}{19}$

c. ____________ $\frac{28}{9}-\frac{2}{3}$

d. ____________ $\frac{18}{17}+\frac{291}{207}$

e. ____________ $\frac{4}{5}-1$

f. ____________ $\frac{5}{4}-1$

g. ____________ $\frac{5}{4}+\frac{4}{3}$

h. ____________ $\frac{2}{9}+2$

i. ____________ $\frac{7}{9}+\frac{11}{12}$

j. ____________ $\frac{15}{14}-\frac{2}{3}$

Section 4-8 *Estimation and Practice--Arithmetic with Fractions*

This section consists of three sets of exercises designed to emphasize what a change in the denominator does to a fraction, to practice arithmetic involving fractions, and to use the calculator to approximate answers in decimal form.

Exercise 1--The Size of the Denominator vs. the Size of the Fraction

Purpose of the Exercise

- To generalize what we have learned about division of fractions

For each of the following pairs, circle the larger number.

1. 1 $\qquad \frac{1}{2}$
2. 1 $\qquad \frac{1}{1/2}$
3. $\frac{3}{2/3}$ $\qquad \frac{3}{1/3}$
4. $\frac{48}{4}$ $\qquad \frac{48}{1/4}$
5. $\frac{1}{2}$ $\qquad \frac{1}{3}$
6. $\frac{48}{1/5}$ $\qquad \frac{48}{1/4}$
7. $\frac{3}{1/3}$ $\qquad \frac{3}{1/2}$
8. $\frac{2897}{1/9}$ $\qquad \frac{2897}{5}$

For each of the following pairs, write the larger denominator in the blank and circle the larger fraction. Each letter represents a particular positive number. If you are having difficulty determining which fraction is larger, pick a positive number to substitute into the two fractions.

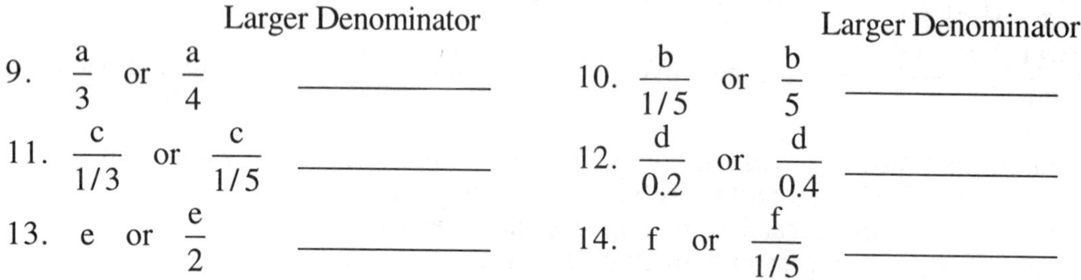

		Larger Denominator			Larger Denominator
9.	$\frac{a}{3}$ or $\frac{a}{4}$	__________	10.	$\frac{b}{1/5}$ or $\frac{b}{5}$	__________
11.	$\frac{c}{1/3}$ or $\frac{c}{1/5}$	__________	12.	$\frac{d}{0.2}$ or $\frac{d}{0.4}$	__________
13.	e or $\frac{e}{2}$	__________	14.	f or $\frac{f}{1/5}$	__________

When using mathematics in real life, we will generally use a calculator for the tedious work. It is important that we recognize reasonable answers whether our work is with pencil and paper or using a calculator, because it is easy to make a mistake by either calculation means.

The following exercises are a check on the concepts learned so far. The first problems check your ability to estimate a reasonable answer visually. Remember when the numerator is larger than the denominator, the fraction is greater than 1. If the numerator is more than twice the denominator, then the fraction is greater than 2. The pattern continues for the numerator being more than 3 times the denominator, 4 times the denominator, and so on. For some of the sums and differences, a number line might be helpful in determining the answer.

Exercise 2--Number Sense

Purpose of the Exercise

- To reinforce the visual sense of fractions
- To reinforce the number sense of addition, subtraction, and multiplication of positive fractions

For each of the following, circle the fraction that best represents the shaded portion of the rectangle.

1. 1/2 3/4 7/8

2. 1/3 1/2 2/3

3. 2/3 3/4 5/6

4. 1/2 2/3 3/4

5. 1/4 1/3 1/2

Without a calculator determine whether each of the following fractions is
(a) greater than 0 but less than or equal to 1 (b) greater than 1 but less than or equal to 2
(c) greater than 2 but less than or equal to 3 (d) none of the above

The first one is done as an example.

ANSWER			ANSWER		
a	6.	$\frac{99}{100}$	______	7.	$\frac{7}{8}$
______	8.	$\frac{2}{3}$	______	9.	$\frac{107}{99}$
______	10.	$\frac{27}{19}$	______	11.	$\frac{107}{10}$
______	12.	$\frac{17}{3}$	______	13.	$\frac{3}{17}$
______	14.	$\frac{2001}{1900}$	______	15.	$\frac{6}{2}$
______	16.	$\frac{15}{7}$	______	17.	$\frac{289}{290}$
______	18.	$\frac{7}{9}$	______	19.	$\frac{10}{7}$
______	20.	$\frac{1}{2} + \frac{1}{4}$	______	21.	$\frac{1}{2} + \frac{1}{3}$

______	22. $\frac{3}{4} + \frac{2}{3}$	______	23. $\frac{1}{3} + \frac{3}{4}$
______	24. $\frac{1}{4} + \frac{2}{3}$	______	25. $\frac{2}{3} - \frac{1}{4}$
______	26. $\frac{1}{4} - \frac{2}{3}$	______	27. $\frac{5}{6} - \frac{3}{5}$
______	28. $1\frac{1}{3} - \frac{1}{4}$	______	29. $\frac{10}{3} - \frac{3}{5}$
______	30. $\frac{10}{7} + \frac{3}{5}$	______	31. $\frac{1}{2} + \frac{2}{4}$

Exercise 3--Practice with Fractions

Purpose of the Exercise

- To practice arithmetic involving fractions
- To reinforce rules of signs
- To recognize the symbols of arithmetic

In each of the following problems, write whether the arithmetic problem is
addition/subtraction of fractions with the same denominator
addition/subtraction of fractions with different denominators
multiplication of fractions
division of fractions

Compute the arithmetic problems, showing work where appropriate. Reduce answers, if needed. The first two are done as an examples.

	Type of Problem	Pos/Neg?	Computation and Answer
1. $\left(\frac{25}{14}\right)\left(\frac{2}{17}\right)$	Multiplication	Positive	$\left(\frac{25}{14}\right)\left(\frac{2}{17}\right)=\left(\frac{25}{7\cdot 2}\right)\left(\frac{2}{17}\right)=\frac{25}{119}$
2. $\frac{14}{3}-\frac{2}{5}$	Subtraction of fractions with different denominators	Positive	$\frac{14}{3}\left(\frac{5}{5}\right)-\frac{2}{5}\left(\frac{3}{3}\right)=\frac{70}{15}-\frac{6}{15}=\frac{64}{15}$
3. $\frac{5}{7}+\frac{1}{7}$			
4. $\left(\frac{5}{7}\right)\left(\frac{1}{7}\right)$			
5. $\left(\frac{5}{7}\right)-\left(\frac{1}{7}\right)$			
6. $3-\frac{5}{9}$			
7. $\frac{5}{9}\div\frac{2}{3}$			
8. $\left(\frac{5}{9}\right)\left(\frac{2}{3}\right)$			
9. $\frac{5}{9}-\frac{2}{3}$			
10. $\left(\frac{5}{9}\right)-\left(-\frac{2}{3}\right)$			

11. $10 \div \frac{7}{10}$			
12. $10 - \frac{7}{10}$			

Exercise 4--Using the Calculator to Approximate Answers

Purpose of the Exercise

- To practice arithmetic involving fractions by converting the fractions to decimal form

Materials Needed

- Calculator

Using a calculator as needed, convert each of the following fractions to decimal approximations, rounded to 4 decimal places, if needed. Then compute the arithmetic. Round the answers to 4 decimal places. The problems are the same as in Exercise 3. Convert the answers from Exercise 3 to decimal form and compare the answers. If the answers are not the same (up to 3-decimal place accuracy), check your work with fractions as well as the work on the calculator to find the discrepancy.

	Problem Rewritten	Answer	Above Answer in Decimal Form
1. $\left(\frac{25}{14}\right)\left(\frac{2}{17}\right)$	(1.7857)(0.1176)	0.2100	$\frac{25}{119}$ 0.2101
2. $\frac{14}{3} - \frac{2}{5}$	4.6667 - 0.4	4.2667	$\frac{64}{15}$ 4.2667
3. $\frac{5}{7} + \frac{1}{7}$			
4. $\left(\frac{5}{7}\right)\left(\frac{1}{7}\right)$			
5. $\left(\frac{5}{7}\right) - \left(\frac{1}{7}\right)$			
6. $3 - \frac{5}{9}$			
7. $\frac{5}{9} \div \frac{2}{3}$			
8. $\left(\frac{5}{9}\right)\left(\frac{2}{3}\right)$			
9. $\frac{5}{9} - \frac{2}{3}$			
10. $\left(\frac{5}{9}\right) - \left(-\frac{2}{3}\right)$			
11. $10 \div \frac{7}{10}$			
12. $10 - \frac{7}{10}$			

Section 4-9 *Solving Equations Involving Fractions*

When solving equations involving fractions, the rules of equations do not change. The order in which we apply the rules, however, may change. In the past, we usually started to solve an equation by combining like terms on the same side of the equation. When there are fractions in the equation, it generally makes our work easier if we first multiply both sides of the equation by an appropriate number and then apply the distributive law to "clear out" the fractions. Combining like terms is easier when there are no fractions than when fractions are present. Before working some examples, we will review rules of equations and the process of finding a least common denominator.

Rule 1 for Solving Equations
We can add the same value to both sides of the equation without changing the balance in the equation.

Rule 2 for Solving Equations
We can subtract the same value from both sides of the equation without changing the balance in the equation.

Rule 3 for Solving Equations
We can multiply both sides of the equation by the same value without changing the balance in the equation.

Rule 4 for Solving Equations
We can divide both sides of the equation by the same nonzero value without changing the balance in the equation.

Since fractions many times give us problems, we will first clear out the fractions by multiplying both sides of the equation by the least common denominator of all fractions on both sides of the equation. After that, we will solve equations the same as we have before.

To **solve equations** involving addition/subtraction and multiplication/division in the same equation, it is helpful to

First: Clear out the fractions by multiplying both sides of the equation by the least common denominator of all fractions in the equation. Use the distributive law as needed on each side of the equation for the multiplication.

Second: Use the distributive law to do multiplications on each side of the equation.

Third: Combine any like terms on each side of the equation separately.

Fourth: As we have done previously, add or subtract to get like terms together on the same side of the equation. Combine like terms in the new equation.

Fifth: Use the multiplication and division rules to solve for the variable.

As with other equations, it is important to check the solution by substituting the solution into the original equation to determine if the left side equals the right side. When checking equations that have fractions in them, it is often tedious to check the answer by pencil-and-paper methods. Another way to check the answer is to work with a calculator. If your calculator has fraction capabilities, you can input the fraction in exact form and evaluate the left and right sides of the equation to determine if they are equal. An alternative is to convert all fractions to decimal form before evaluating the left and right sides of the equation. When checking a solution by using decimal forms of fractions, it is generally a good idea to round the fractions to several decimal places earlier in the computations and save the more severe rounding for the final answer. When we use decimal approximations for fractions the equations may not check out exactly; however, even with rounding errors, checking determines if the answer is reasonable.

Remember, when solving an equation for a variable, the objective is to use the rules of equations to transform the equation into the form, "the variable = a number."

Example 1: Solve $\frac{5x}{7} = \frac{2}{3}$ for x.

First: Multiply both sides of the equation by the least common denominator of the two fractions.

$$21\left(\frac{5x}{7}\right) = 21\left(\frac{2}{3}\right)$$ Multiply both sides by 21.

$$\frac{7 \cdot 3}{1}\left(\frac{5x}{7}\right) = \frac{7 \cdot 3}{1}\left(\frac{2}{3}\right)$$ Factor 21 to divide out common factors.

$$\left(\frac{7}{7}\right)\frac{3 \cdot 5x}{1} = \left(\frac{3}{3}\right)\frac{7 \cdot 2}{1}$$

$$15x = 14$$ Divide out the common factors on each side.

Since there is only one term per side, there is nothing to do in Steps 2, 3, and 4.

Fifth: Divide both sides of the equation by 15.

$$\frac{15x}{15} = \frac{14}{15}$$

$$x = \frac{14}{15} \approx 0.933$$

To check, substitute x = 0.933 into the original equation, $\frac{5x}{7} = \frac{2}{3}$.

$$\frac{5(0.933)}{7} = \frac{2}{3}$$

$$0.6667 \approx 0.6667$$

The two sides are equal or approximately equal when the rounded answer is substituted for x.

Example 2: Solve $\frac{3}{4}t - \frac{1}{3} = 4 - \frac{t}{2}$ for t.

First: The three fractions in the equation have denominators, 4, 3, and 2. The least common denominator is 12. We will multiply both sides of the equation by 12.

$$12\left(\frac{3}{4}t - \frac{1}{3}\right) = 12\left(4 - \frac{t}{2}\right)$$

Second: There are two terms per side, so use the distributive law to multiply both sides of the equation by the least common denominator.

$$12\left(\frac{3}{4}t\right) - 12\left(\frac{1}{3}\right) = 12(4) - 12\left(\frac{t}{2}\right)$$

$$9t - 4 = 48 - 6t$$

Third: There are no like terms to combine.

Fourth: $9t - 4 + \mathbf{6t} + \mathbf{4} = 48 - 6t + \mathbf{6t} + \mathbf{4}$ Add 6t and 4 to both sides of the equation.

$$15t = 52$$ Combine like terms.

Fifth:

$$\frac{15t}{15} = \frac{52}{15}$$

$$t = \frac{52}{15}$$

$$\approx 3.467$$

To check the solution, substitute 3.467 for t into the equation,

$$\frac{3}{4}t-\frac{1}{3}=4-\frac{t}{2}$$

$$\frac{3}{4}(3.467)-\frac{1}{3}=4-\frac{3.467}{2}$$

$$2.2669 \approx 2.2665$$

The left and right sides did not come out exactly the same. That is because of the round-off error in converting $\frac{52}{15}$ to decimal form.

Exercise 1--Solving Equations

Purpose of the Exercise

- To practice solving equations involving fractions

Materials Needed

- A calculator

In each of the following equations, record the number of terms per each side. On a separate sheet of paper solve each equation, clearly showing the equation and your work. Describe in words what you did in each step. Be careful with the rules of signs and rules of fractions. Use a calculator as needed and when appropriate put your solution in decimal form rounded to 2 decimal places. Check your work by substituting your solution into the original equation. (If you are rounding off answers or other fractions, your sides may not be exactly the same.)

1. $\frac{t}{5}=\frac{3}{4}$
   no. of terms left side ________
   no. of terms right side ________

2. $\frac{-m}{11}=\frac{3}{22}$
   no. of terms left side ________
   no. of terms right side ________

3. $\frac{-m}{11}=\frac{3}{22}-2$
   no. of terms left side ________
   no. of terms right side ________

4. $\frac{-7}{5}+6x=4-\frac{1}{5}$
   no. of terms left side ________
   no. of terms right side ________

5. $3\left(\frac{2}{3}-y\right)=4$
   no. of terms left side ________
   no. of terms right side ________

6. $\frac{1}{9}=2+\frac{x}{3}$
   no. of terms left side ________
   no. of terms right side ________

7. $\frac{3a}{500}=-500$
   no. of terms left side ________
   no. of terms right side ________

8. $\frac{1}{5}x=5$
   no. of terms left side ________
   no. of terms right side ________

9. $\frac{x}{5} - \frac{3}{5} = 4$

no. of terms left side ________

no. of terms right side ________

10. $\frac{x}{5} - \frac{3}{5} = 4x$

no. of terms left side ________

no. of terms right side ________

11. $\frac{1}{99} = 2x$

no. of terms left side ________

no. of terms right side ________

12. $\frac{1}{99} = 2 + x$

no. of terms left side ________

no. of terms right side ________

13. $\frac{t}{4} - \frac{5}{2} = \frac{1}{2}$

no. of terms left side ________

no. of terms right side ________

14. $\frac{t}{3} + \frac{t}{6} = \frac{1}{2}$

no. of terms left side ________

no. of terms right side ________

15. $\frac{7}{9} = \frac{5x}{4}$

no. of terms left side ________

no. of terms right side ________

16. $3 + \frac{2x}{3} = 5$

no. of terms left side ________

no. of terms right side ________

17. $x - \frac{3}{4} = 1$

no. of terms left side ________

no. of terms right side ________

18. $1 - \frac{3}{4} = x$

no. of terms left side ________

no. of terms right side ________

Review of Terms

- The **numerator** of a fraction is the top number of the fraction. The numerator is the dividend. (page 241)
- The **denominator** of a fraction is the bottom number of the fraction. The denominator is the divisor. (page 241)
- An **improper fraction** is one in which the numerator is larger than or equal to the denominator. An improper fraction is greater than or equal to 1. (pages 241, 246)
- A **mixed number** consists of a whole number and a fraction. (pages 241, 246)
- A fraction is in **reduced form** when it is written with the smallest possible numerator and denominator. A fraction is in reduced form when 1 is the greatest common factor of the numerator and denominator. (page 246)
- **Equivalent fractions** are two fractions that represent the same amount. (page 246)
- When a fraction is converted to decimal form and a particular pattern of digits after the decimal point repeats no matter how far we carry out the division, the number is a **repeating decimal.** (page 253)
- When a fraction is converted to decimal form and after a certain number of decimal places, the remainder is zero, we say the number is a **terminating decimal.** (page 253)
- A **rational number** is any number that can be written with a repeating decimal pattern or with a last (terminal) decimal place. Any rational number can also be written as a fraction in which the numerator and denominator are integers. (page 255)
- An **irrational number** is not rational. (page 255)
- A **common denominator** of several fractions is a number that all of the denominators will divide evenly into. (page 282)
- A **least common denominator (LCD)** of several fractions is the least common multiple of the denominators of the fractions. (page 282)
- The **reciprocal** of a fraction is another fraction with the numerator and denominator reversed. (page 292)

Processes to Review

Reducing a Fraction (page 251)

1. Factor the numerator and denominator of the fraction.
2. Find the greatest common factor (GCF) of the numerator and denominator.
3. Divide the numerator and denominator by the GCF.
4. Multiply the remaining factors in the numerator.
5. Multiply the remaining factors in the denominator.

To **build up a fraction**, multiply both the numerator and the denominator by the same number. (page 252)

To **convert a fraction to a decimal**, divide the numerator by the denominator using either a calculator or a division algorithm. (page 253)

Multiplying Fractions (page 260)

1. Factor each numerator and each denominator.
2. Divide factors that are common to the numerators and the denominators (of either fraction--or you can divide across fractions).
3. Multiply the remaining factors in the numerators.
4. Multiply the remaining factors in the denominators.

To **add or subtract fractions with the same denominator**, add or subtract the numerators, keeping the denominator the same. (page 270)

To **add/subtract a fraction to/from a whole number**, first change the whole number to fraction form, selecting the denominator to be the same as that of the other fraction; then add/subtract fractions with the same denominator. (page 270)

To **convert a mixed number to an improper fraction**, rewrite the mixed number as a whole number plus a fraction and follow the process for adding a whole number and a fraction. (page 271)

To **convert an improper fraction to a mixed number**, use an algorithm to divide the denominator into the numerator. The quotient is the whole number part of the mixed number and the remainder is the numerator of the fractional part. The denominator remain the same. (page 271)

Finding the Least Common Denominator (page 282)

1. Factor each denominator into prime factors.
2. If a factor appears more than once in any one denominator, write that factor in exponential notation.
3. The least common denominator is the product of the factors that appear in all of the denominators. If a factor has an exponent larger than one in at least one of the denominators, include that factor to the highest power that appears in any one denominator.

Adding and Subtracting Fractions with Different Denominators (page 282)

1. Find the least common denominator of all of the fractions.
2. Find an equivalent fraction for each fraction so that all fractions have the same denominator.
3. Follow the process and rules of signs for adding and subtracting fractions with the same denominators.
4. Reduce the answer if possible.

To **add or subtract mixed numbers,** change each mixed number to an improper fraction; then add or subtract. (page 283)

To **divide fractions**, find the reciprocal of the divisor then restate the division as multiplication of fractions and follow the procedure for multiplying fractions. (page 293)

Summary Test--Chapter 4

Part 1: Do not use a calculator.

1. In the following, do the arithmetic as indicated. Reduce answers where appropriate. Show work.

a. $\frac{6}{7} - \frac{3}{4}$ b. $\frac{5}{7} - \frac{-5}{7}$

c. $(9)\left(\frac{2}{3}\right)$ d. $\left(\frac{3}{5}\right)\left(\frac{5}{6}\right)$

e. $\left(\frac{3}{5}\right) \div \left(\frac{5}{6}\right)$ f. $\left(\frac{3}{5}\right) - \left(\frac{5}{6}\right)$

2. Determine if each of the following fractions are

a. between 0 and 1 b. equal to 1
c. between 1 and 2 d. equal to 2
e. between 2 and 3 f. equal to 3

_______ $\frac{17}{12}$ _______ $\frac{245}{101}$ _______ $\frac{87}{98}$

_______ $\frac{300}{100}$ _______ $\frac{300}{111}$ _______ $\frac{3087}{3187}$

3. In each of the following, determine if the answer is positive or negative.

a. $\left(\frac{-15}{9}\right)-\left(\frac{-12}{9}\right)$

b. $(-8)\left(\frac{7}{29}\right)$

c. $6-\frac{17}{2}$

d. $-\frac{9}{23}+\frac{11}{12}$

4. In each of the following, circle the answer that is closest to correct.

a. $4\frac{9}{13}-1\frac{3}{15}$ 3.5 5 6

b. $\frac{7}{9}-\frac{21}{55}$ 1.2 0.9 0.3

5. Write each of the following fractions in decimal form.

a. $\frac{3}{4}$ = ________ b. $\frac{18}{100}$ = ________ c. $\frac{7}{5}$ = ________

6. Solve the following equations for x.

a. $\frac{3}{4}x - 5 = 2x + 1$

b. $x-\frac{3}{5}=\frac{1}{2}x+3$

Part 2--Use a calculator as needed.

7. Convert the following fractions to decimal form, rounded to 3 decimal places.

a. $\frac{88}{93} \approx$ ____________ b. $\frac{3567}{146} \approx$ ____________ c. $\frac{1298}{6754} \approx$ ____________

8. In each of the following, circle the fraction that is larger.

a. $\frac{98}{103}$ or $\frac{79}{81}$ b. $\frac{256}{1067}$ or $\frac{65}{281}$ c. $\frac{316}{604}$ or $\frac{7089}{13999}$

9. Do the arithmetic as indicated on the following fractions. Leave the answer in decimal form, rounded to 2 decimal places.

a. $\frac{783}{913}-\frac{354}{675}$

b. $\dfrac{\frac{4}{15}-\frac{3}{8}}{\frac{7}{9}-\frac{-9}{14}}$

10. Trish, Jody, Jesse, and Terry planned and produced a dance that cleared $1250 after expenses were paid. The four producers agreed to donate 1/3 of the profit to a charitable organization and to split the rest of the profit among the four of them. Answer the following questions. Show your work.

a. How much was donated to the charity?

b. How much was each producer's share of the profit?

Chapter 5 Ratios, Proportions, Unit Conversions and Rates

Key Topics

- Ratios and proportions
- Unit conversions--linear, time, metric, English, square
- Rates and rate conversions--miles per hour, miles per gallon, cost per person
- Scaling
- Slopes and angles

Materials Needed for the Exercises

- Large jar of red beans and white beans
- Small containers or bags
- 20-30 each, red and white beans
- Line level
- String
- Measuring tape, yardstick or meter stick, or ruler.
- Metric ruler
- Calculator
- Protractor
- Graph paper
- Map

Section 5-1 *Ratios and Proportions*

Mathematically speaking, a **ratio** of one number to another is a fraction in which the first number is the numerator and the second number is the denominator. We will define the more intuitive notion of a ratio of one object to another with the following examples.

Example 1:

The picture above represents 2 red beans and 3 white beans. Notice that there is a total of 5 beans.

The **ratio** of red beans to white beans is 2 to 3 (2 red beans to 3 white beans). This ratio can also be written as a fraction, 2/3. We also use the notation 2:3 to represent a 2 to 3 ratio.
The **ratio** of white beans to red beans is 3 to 2. This ratio can be written as 3/2 or 3:2.
The **ratio** of red beans to all beans is 2 to 5, which can be written as 2/5 or 2:5.
The **ratio** of white beans to all beans is 3 to 5, which can be written as 3/5 or 3:5.

Example 2: If there are 13 women and 9 men employed at a company, there is a total of 22 employees. We get the following ratios:
The ratio of women to men is 13 to 9, also written 13:9 or 13/9.
The ratio of men to women is 9 to 13.
The ratio of women to total number of employees is 13 to 22.
The ratio of men to total number of employees is 9 to 22.

Example 3:
Since $3/4 = 0.75$, the ratio 3:4, can also be expressed as 0.75:1
Since $7/9 \approx 0.778$, the ratio 7:9 can also be expressed as 0.778:1.

Exercise 1--Ratios

Purpose of the Exercise
- To understand ratios as fractions
- To understand that same ratios are expressed by equivalent fractions

Materials Needed
- Beans -- optional

Write the following ratios in fraction form, first not reduced, and then reduced, and then in decimal form. Compare the reduced form to the ratios listed in Example 1.

1. Double the number of red and white beans in Example 1.

	not reduced	reduced	decimal to one ratio
a. The ratio of red beans to white beans =	________	________	________
b. The ratio of white beans to red beans =	________	________	________
c. The ratio of red beans to total beans =	________	________	________
d. The ratio of white beans to total beans =	________	________	________

2. Triple the number of red and white beans in Example 1.

    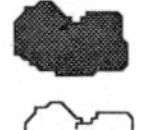  

	not reduced	reduced	decimal to one ratio
a. The ratio of red beans to white beans =	__________	__________	__________
b. The ratio of white beans to red beans =	__________	__________	__________
c. The ratio of red beans to total beans =	__________	__________	__________
d. The ratio of white beans to total beans =	__________	__________	__________

3. How do the reduced answers in Problems 1 and 2 compare to the ratios from Example 1?

____

If a ratio of a few red beans to white beans represents a ratio in a larger pot of beans, we can use the ratio to predict the number of red and white beans in the larger "pot." To do this we set up a proportion (an equation of ratios) and solve the proportion. We will see in the following examples how to solve the proportion and how to use the solution to make predictions.

> **Vocabulary**
> A **proportion** is an equation of the form: ratio 1 = ratio 2.

Example 4: If we assume the ratio of red to white beans in a large container of beans is 2 to 3, and we know there are 520 red beans in the container, we can determine (or estimate) the number of white beans in the container in the following way.

First: Set up the proportion: $\dfrac{2\text{ red beans}}{3\text{ white beans}} = \dfrac{520\text{ red beans}}{w\text{ white beans}}$

If the number of red beans is in the numerator on one side of the equation, they must also be in the numerator on the other side of the equation. We choose a letter to represent the unknown number of white beans in the container. In this case, we are using w for the unknown number of white beans.

After setting up the proportion including the units to assure proper placement of the numbers and variable, we rewrite the equation without the units before solving.

$$\frac{2}{3} = \frac{520}{w}$$

Second: The answer to the question is the value of w that makes the two fractions equivalent.

We sill discuss two strategies to find the value of w in the proportion above.

Strategy 1: Educated Guesses and Checks

1. Guess at the value of the variable.
2. Record the guess.
3. Substitute the value of the variable into the equation.
4. Determine whether the fractions on the left and right are the same (or very close.)
   If the two fractions are the same, then the value of the variable is the right answer, so the problem is solved.
   If the two fractions are different, determine if the value of the variable is too large or too small. Make another educated guess for the value of the variable and repeat the process. **Each wrong answer helps us to make a better guess the next time.**
   Continue this process of guessing and testing until you get the right value for the variable.

Example 5: Solve $\frac{2}{3} = \frac{520}{w}$ for w using strategy 1.

1. Guess a value for w that makes the two ratios the same. Since the denominator on the left is larger than the numerator but not quite twice double the numerator, our first guess will be a number larger than 520, but not as large as 1140 (which is double 520.)
2. **First guess for w: w = 600.** (The guess is recorded.)
3. Substitute w = 600 into the equation and see if the fractions are equivalent.
4. Does $\frac{2}{3} = \frac{520}{600}$?

An easy way to check is to use a calculator to convert both fractions to decimal.

$\frac{2}{3} \approx 0.666666667$ and $\frac{520}{600} \approx 0.86666667$ (The number of decimal places is different on different calculators.)

Comparing the decimal equivalents to the fractions, $\frac{520}{600}$ is too big. To make the fraction smaller, we need a larger denominator, so we will repeat the process with a larger value for w.

1. Make a second guess for the value of w. This time we will pick a number larger than 600.
2. **Second guess for w: w = 800.** (The guess is recorded.)
3. Substitute w = 800 into the equation. Does $\frac{2}{3} = \frac{520}{800}$?
4. $\frac{2}{3} \approx 0.666666667$ and $\frac{520}{800}$ 0.65

Now the fraction is too small but is getting close. We will pick a value of w between 600 and 800 but closer to 800 (since $\frac{520}{800}$ is closer to $\frac{2}{3}$ than $\frac{520}{600}$).

This is the idea of how to make and adjust educated guesses. Continuing in this process using a calculator, we quickly find that $\frac{2}{3} \approx \frac{520}{780}$, so there are approximately 780 white beans in the container.

Strategy 2: Solve an Algebraic Equation

Use algebra to solve the equation $\frac{2}{3} = \frac{520}{w}$.

As discussed before, the **solution of an algebraic equation** is the value for the variable that makes the equation "balanced"--that is, the value of the variable that makes both sides of the equation equal. There are several steps one can take in changing the look of the equation while still maintaining the balance. Let's review the rules of equations.

Rules of Equations

1. We can add the same number to both sides of the equation.
2. We can subtract the same number from both sides of the equation.
3. We can multiply both sides of the equation by the same number.
4. We can divide both sides of the equation by the same non-zero number.

Example 6: Solve the equation, $\frac{2}{3} = \frac{520}{w}$ using the rules of equations.

Multiply both sides of the equation by 3w. 3w was chosen as a multiplier because it is the least common denominator of the two fractions in the problem. Multiplying both fractions by the least common denominator "clears out" the fractions. See below and review fractions if necessary.

$\frac{2}{3}(3w) = \frac{520}{w}(3w)$ Multiply both sides by 3w.

$\left(\frac{2}{3}\right)\left(\frac{3w}{1}\right)=\left(\frac{520}{w}\right)\left(\frac{3w}{1}\right)$ The problem restated.

$\left(\frac{2w}{1}\right)\left(\frac{3}{3}\right)=\left(\frac{520\cdot 3}{1}\right)\left(\frac{w}{w}\right)$ Factors of the numerator and denominator rearranged.

$\left(\frac{2w}{1}\right)(1)=\left(\frac{520\cdot 3}{1}\right)(1)$ 3/3 = 1 and w/w = 1.

$(2)(w) = (520)(3)$ After the above divisions, there are no fractions left.

$2w = 1560$ To isolate w, divide both sides of the equation by 2.

$$\frac{2w}{2} = \frac{1560}{2}$$

$w = 780$ Solution after dividing.

Important last step: Make sure the process was done correctly by checking the answer. Substitute w equals 780 into the original equation to see if the left side equals the right side. An easy way to make the check is with the calculator. Both fractions come out the same on the calculator, so the answer is correct. The calculator may give a rounded answer, but not to worry--it rounds off to more significant digits than we usually need for the precision required.

$$\frac{2}{3} \approx 0.666666667 \text{ and } \frac{520}{780} \approx 0.666666667$$

Ratios are also commonly expressed as a fraction where one of the numbers is a decimal. We use the same strategies for solving these ratio problems.

Example 7--Using Strategy 1:

If the ratio of females to males in a particular college is 2.1 to 1 and there are 4352 males in the college, we can determine the number of females using either strategy.

We will use the "guesstimate" method first.

$$\frac{2.1 \text{ females}}{1 \text{ male}} = \frac{x \text{ females}}{4352 \text{ males}}$$

is the proportion. Now write the proportion with just the numbers and the variable.

$$\frac{2.1}{1} = \frac{x}{4352}$$

Since the ratio is 2.1 to 1, the value for x will be a little more than double 4352. We will take 8800 as the first estimate.

$$\frac{2.1}{1} = 2.1 \qquad \frac{8800}{4352} \approx 2.022058824$$

8800 is too small, so we will make another estimate.

After substituting several values for x on the calculator, we find that

$$\frac{9140}{4352} \approx 2.100183824 \quad \text{and} \quad \frac{9139}{4352} \approx 2.099954044.$$

The last two ratios round to 2.1, so there are either 9139 or 9140 female students at the college. Since students do not come in parts, we choose either of the numbers for the answer. There are about 9140 female students at the college.

Example 8--Using Strategy 2:

Now we will solve the same ratio/proportion problem by using an algebraic equation.

$$\frac{2.1}{1} = \frac{x}{4352}$$

Multiply both sides of the equation by 4352. Get your calculator handy to help multiply.

$$\frac{2.1}{1}(4352) = \frac{x}{4352}(4352)$$

Multiply on the left side of the equation and divide common factors on the right.

$$9139.2 = x.$$

Students do not come in parts of students, so we conclude there are 9139 female students at the college.

Example 9--Using Strategy 1:

If the ratio of females to males in a particular college is 2.1 to 1 and there are 4352 students in the college, we can determine the number of males and females using either strategy.

Ratio of females to males: 2.1 to 1. ▢▢▯ to ▢

females males

Ratio of females to total number of students: 2.1 to 3.1.

$$\frac{2.1 \text{ females}}{3.1 \text{ students}} = \frac{x \text{ females}}{4352 \text{ students}}$$

is the proportion. Now write the proportion without the words.

$$\frac{2.1}{3.1} = \frac{x}{4352}$$

Since the ratio is 2.1 to 3.1, the value for x will be less than 4352 but more than half of 4352. We will take 3000 as the first estimate.

$$\frac{2.1}{3.1} \approx 0.677 \qquad \frac{3000}{4352} \approx 0.689$$

3000 is too large, so we will make another estimate.

Using the calculator to make further estimations, we find that $\frac{2948}{4352} \approx 0.677$.

There are about 2948 female students. Since the rest of the students are male, there are about 1404 male students.

Example 10--Using Strategy 2:

Now we will solve the same ratio/proportion problem by using an algebraic equation.

$$\frac{2.1}{3.1} = \frac{x}{4352}$$

Multiply both sides of the equation by (4352)(3.1). Get your calculator handy.

$$\frac{2.1}{3.1}(4352)(3.1) = \frac{x}{4352}(4352)(3.1)$$

Divide common factors and then multiply the remaining factors on both sides.

$$9139.2 = 3.1\,x.$$

Divide both sides of the equation by 3.1.

$$\frac{9139.2}{3.1} = \frac{3.1x}{3.1}$$

Use a calculator to divide the left side of the equation. On the right side, divide the common factors.

$$2948.129 \approx x$$

x represents the number of females, so there are 2948 females. This leaves 1404 male students.

For the following exercises you may need to review how to compute an average (Chapter 1).

Exercise 2--Beans in a Jar (Group)

Purpose of the Exercise

- To practice making predictions about large populations from a sample
- To learn about samples

Materials Needed

- One large jar filled with red and white beans for the whole class
- Small containers to hold beans, one for each group
- Calculator

The instructor tells the students the total number of beans in the jar.

1. Each group takes out a small scoop or handful of beans, counts the number of red beans and white beans in their sample, and writes the ratio of red to white beans as a fraction and then as a decimal. Each group will set up a proportion and estimate the number of red beans and the number of white beans in the large container. They should then write their ratios and proportion on the board along with their estimates of red beans and white beans in the jar.

2. The entire class will average (in decimal form) all the group answers for the numbers of red beans and white beans and compute the ratio of red to white beans.

3. Find out from the instructor the actual mix of red to white beans in the jar and compute the actual ratio in decimal form and compare answers and discuss the differences in the answers.

4. Make the following estimates.

If the ratio of red to white beans stays the same,

a. Estimate the number of red beans and white beans in a container that has 2357 beans.

b. Estimate the number of red beans and white beans in a container that has 19,467 beans.

c. Estimate the number of red beans and white beans in a container that has 145,768 beans.

d. Estimate the number of red beans and white beans in a container that has 378,926 beans.

5. Discussion questions
   a. Why did different groups get different estimates?
   b. Would you expect a better estimate from a smaller sample or a larger sample?
   c. Would you expect very accurate results if the beans were not very well mixed?
   d. When conducting opinion polls, would you expect the opinions you collected to reflect the greater population if you polled only your closest friends and relatives?
   e. When conducting opinion polls, would you expect the opinions you collect to reflect the greater population if a computer selected the sample by selecting phone numbers at random from a phone book?

For the following exercises, you may need to review these formulas.

Area of a rectangle: $A = LW$

Area of a circle: $A = \pi r^2$

Exercise 3--Comfortable Number of People in a Room (Group)

Purpose of the Exercise

- To use ratios to estimate room capacity

Materials Needed

- Measuring tape, yardstick, or meter stick, or ruler
- Calculator

Have the members of your group stand in a part of the room spacing themselves at a comfortable but easy conversational distance from each other. Measure the floor space you are occupying and compute and record the corresponding number of square feet in the area. Determine the ratio of the number of people to number of square feet of floor space.

Measure the room, compute the floor space of the entire room and set up a proportion and determine the number of people who would comfortably fit in the room if all of the furniture were removed.

Each group reports their calculations and findings to the entire class.

In the report to the class

1. Include the number of people in your group and the floor space required for all of you to stand comfortably.

2. Show the steps you went through to determine how many people would fit in the room.

3. State the estimated number of people who will fit in the room.

Section 5-2 *Unit Conversions--Metric and English Measurements and Weights*

From our common knowledge of money, we know that four quarters make up a dollar. From this same connection between quarters and dollars, we know that 1 quarter is 1/4 of a dollar. (That is why it is called a quarter--it is one-quarter of a dollar.)

Exercise 1--Quarters to Dollars

Purpose of the Exercise

- To practice money conversions in a familiar setting

Using your knowledge of money, answer the following questions.

1. 12 quarters are how many dollars? ____________________
2. 17 quarters are how many dollars? ____________________
3. 25 quarters are how many dollars? ____________________
4. 33 quarters are how many dollars? ____________________
5. If you get change for $7 in quarters, how many quarters do you get? ____________
6. If you change $7.50 into quarters, how many quarters do you get? ____________
7. If you change $12.75 into quarters, how many quarters do you get? ____________
8. If you change $152.25 into quarters, how many quarters do you get? ____________
9. 80 quarters are how many dollars? ____________________
10. 100 quarters are how many dollars? ____________________
11. If you change $123 into quarters, how many quarters do you get? ____________

From the correspondence between quarters and dollars we can write the following ratio:

$$1 = \frac{4 \text{ quarters}}{4 \text{ quarters}} = \frac{4 \text{ quarters}}{1 \text{ dollar}} = \frac{1 \text{ dollar}}{4 \text{ quarters}}$$

Since each of these ratios is equivalent to 1, we can use the ratio as a multiplier without changing the value of the original money amount. A ratio used in such a way is called a **conversion factor.**

An algebraic way to make the money conversions such as the ones above is to

1. Multiply the original amount by the appropriate conversion factor. Make sure the unit in the denominator of the conversion factor is the same unit as in the original quantity.
2. Cancel common units in the numerator and denominator just as you would divide common factors in a fraction.

This method is valuable for unit conversions that are not as intuitive as quarters to dollars. The algebraic method for making money conversions is shown in the examples below.

When changing quarters to dollars, the quarters must cancel out so the proper unit conversion factor will have quarters in the denominator of the fraction. When changing dollars to quarters, the dollars must cancel out, so the dollars must be in the denominator of the conversion factor.

Example 1: Convert 12 quarters to dollars using the algebraic method.
First, from our common knowledge of money, we know that 4 quarters equals 1 dollar, so 12 quarters = 3 dollars. Now let us set up the problem in algebraically.

GIVEN: 12 quarters **WANTED**: dollars **PATH**: quarters to dollars

$\frac{1 \text{ dollar}}{4 \text{ quarters}}$ and $\frac{4 \text{ quarters}}{1 \text{ dollar}}$ are both ratios equal to 1. We choose the first ratio as a multiplier so that the unit, quarters, will cancel.

Algebraic process: $12 \text{ quarters} \times \frac{1 \text{ dollar}}{4 \text{ quarters}} = \frac{12 \cancel{\text{quarters}}}{1} \times \frac{1 \text{ dollar}}{4 \cancel{\text{quarters}}}$

$= 3 \text{ dollars}$

Cancel the unit, quarters, and then multiply 12 by 1/4.
The unit left is dollars and 12 times 1/4 is 3. This leaves 3 dollars.

Example 2: Convert 12 dollars to quarters using the algebraic method.

GIVEN: 12 dollars **WANTED**: quarters **PATH**: dollars to quarters

Algebraic process: $12 \text{ dollars} \times \frac{4 \text{ quarters}}{1 \text{ dollar}} = \frac{12 \cancel{\text{dollars}}}{1} \times \frac{4 \text{ quarters}}{1 \cancel{\text{dollar}}}$

$= 48 \text{ dollars}$

Cancel the unit, dollars, and then multiply 12 by 4.
The unit left is dollars and 12 times 4 is 48. This leaves 48 dollars.

Exercise 2--Quarters/Dollar Conversions Using the Algebra Process

Purpose of the Exercise
- To use conversion factors

Do the conversions in Problems 1 through 10 by using the appropriate conversion factor.

1. 17 quarters are how many dollars? ____________________

   Given: __________ Wanted: __________ Path: ____________________
   Algebra process:

2. 25 quarters are how many dollars? ____________________

   Given: __________ Wanted: __________ Path: ____________________
   Algebra process:

3. 33 quarters are how many dollars? ____________________

   Given: __________ Wanted: __________ Path: ____________________
   Algebra process:

4. If you get change for $7 in quarters, how many quarters do you get? ___________

Given: ___________ Wanted: ___________ Path: ________________________
Algebra process:

5. If you change $7.50 into quarters, how many quarters do you get? ___________

Given: ___________ Wanted: ___________ Path: ________________________
Algebra process:

6. If you change $12.75 into quarters, how many quarters do you get? ___________

Given: ___________ Wanted: ___________ Path: ________________________
Algebra process:

7. If you change $152.25 into quarters, how many quarters do you get? ___________

Given: ___________ Wanted: ___________ Path: ________________________
Algebra process:

8. 80 quarters are how many dollars? ____________________

Given: ___________ Wanted: ___________ Path: ________________________
Algebra process:

9. 100 quarters are how many dollars? ____________________

Given: ___________ Wanted: ___________ Path: ________________________
Algebra process:

10. If you change $123 into quarters, how many quarters do you get? ___________

Given: ___________ Wanted: ___________ Path: ________________________
Algebra process:

Unit Conversions--Weights and Linear Units

1 pound (lb) = 454 grams (g) From this correspondence, we can set up two conversion factors:

$$1 = \frac{1\ lb}{454g} = \frac{454g}{1\ lb}$$

To convert from **grams to pounds**, grams have to cancel, so choose the conversion factor with grams in the denominator.

To convert from **pounds to grams**, choose the conversion factor with pounds in the denominator.

Example 3: Convert 17 pounds to grams.

To get an idea of what a reasonable answer is:

1. Decide if the number of grams will be more than 17 or fewer than 17. More or less? More because it takes a lot of grams to make up one pound.
2. Will the number of grams be a lot more or a lot less than 17? A lot more.

Now to compute: GIVEN: 17 lb WANTED: grams PATH: pounds to grams

$$17\,\cancel{lb} \times \frac{454g}{1\cancel{lb}} = 17 \times \frac{454}{1}\,g = 7718\ g$$

Answer: 17 pounds equals 7718 grams.

Example 4: Convert 392 grams to pounds.

1. Will the number of pounds be more or less than the number of grams? less, because a pound is larger than a gram and therfore it take fewer pounds to weigh the same as a given number of grams.
2. Will the number of pounds be more or less than 1? less

Now to compute: GIVEN: 392 grams WANTED: pounds PATH: grams to pounds

$$392\cancel{g} \times \frac{1lb}{454\cancel{g}} = 392/454\ \text{lb} = 0.863\ \text{lb (rounded)}$$

Answer: 392 grams equals approximately 0.863 pounds.

Sometimes there is more than one possible correct conversion factor to use. From the chart at the back of the chapter, we find that

1 inch = 2.54 centimeters and that **1 centimeter = 0.39 inches.**

This gives us two sets of conversion factors:

$$1 = \frac{1\text{ in.}}{2.54\text{ cm}} = \frac{2.54\text{ cm}}{1\text{ in.}} \quad \text{and} \quad 1 = \frac{0.39\text{ in.}}{1\text{ cm}} = \frac{1\text{ cm}}{0.39\text{ in.}}$$

Example 5: Convert 22 inches to centimeters:

Will the number of centimeters be more or less than 22? more, because a centimeter is smaller than an inch. Therefore it will take more of them to cover the same distance as inches.

GIVEN: 22 inches WANTED: centimeters PATH: inches to centimeters

$$(22\cancel{in.})\left(\frac{1\text{ cm}}{0.39\cancel{in.}}\right) = \frac{(22)(1)}{0.39}\text{ cm} \quad \underline{56.4\text{ cm}} \quad \underline{56\text{ cm}}\text{ or:}$$

$$(22\cancel{in.})\,\frac{2.54\text{ cm}}{1\cancel{in.}} = (22)\left(\frac{2.54}{1}\right)\text{cm} \quad \underline{55.8\text{ cm}} \approx \underline{56\text{ cm}}$$

Answer: 22 inches is approximately 56 centimeters.

Notice that the two answers in Example 5 are close but not exactly the same. The difference is because the original conversions use rounded numbers.

Exercise 3--Unit Conversions

Purpose of the Exercise

- To practice unit conversions

Materials Needed

- Conversion Chart (page 348)
- Calculator

For each of the conversions, fill out the data given, unit wanted, and the path. Decide whether the answer will be larger or smaller than the number given. Do the conversion, showing your algebraic work. Write your answer in a complete sentence.

1. 25 quarts to liters

GIVEN ____________ WANTED __________ PATH____________

Answer larger or smaller than 25? ______________________________

Conversion: ____

Answer: ____

2. 35 meters to yards

GIVEN ____________ WANTED __________ PATH____________

Answer larger or smaller than 35? ______________________________

Conversion: ____

Answer: ____

3. 293 miles to kilometers

GIVEN ____________ WANTED __________ PATH____________

Answer larger or smaller than 293? _____________________________

Conversion: ____

Answer: ____

4. 88 kilograms to pounds

GIVEN ____________ WANTED __________ PATH____________

Answer larger or smaller than 88? ______________________________

Conversion: ____

Answer: ____

5. 88 kilograms to ounces (Hint: There are two conversions.)

GIVEN ____________ WANTED __________ PATH____________

Answer larger or smaller than 88? ______________________________

Conversion: ____

Answer: ____

6. 7 teaspoons to tablespoons

GIVEN ____________ WANTED __________ PATH____________

Answer larger or smaller than 88? ______________________________

Conversion: ____

Answer: ____

7. 7 feet to centimeters

GIVEN ____________ WANTED __________ PATH____________

Answer larger or smaller than 7? _______________________________

Conversion: ____

Answer: ____

8. 89 inches to centimeters

GIVEN ____________ WANTED __________ PATH____________

Answer larger or smaller than 89? ______________________

Conversion: ______________________

Answer: ______________________

9. 89 grams to ounces

GIVEN ____________ WANTED __________ PATH____________

Answer larger or smaller than 89? ______________________

Conversion: ______________________

Answer: ______________________

10. 125 ounces to pounds

GIVEN ____________ WANTED __________ PATH____________

Answer larger or smaller than 125? ______________________

Conversion: ______________________

Answer: ______________________

11. 319 yards to centimeters

GIVEN ____________ WANTED __________ PATH____________

Answer larger or smaller than 319? ______________________

Conversion: ______________________

Answer: ______________________

12. 29 centimeters to millimeters

GIVEN ____________ WANTED __________ PATH____________

Answer larger or smaller than 29? ______________________

Conversion: ______________________

Answer: ______________________

13. 19 quarts to liters

GIVEN ____________ WANTED __________ PATH____________

Answer larger or smaller than 19? ______________________

Conversion: ______________________

Answer: ______________________

14. 195 days to years

GIVEN ____________ WANTED __________ PATH____________

Answer larger or smaller than 195? ______________________

Conversion: ______________________

Answer: ______________________

15. Make up and solve five other unit conversion problems with units not used above.

Section 5-3 *Unit Conversion--Linear vs. Square Units*

We have already worked with area and perimeter formulas for rectangles and other shapes and with different units of measurement. In this section, we will work more with the formulas and converting from one unit to another. As before, we must be aware of the type of unit we are working with, for example, inches vs. centimeters, but also linear units vs. square units.

Review of area formulas for rectangles and squares

Area of a rectangle: $\mathbf{A = LW}$
Area of a square is $\mathbf{A = s^2}$
Perimeter of a rectangle: $\mathbf{P = 2L + 2W}$
Perimeter of a square: $\mathbf{P = 4s}$

L and W are the length and width of the rectangle and s is the side of the square. L, W, s, and P are all linear measures, and A is measured in square units.

Ratios can be used to compare one linear unit to another linear unit, one square unit to another square unit, or one cubic unit to another cubic unit. To get the idea of how we do this, let's look at a diagram. We know that 3 feet = 1 yard. So if we draw (to scale) 1 foot and 1 yard, we can compare visually and determine the ratio of feet to yards .

To determine the ratio of feet to yards visually, look at the diagram. We can see that the ratio of feet to yards is 1 to 3, or $\frac{1}{3}$.

Example 1: Compute the ratio of feet to yards algebraically.

1. Write the ratio of 1 foot to 1 yard keeping the different units.
$$\frac{1 \text{ ft}}{1 \text{ yd}}$$
2. Write the ratio in the same units by substituting 3 feet in the place of 1 yard.
$$\frac{1 \text{ ft}}{3 \text{ ft}}$$
3. Cancel the unit. We are left with the ratio of 1 foot to 1 yard as 1 to 3.
$$\frac{1 \cancel{\text{ft}}}{3 \cancel{\text{ft}}} = \frac{1}{3}$$

1 foot = 12 inches. To determine the number of square inches in a square foot, look at the following drawing representing a square foot. The smaller squares represent square inches.

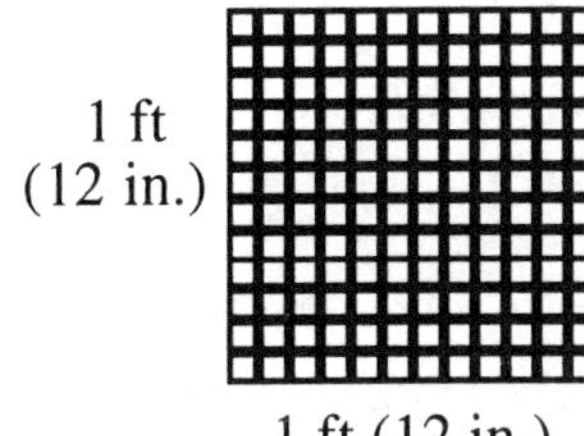

Dimensions in feet	Dimensions in inches
s = 1 ft	s = 12 in.
$A = 1 \text{ ft}^2$	$A = 144 \text{ in.}^2$

Example 2: Write the ratio of feet to inches.

1. Write the ratio as 1 foot to 1 inch. $\frac{1\text{ ft}}{1\text{ in.}}$
2. Convert the 1 foot to 12 inches. $\frac{12\text{ in.}}{1\text{ in.}}$
3. Cancel the unit, inch. $\frac{12\,\cancel{\text{in.}}}{1\,\cancel{\text{in.}}}$

Answer: The ratio of feet to inches is 12 to 1.
The ratio of inches to feet is 1 to 12. (If in doubt which way to write the ratio, 1 to 12 or 12 to 1, remember an inch is smaller than a foot, so the ratio of inches to feet would put the smaller number first.)

Example 3: Write the ratio of square feet to square inches.

1. Write the ratio as 1 square foot to 1 square inch. $\frac{1\text{ ft.}^2}{1\text{ in.}^2}$
2. Convert the 1 square foot to 144 square inches. $\frac{144\text{ in.}^2}{1\text{ in.}^2}$
3. Cancel the unit, square inches. $\frac{144\,\cancel{\text{in.}^2}}{1\,\cancel{\text{in.}^2}}$

Answer to the question: The ratio of square feet to square inches is 144 to 1.
The ratio of square inches to square feet is 1 to 144.

Exercise 1--Conversion of Linear and Square Units

Purpose of the Exercise

- To practice linear unit conversions and square unit conversions

Materials Needed

- Calculator

Answer the following questions, including the proper unit with each answer. Use the conversion chart at the end of the chapter when needed.

1. (Square: 1 yd by 1 yd)

Dimensions in feet	Dimensions in inches	Dimensions in meters
s = ______ ft	s = ______ in.	s = ______ m
A = ______ ft^2	A = ______ in.^2	A = ______ m^2

Compute the following ratios. Show the setup and work.

a. Ratio of feet to yards = ________ b. Ratio of square feet to square yards = ________

c. Ratio of yards to feet = ________ d. Ratio of square yards to square feet = ________

e. Ratio of inches to yards = ______ f. Ratio of square inches to square yards = ______

g. Ratio of yards to inches = ______ h. Ratio of square yards to square inches = ______

i. Ratio of meters to yards = ______ j. Ratio of square meters to square yards = ______

k. Ratio yards to meters = ________ l. Ratio of square yards to square meters = ______

2. For each of the following, use the conversion charts at the end of the chapter to find the ratios (m is an abbreviation for meter and cm is an abbreviation for centimeter).

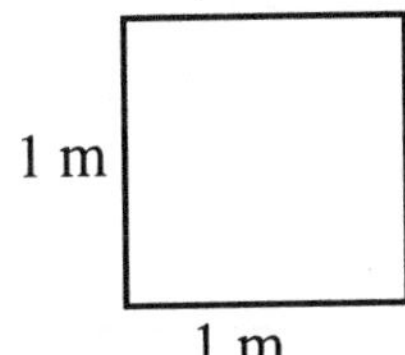

Dimensions in centimeters	Dimensions in inches	Dimensions in yards
s = _______ cm	s = _______ in.	s = _______ yd
A = _______ cm^2	A = _______ $in.^2$	A = _______ yd^2

a. Ratio of centimeters to meters = ____________________

b. Ratio of square centimeters to square meters is __________

c. Ratio of meters to centimeters is ____________________

d. Ratio of square meters to square centimeters is __________

3.

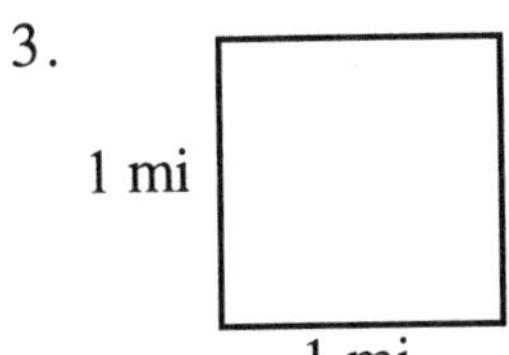

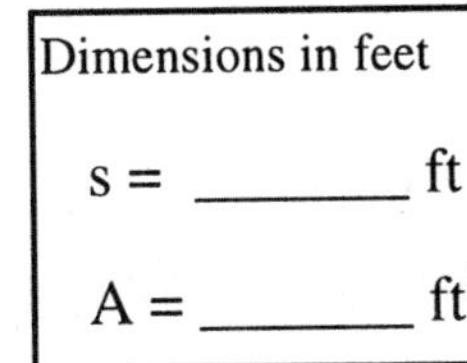

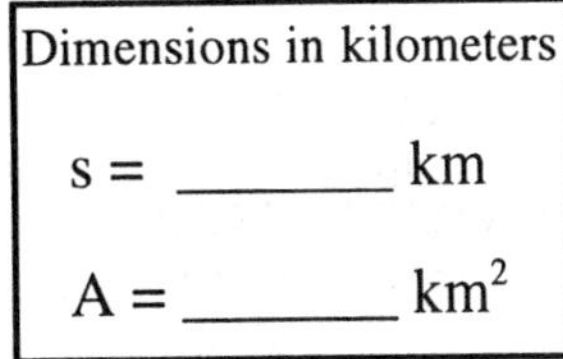

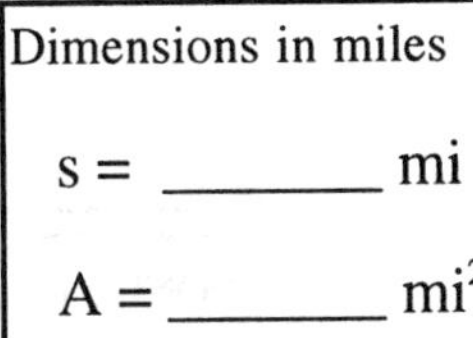

a. What is the ratio of miles to feet? _______

b. What is the ratio of square miles to square feet? _________

c. What is the ratio of feet to miles? _______

d. What is the ratio of square feet to square miles? _________

4. For each of the following rectangles, write the length, width and area in both of the units.

a.

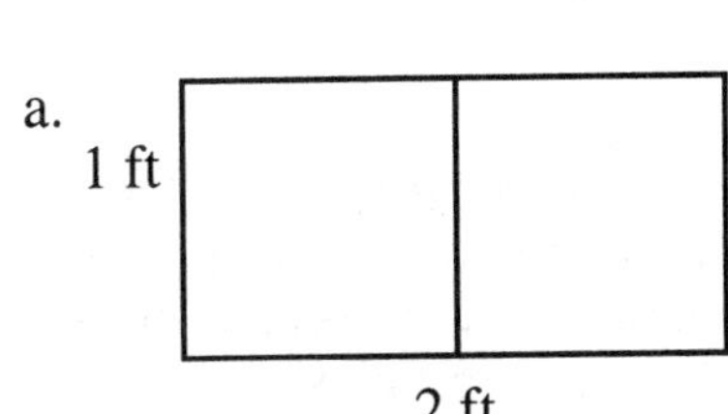

In feet
L = ________ ft
W = _______ ft
A = _______ ft^2

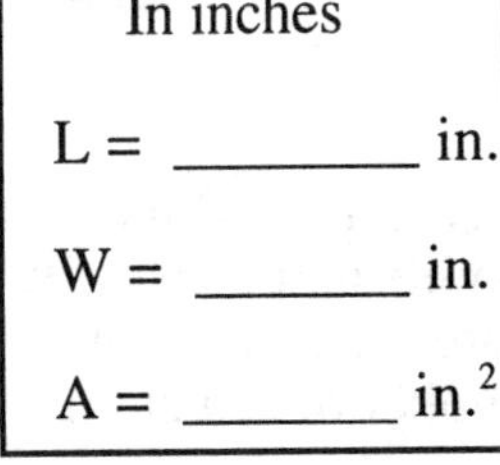

b.

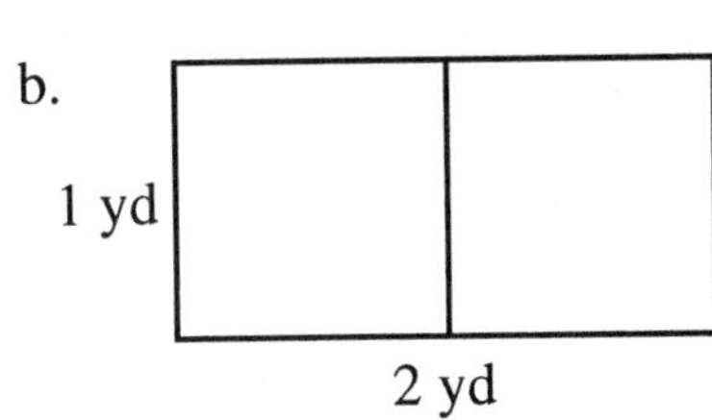

In feet
L = ________ ft
W = _______ ft
A = _______ ft^2

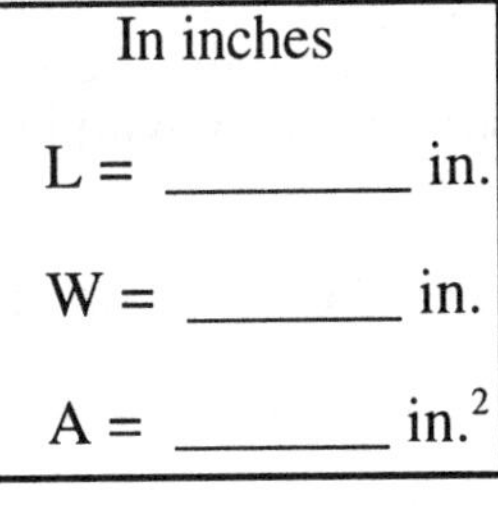

c.

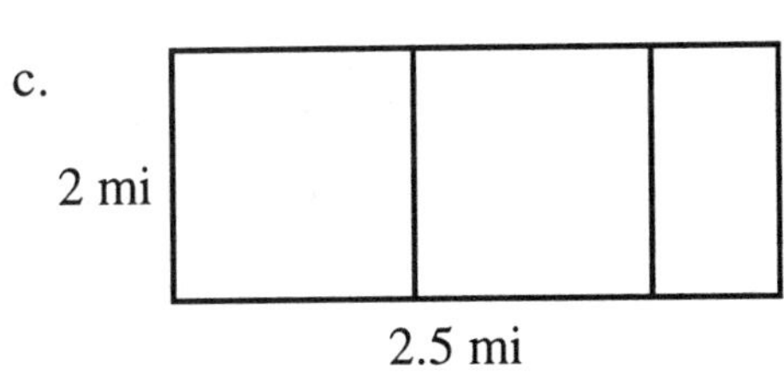

In feet	In kilometers
L = ________ ft	L = ________ km
W = _______ ft	W = _______ km
A = _______ ft^2	A = _______ km^2

d.

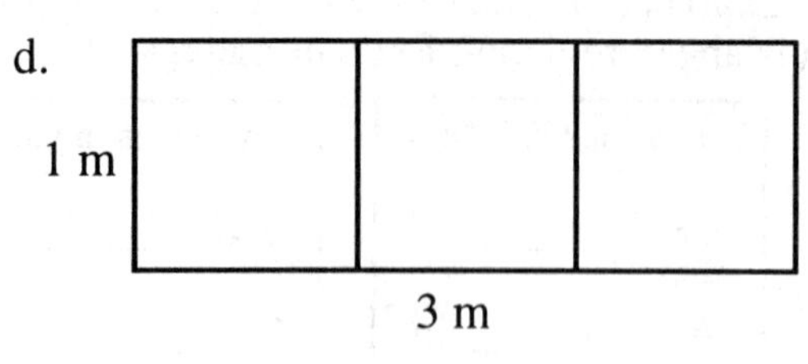

In centimeters	In meters
L = ________ cm	L = ________ m
W = ________ cm	W = ________ m
A = ________ cm^2	A = ________ m^2

Maps and Other Drawings to Scale

To help interpret a map, a scale or legend is shown that helps us estimate distances on the map. A legend may look something like the following:

1:1,000,000

0 10 20 30 Scale in Miles

ONE INCH EQUALS APPROXIMATELY 16 MILES OR 25.4 KILOMETERS

The ratio 1:1,000,000 means that one inch on the map represents 1,000,000 actual inches that is, the distance between locations on the map is one millionth of the actual distance between the locations. The ruler shown or a standard ruler can be used as a measuring tool.

Example 4: Use the scale above to estimate the distance between the following fictional towns.

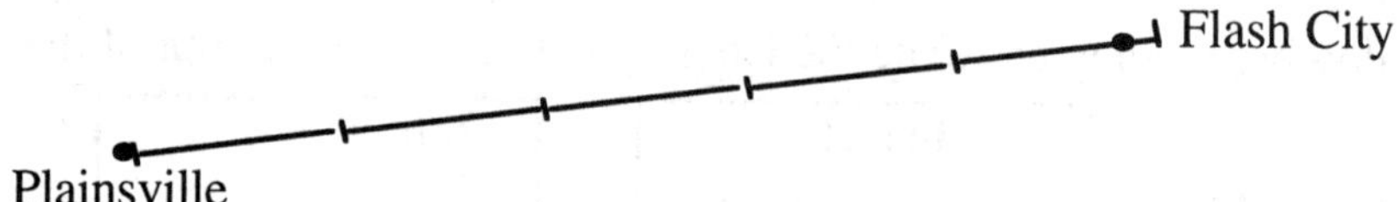

Since each unit on the ruler shown in the scale represents 10 miles, it appears that Plainsville and Flash City are about 48 miles apart. You could also use a standard ruler, measure the inches between the towns, and multiply the number of inches by 16 to get the number of actual miles (since one inch equals approximately 16 miles).

Other drawings to scale, such as floor plans, are done similarly. To make a drawing to scale, first measure the room (or object) and determine a scale so that the diagram will be large enough to work with but will also fit on the paper.

Example 5: On an $8\frac{1}{2}$ in. x 11 in. sheet of paper, make a scale drawing of a room that is 18 ft by 27 ft and has a 4-foot-wide doorway one foot from a corner of the room.

If we choose the scale 1 inch equals 3 feet, the drawing would be 6 in. by 9 in. A 6-in. x 9-in. rectangle would be a representation that fits on the paper.

To represent 1 foot on the drawing, solve the proportion, $\frac{1 \text{ in.}}{3 \text{ ft}} = \frac{x \text{ in.}}{1 \text{ ft}}$. x = 1/3 in.

To represent 4 feet on the drawing, solve the proportion, $\frac{1 \text{ in.}}{3 \text{ ft}} = \frac{x \text{ in.}}{4 \text{ ft}}$. x = 4/3 in.

Exercise 2--Maps and Other Scale Drawings (Group)

Purpose of the Exercise

- To practicing using the scale on a map
- To practice creating a drawing to scale
- To practice reading a drawing to scale

Materials Needed

- Yardstick, meter stick, or tape measure
- Map with scale shown

1. On your map pick several locations and, using the scale, approximate actual distances between locations. If the map is a road map with distances along a particular highway marked, compare the shortest distance "as the crow flies" to the distance between the locations along the road.

2. With the measuring tool, measure the room, including all doorways and closets or other irregularities. Choose a scale and make a scale drawing of the room, labeling the scale and significant parts of the room (North wall, doorway, closet, etc.).

3. Using the following scale diagram, determine the dimensions in feet as well as yards of each room and hallway. Compute the floor space in each room in square feet as well as in square yards.

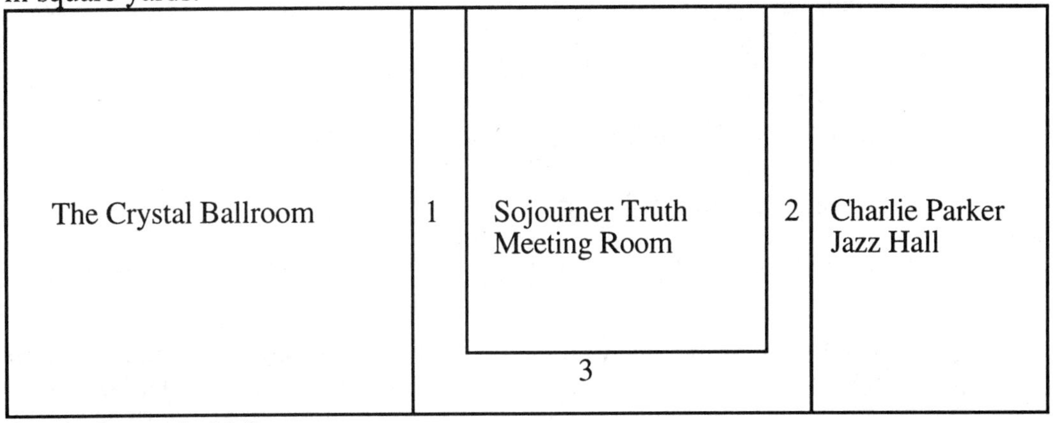

The Crystal Ballroom

1

Sojourner Truth Meeting Room

2

Charlie Parker Jazz Hall

3

One inch equals 30 feet

The Crystal Ballroom: dimensions in feet ________________

square feet ______________________ square yards ______________________

Sojourner Truth Meeting Room: dimensions in feet ________________

square feet ______________________ square yards ______________________

Charley Parker Jazz Hall: dimensions in feet ________________

square feet ______________________ square yards ______________________

Hallway 1: dimensions in feet ________________

square feet ______________________ square yards ______________________

Hallway 2: dimensions in feet ________________

square feet ______________________ square yards ______________________

Hallway 3: dimensions in feet ________________

square feet ______________________ square yards ______________________

Section 5-4 *Batting Average, Rates, and Rate Conversions*

A rate is how fast something changes or is being consumed. For example, the speed we travel is a rate because it measures how fast the distance from a starting point changes. The faster we move away from the starting point, the faster our speed. If we use 1374 kilowatts of electricity in a month (30 days), the daily rate of electricity use is the average number of kilowatts consumed in one day which is the number of kilowatts divided by the number of days. The daily rate of electricity consumption = 1374 kilowatts/30 days = 45.8 kilowatts per day. If the electric bill for that month is $121.93, the daily rate, or cost per day, is $121.93/30 days $4.06 per day. If we drive 455 miles using 26 gallons of gas, the average number of miles per a gallon (the rate of gas consumption) is 455 mi/26 gal = 17.5 miles per gallon. In this section we will learn how to compute rates, how to use rates to make predictions, and how to convert from one rate to another.

Computing rates, such as miles per hour or miles per gallon, is easy if we understand units. To begin with, we will write miles per hour as "miles/hour" --taking out the word per and putting in a slash. This tells us how to set up the division problem to compute miles per hour--we put miles in the numerator of the fraction and hours in the denominator.

Example 1: Compute the average speed in miles per hour if Opal travels 458 miles in 6 hours and 22 minutes. At the same speed, how far would Opal travel in 9.5 hours?

First we need to convert the 6 hours and 22 minutes to hours.

$$22 \text{ minutes} = (22 \cancel{\text{min}})\left(\frac{1 \text{ hr}}{60 \cancel{\text{min}}}\right) = \frac{22}{60} \text{ hr} \approx 0.367 \text{ hr}$$

6 hr and 22 min ≈ 6.367 hr.

458 miles traveled in 6.367 hours is $\dfrac{458 \text{ mi}}{6.367 \text{ hr}} \approx 71.9$ miles per hour (mph)

To compute how far Opal would travel in 9.5 hours, we set up a product with the units so that the hours cancel out and the unit left is miles. 71.9 miles per hour written as a ratio is $\dfrac{71.9 \text{ mi}}{1 \text{ hr}}$. Multiply this by the number of hours and cancel the unit, hours, leaving us with miles.

$$\frac{71.9 \text{ mi}}{1 \cancel{\text{hr}}} \text{ x } 9.5 \cancel{\text{hr}} = 683.05 \text{ mi.}$$

In 9.5 hours, Opal travels about 683 miles.

When working with rates, we can set up algebraic equations to use the rates or other related ratios to make predictions. As we noticed in Example 1, to compute the number of miles traveled in a particular time period, we multiplied the speed (rate) times the time to get the distance. Written as a formula, rt = d, where r represents rate, t represents time, and d represents distance. This is a formula we frequently use without thinking about it. For example, if we travel 50 miles per hour for two hours, common sense tells us that we traveled 50 miles the first hour and 50 miles the second hour for a total of 100 miles. The distance traveled is (50 miles per hour)(2 hours) = 100 miles. Algebraic formulas involving other rates are similar. Stating rates as a ratio helps us correctly match the units.

Formulas

d = rt distance = rate times time. (If the rate is in miles per hour, distance is measured in miles and time in hours.)

Other formulas involving rates are easier to understand expressed in words. Here are a few examples.

Total cost = (cost per day) times (the number of days)

Total number of miles a car will travel on a particular number of gallons of gas = (miles per gallon) times (the number of gallons)

Total price = (price per unit) times (number of units)

Exercise 1--Walking Speed (Group)

Purpose of the Exercise

- To compute walking speed
- To use speed to make predictions

Materials Needed

- Stopwatch or watch with a second hand
- Yardstick, meter stick or measuring tape

1. Measure the distance from one wall to the other in the room (or the length of a hallway). Have one group member walk the distance measured and clock the time it takes. Compute his/her speed in feet per second or meters per second.

2. Convert the walking speed from feet per second (or meters per second) to miles per hour. To do this, set up the speed as a ratio, then multiply by ratios equal to 1 to cancel feet and introduce miles and to cancel seconds and introduce hours. For example, if the speed were 4 feet per second,

$$\left(\frac{4\,\cancel{\text{ft}}}{\cancel{\text{sec}}}\right)\left(\frac{3600\,\cancel{\text{sec}}}{1\text{ hr}}\right)\left(\frac{1\text{ mi}}{5280\,\cancel{\text{ft}}}\right) = \frac{(4)(3600)(1\text{ mi})}{(1\text{ hr})(5280)} \quad \text{After the units are canceled}$$

$$\approx 2.73 \text{ miles per hour}$$

3. Assuming you could maintain the same walking speed for several hours, compute how long it would take you to walk in a 5-mile marathon. How long would it take you to complete a 10-kilometer walk? (Hint: Either convert the 10 kilometers to miles, or convert your walking speed from miles per hour to kilometers per hour.)

Example 2: If Sue worked a total of 493 days in 2 years, what was the average number of days per month that she worked? At that same average, how many days would Sue work in 3.5 months?

First convert 2 years to months since the rate is days per month.

2 years = 24 months

493 days worked in 24 months is $\frac{493\text{ days}}{24\text{ month}} \approx 20.5$ days per month.

To compute the number of days Sue would work in 3.5 months, set up the equation, total number of days worked = (days per month) times (number of months).

$$\frac{20.5\text{ days}}{1\,\cancel{\text{mo}}} \times 3.5\,\cancel{\text{mo}} = 71.75\text{ days}$$

Including the units assures us that the setup is right.

Example 3: Compute the average number of miles per gallon that Nathan's car got if it used 82 gallons of gas for a 2467 mile trip. How much did he pay for gasoline if gasoline cost an average \$1.27 per gallon? If Nathan took a 550-mile trip in the same vehicle, about how many gallons of gas would the trip take?

2467 miles per 82 gallons gives $\frac{2467 \text{ mi}}{82 \text{ gal}} \approx 30$ miles to the gallon.

How much did he pay for gasoline if gasoline cost an average \$1.27 per gallon?

(price per gallon) times (number of gallons) = Total price

(\$1.27 per gallon) x (82 gallons) ≈ \$104.14

If Nathan took a 550-mile trip in the same vehicle, about how many gallons of gas would the trip take?

(number of gallons) times (miles per gallon) = number of miles

(x gallons)(30 miles per gallon) = 550 miles Rewriting without units.

$30x = 550$ Divide both sides by 30.

$x \approx 18.3$ gallons

Batting Average

In baseball, a player's batting average is not a true average. It is more like miles per gallon or price per pound because it is the number of hits per the number of times at bat. Typically, the batting average is rounded to three significant digits and is written as a decimal. The phrase "batting 300" means the player has gotten 300 hits out of 1000 times at bat. Notice that the batting average is $\frac{300}{1000} = .300$.

$$\text{Batting average} = \frac{\text{number of hits}}{\text{number of times at bat}}$$

Using AVG to represent batting average, H for number of hits and AB for the number of times at bat, the formula becomes,

$$AVG = \frac{H}{AB}$$

Example 4: Jerry was at bat 136 times and had a batting average of .287 (is batting 287). How many hits did Jerry have?

Using the formula, $AVG = \frac{H}{AB}$, we get the equation,

$$.287 = \frac{H}{136}$$ To solve for H, multiply both sides of the equation by 136.

$$(136)(.287) = \left(\frac{H}{136}\right)(136)$$

$$38.9 = H$$

Since Jerry cannot be at bat 38.9 times, we round to the nearest whole number. Jerry was at bat 39 times.

Check: $\frac{H}{AB} = \frac{39}{136} \approx .287$

Exercise 2--Batting Average, Price per Pound, and Other Rates

Purpose of the Exercise

- To practice computing batting averages and work with batting average as a ratio
- To practice computing rates
- To review setting up and solving equations involving ratios

Materials Needed

- Calculator

Compute the batting average for each of the following American League Best Batters for the 1997 season through April 30, 1997. The first one is done as an example.

Player	AB	H	AVG
1. Bib Roberts	86	34	$\frac{34}{86}$ = .395 (rounded to 3 places.)
2. Sandy Alomar	75	29	
3. David Justice	83	32	
4. Brady Anderson	79	30	
5. Paul Oneill	91	32	
6. Geronimo Berroa	80	28	
7. Dan Wilson	86	30	
8. Jim Leyritz	69	24	
9. Damon Mashore	75	26	
10. Bernie Williams	110	38	
11. Ken Griffey, Jr.	103	35	
12. Troy Oleary	74	25	
13. Jim Thome	80	27	
14. Russ Davis	86	29	
15. Garret Anderson	101	34	
16. Marino Duncan	66	22	
17. Julio Franco	81	27	
18. Alex Rodriquez	120	40	
19. Nomar Garciaparra	116	38	
20. Tino Martinez	113	37	

Source: YAHOO scoreboard, May 1, 1997.

21. Who was at bat the most number of times? ____________________

    How many times? __________

22. Who had the most number of hits? ____________________

    How many hits? __________

23. Who had the highest batting average? ______________________________

Highest average = ____________

In each of the following, use the formula for batting average, fill in the numbers given in their appropriate places in the formula, and then solve the equation (if necessary) to find the missing piece of information. Show the setup and your work. Check your answers.

$$\text{AVG} = \frac{\text{H}}{\text{AB}}$$

24. Justin had 57 hits and a batting average of .127. How many times was he at bat?

25. Jane had 33 hits out of 72 times at bat. What is Jane's batting average?

26. Skip was at bat 84 times and had a batting average of .298. How many hits did Skip get?

27. If Julie's batting average is .242, how many times has she hit the ball out of 33 times at bat? ____________ If she continued with the same batting average, how many times would she hit the ball out of 1000 times at bat? ____________ Out of 87 times at bat? ____________________

Discussion Questions

28. Suppose a car salesman says to his manager, "I've dealt with three customers today and am batting a thousand." How many cars has he sold?

29. A senator talks to 18 other senators, trying to persuade them to vote for her piece of legislation. In an interview, she says that she is "batting 500." How many senators has she convinced to vote for her legislation?

In each of the following problems, set up a appropriate mathematical expression or equation, simplify the expression or solve the equation, and answer the question.

30. Braeburn apples cost \$1.19 per pound.

    a. Carlton bought 3.2 pounds of apples. How much did the apples cost him?

    b. Cheryl paid \$7.53 for a bag of apples. How many pounds of apples did she buy?

31. Wayne drove 367 miles on 13.5 gallons of gasoline.

    a. How many miles per gallon did Wayne get on his trip?

    b. If Wayne took a 2378 mile trip in the same vehicle, about how many gallons of gas would the trip take?

32. Natalie filled 52 crates in one hour.

    a. Approximately how many crates would Natalie fill in an 8-hour work day?

    b. Explain why the number of crates she could fill in a day might not reflect exactly the same rate as she could do in an hour.

33. Challenge problem: Perry stored 10 chairs in a 3-ft by 5 ft area. About how many chairs could he store in room that is 15 ft by 20 ft.

Converting from one rate to another is very similar to converting units as we did in Sections 5-2 and 5-3.

Example 5: Using individual conversions.

Suppose a car gets 11 kilometers (km) per liter of gas and we want to know whether that is good or bad mileage, but we are accustomed to gas mileage in miles per gallons. We will convert 11 kilometers per liter to miles per gallon by multiplying by proper conversion factors.

From the conversion chart, 1 kilometer ≈ 0.62 mile and 1 liter = 1.06 quart

First conversion: $1 \text{ liter} \approx (1.06 \text{ qt}) \dfrac{1 \text{ gal}}{4 \text{ qt}} \approx \dfrac{1.06}{4} \text{gal} = 0.265 \text{ gallons}$

Second conversion: $(11 \cancel{\text{km}}) \left(\dfrac{0.62 \text{ mi}}{1 \cancel{\text{km}}} \right) = 6.82 \text{ miles.}$

Traveling 6.82 miles on 0.265 gallons give a rate of $\dfrac{6.82 \text{ mi}}{0.265 \text{ gal}} = 25.7$ miles per gallon.

11 kilometers per liter is about 25.7 miles per gallon.

Example 6: Here is a quicker way to do the same conversion as in Example 4.

We will convert 11 kilometers per liter to miles per gallon by multiplying by proper conversion factors. (This method was used in Exercise 1.)

1 kilometer ≈ 0.62 mile 1 liter ≈ 1.06 quart 4 quarts = 1 gallon

For each conversion factor, place the unit to cancel the desired unit. Since we want to replace liters with quarts, then gallons, we put liters in the numerator of the conversion factor to cancel the liter in the denominator. To cancel kilometers, we put kilometers in the denominator of the conversion factor.

$$\left(\frac{11 \cancel{\text{km}}}{1 \cancel{\text{liter}}} \right) \left(\frac{1 \cancel{\text{liter}}}{1.06 \cancel{\text{qt}}} \right) \left(\frac{4 \cancel{\text{qt}}}{1 \text{ gal}} \right) \left(\frac{0.62 \text{ mi}}{1 \cancel{\text{km}}} \right) = \frac{(11)(4)(0.62) \text{ mi}}{1.06 \text{ gal}} = 25.7 \text{ miles per gallon}$$

11 kilometers per liter is about 25.7 miles per gallon.

Example 7: Price per unit also works like miles per hour or other rates. If 1.5 pounds of coffee cost $11.35, to set up the cost per pound, we would set up a ratio of cost to pounds. The ratio is $\dfrac{11.3 \text{ dollars}}{1.5 \text{ pounds}}$. If we want cost per pound, not cost per 1.5 pounds, we merely divide 11.3 by 1.5 to convert the ratio to decimal form.

$$\frac{11.3 \text{ dollars}}{1.5 \text{ pounds}} \approx \frac{7.53 \text{ dollars}}{1 \text{ pound}}$$

The coffee costs about $7.53 dollars per pound.

Exercise 3--Rates and Rate Conversions

Purpose of the Exercise

- To practice computing rates
- To practice converting rates from one unit to another

Materials Needed

- Calculator

Compute the following rates. Include the appropriate units with the answer.

1. A bicyclist covers 52 miles in $7\frac{1}{2}$ hours. Her bicycling rate is ________________
2. A secretary types 397 words in 7 minutes. His typing speed is ________________
3. A coffee connoisseur drinks 8956 cups of coffee in 4 years. How many cups of coffee per day does she average? ________________________
4. A European train traveled 87 kilometers in 45 minutes. How fast did it travel per hour? ________________
5. A runner ran 1 mile in 5 minutes. What was his speed in miles per hour? ________
6. It took 2 minutes to fill a 9-gallon tank with gasoline. How many gallons per minute were being pumped? ________________________________
7. Marie was paid $698 (before deductions) for working 73 hours. How much money did she make per hour? ________________________________
8. Jerry bought a 12.5 pound turkey for $8.93. How much per pound did the turkey cost?
9. Gourmet jelly beans cost $3.75 a pound. How much will 5 pounds of jelly beans cost?
10. Convert the following rates as indicated.
    a. 75 feet per second = ____________________ miles per hour.
    b. 69 kilometers per hour = ________________ miles per hour.
    c. 278 days worked per year = ______________ days worked per month.
    d. 876 gallons per hour = _________________ liters per hour.
11. Challenge problem: In each of the following, on graph paper draw a diagram showing the starting point and direction of the planes as well as how far they travel over the time given. To answer one of the questions, you may need to review the Pythagorean theorem. In each case, the speed of one plane is 372 miles per hour and the second plane flies at 428 miles per hour.
    a. The two planes start at the same place at the same time, but one flies North while the other flies South. How far apart are the two planes after 2.5 hours in flight?
    b. The two planes start 2563 miles apart and fly towards each other. How far apart are the two planes after flying 1 hour and 20 minutes?
    c. The two planes start at the same point, but one flies East and the other flies South. How far apart are the two planes after 3 hours and 45 minutes in flight?

Select one of the Following Projects.

1. Laundromat vs. buying a washer and drier -- An appropriate project if you do your laundry in a laundromat
   - If you do laundry at a laundromat, estimate how much it costs you per month to do laundry. Check with a store that sells washers and dryers and find out how much it would cost to buy a washer and drier suitable to your needs. (Don't forget to figure in taxes.)
   - Calculate how many months it would take you to recover the cost of the washer and drier in saved laundromat costs.
   - Write a report giving all the information you collected. Explain how you came up with your answer.
   - Explain why your answer is an estimate and might not be exactly correct.
   - If you bought a washer and drier on credit, how much would it increase your total cost? What would be the difference between your monthly payment and your cost at the laundromat?
2. Conference meals
   - Suppose people at a conference have a choice of a vegetarian, turkey, or roast beef lunch. You are in charge of telling the caterers approximately how many of each type of meal to prepare. You expect 1352 people at the conference. Take a poll of 35 people and ask them which lunch they would prefer. From the result of this poll, use the ratios to predict how many people at the conference will choose each kind of meal.
   - Write a report giving all the information you collected. Explain clearly how you came up with your answer.
   - Explain why your answer is an estimate and might not be exactly correct.
   - What would your prediction be if 293 people were expected to attend the conference?
3. Hourly wage for cooking
   - Cook or bake some food dish. Figure how much money the dish cost to make, how much time it took to make it, and how many servings you got from the dish. If you were selling servings from the dish, how much would you charge per serving if your goal is to make $10.00 an hour?
   - Write a report giving all the information you collected. Explain how you came up with your answer.
   - Explain why your answer is an estimate and might not be exactly correct.
   - How might your figures change if you were cooking for 100 people?
4. Prorating bills according to floor space
   - Measure the rooms in your living space and figure the area of each room in square feet. Suppose one of your rooms is used for a home business and you meet all of the legal requirements for deducting this space as a home business. Total your electric and gas bills you paid last year. What would be the admissible tax deduction for these utilities?
   - Write a report giving all the information you collected. Explain clearly how you came up with your answer.
   - Is the answer an estimate? Explain.
   - How would this change if you had moved your business from one room to another during the year?

5. Mixed nuts and unit pricing (You will need a good scale that measures ounces accurately.)
   - Buy a can or pound of mixed nuts. Separate the nuts into piles of the different kinds. Compute the price per pound of the mixed nuts. Find out how much each type of nut costs per pound. (You may have to check with a store that sells bulk nuts for this.) If you had bought each kind of nut separately, how much would a pound of nuts (of the same mixture) cost?
   - Devise your own preference of mixed nuts. What percent of each kind would you mix in (to your taste, of course), and how much would the mix cost per pound?
   - Write up your results and explain how you did your experiment as well as where and how you got your pricing. Show/explain your computations.
   - If you did the same experiment again, explain why your result might be different.

6. Number of bags of sand to fill a sandbox
   - Determine the approximate shape of your sandbox and decide how deep you want the sand to be in the sandbox. Look up formulas for the volume of the shape. Approximate the volume of sand you will need.
   - Examine a bag of sand. (A lumber yard is a good place to start.) If the volume of sand is not written on the bag, estimate the volume of sand in the bag. This could be done by assuming the bag is in the shape of a rectangular box, or you could devise another means where you empty the sand into a container with which you can measure the volume more accurately. Find how many pounds of sand make up a cubic foot of sand.
   - Figure out how many bags of sand it will take to supply the sandbox and how much the sand will cost.
   - Suppose you have the sand in the sandbox and you decide you need 2 more inches of sand. (Either you changed your mind, or the kids have managed to make some of the sand disappear.) How many bags will you need?
   - Show all of your computations and explain how what you did and how you came to your conclusions. Explain what makes the numbers approximations instead of hard fast figures.

7. Determining the gas mileage and budgeting a trip
   - Fill your tank with gas and record the mileage from the odometer. The next time your fill your tank, record the mileage to determine how far you drove on the tank of gas and compute the miles per gallon that your car gets. For more accurate results, do this for several tanks of gas and compute the mileage using the total number of miles driven and the total number of gallons of gas used.
   - Plan a lengthy trip and determine about how many gallons of gas the trip would take. Estimate the price per gallon of gas and budget the amount of money you would need for gas on the trip.
   - Write up your results, showing the computations and reporting on the results. Explain why the actual number of gallons of gas used and the cost of gasoline are only estimates. What might make the numbers come out differently?

Section 5-5 *Ratios, Angles, and Slopes*

In Section 5-1, we learned about ratios of countable objects, such as the ratio of males to females. In this section we will learn about the slope as a ratio of lengths that measures the steepness of the incline of a line, a mountain, a ramp, or a roof.

The following diagrams represent side views of roofs or ramps. The horizontal distance is called the run and the vertical distance is called the rise.

> **Vocabulary**
> **Slope** is the ratio of rise to run. Stated differently, it is a measure of the change in vertical position divided by a horizontal distance.

Exercise 1--Slope (Group)

Purpose of the Exercise

- To understand slopes
- To understand the arbitrary nature of the units used in ratios

Materials Needed

- 2 long pieces of string and a line level
- A small weight or plumb bob
- A yardstick and meter stick

Find two different ramps or staircases. Measure and record the rise and run in centimeters and then in inches. Record the slope as a fraction and then compute as a decimal.

To use a line level to measure a slope of a ramp, you need a line level, a long piece of string, and a string with a weight or plumb bob attached. Since we cannot usually measure the run below the ramp, we measure it above the ramp. It is helpful to have four or five people to do this.

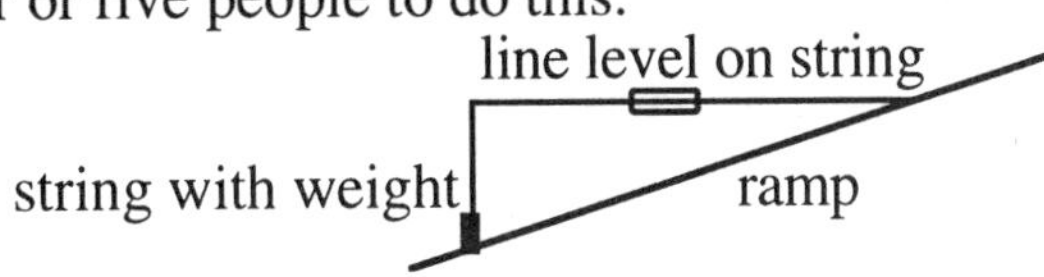

- First person: Hold one end of a string touching one spot on the ramp.
- Second person: Hold the other end of the string.
- Third person: Put the line level on the string and have the first and second persons hold the string taut and level. Read the gauge on the line level and have the second person raise or lower his/her end of the string until the string is level.
- Fourth person: (The second person could also manage this job); Hold the string with the weight attached so that it dangles to the ramp and the top touches the level string.
- Fifth person: Measure and record the rise and the run.

1. a. What were the rise and run, measured in centimeters, of the first ramp or staircase you measured? What is the ratio of rise to run?

rise = ______ cm run = ______ cm slope = ________ = __________
fraction decimal

b. What were the rise and run, measured in inches, of the first ramp or staircase you measured? What is the ratio of rise to run?

rise = ______ in. run = ______ in. slope = ________ = __________
fraction decimal

2. a. What were the rise and run, measured in centimeters, of the second ramp or staircase you measured? What is the ratio of rise to run?

rise = ______ cm run = ______ cm slope = ________ = __________
fraction decimal

b. What were the rise and run, measured in inches, of the second ramp or staircase you measured? What is the ratio of rise to run?

rise = ______ in. run = ______ in. slope = ________ = __________
fraction decimal

Now that we understand what a slope is, we can represent an incline in a diagram. In the following exercises, you will measure rise/run on the drawing of an incline and see that the slope is the same no matter which rise and corresponding run that you pick.

Exercise 2--The Rise-to-Run Ratio of an Angle

Purpose of the Exercise

- To learn about the rise:run ratio

Materials Needed

- Metric ruler
- Calculator

In the following drawings, measure and record the indicated lengths in centimeters accurate to 1 decimal place. Record the ratios indicated in fraction from as well as decimal form correct to 2 decimal places.

1.

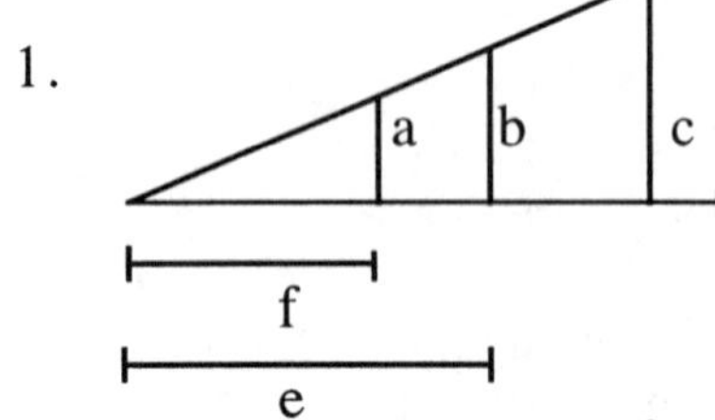

a = ________ b = ________ c = ________ (rise)

d = ________ e = ________ f = ________ (run)

	fraction	decimal		fraction	decimal		fraction	decimal
$\frac{a}{f}$ =	______ =	________	$\frac{b}{e}$ =	______ =	________	$\frac{c}{d}$ =	______ =	________

2.

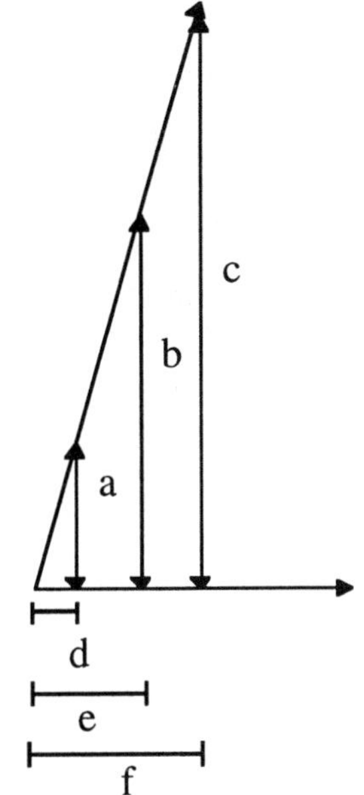

a = ________ b = ________ c = __________ (rise)

d = ________ e = ________ f = __________ (run)

	fraction		decimal
$\frac{a}{d}$ =	______	=	________

	fraction		decimal
$\frac{b}{e}$ =	______	=	_______

	fraction		decimal
$\frac{c}{f}$ =	______	=	_______

In each angle above, notice that the decimal forms of the ratios are about the same. If your measurements could be exactly right, the ratios would be the same. If your ratios are not approximately the same in any one problem, check your measurements and calculations.

3. Which problem had the largest <u>angle</u> and which has the smallest?
   largest angle is in Problem _________ smallest angle is in Problem _________

4. Which problem has the largest <u>ratio</u> and which has the smallest?
   largest ratio is in Problem __________ smallest angle is in Problem _________

As you noticed in the above exercise, for a particular angle (positioned with one side horizontal) the ratio of rise to run is the same ratio regardless of where you measured. For this reason, there is a correspondence between the measurement of an angle and the ratio of rise to run (provided the angle is positioned so that one side is horizontal). This correspondence in an acute angle is called the tangent of the angle and can be obtained from your calculator. But first, we will learn to use a protractor to measure an angle.

Protractors and Angles

Vocabulary review:

A **right angle** is one that is made with two perpendicular lines. The measure of a right angle is 90° .

An **acute angle** is an angle whose measure is less than 90°

An **obtuse angle** is an angle whose measure is greater than 90°.

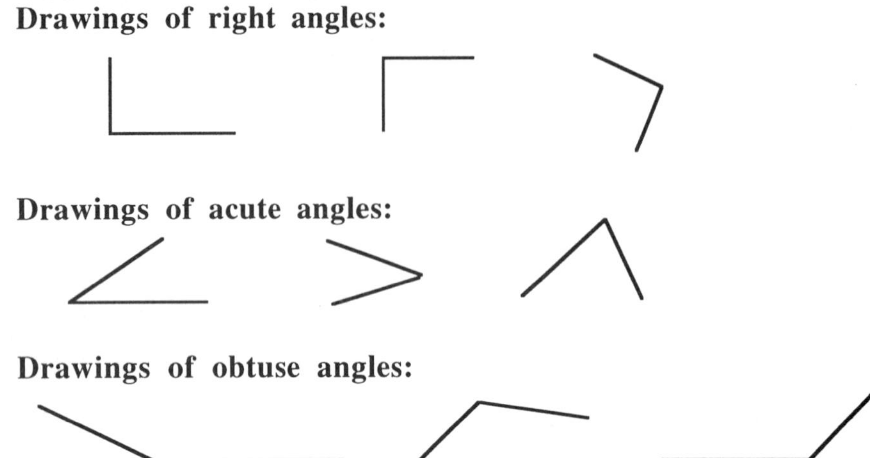

Drawings of right angles:

Drawings of acute angles:

Drawings of obtuse angles:

To measure an angle we can use a protractor. There are two scales on a protractor and we have to determine which of the scales gives the right answer. When measuring a 90° angle, both scales give the same answer, so it does not matter which scale is used. In other angles, we have to determine whether the angle is acute or obtuse and pick the appropriate scale accordingly. We may have to extend one or both of the lines that form the angle to reach the scale on the protractor. The following examples demonstrate how to measure an angle using a protractor.

Example 1:

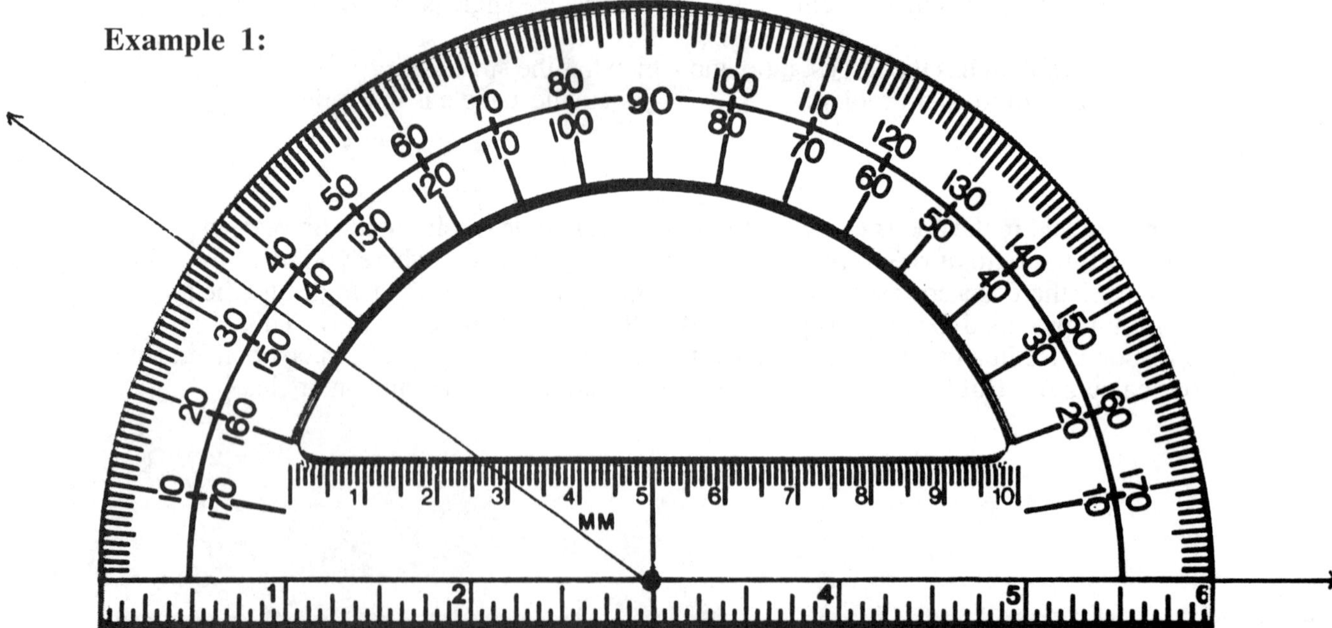

The angle above is clearly obtuse. The two readings on the protractor are about 35° and about 145°.   145° is the measure of an obtuse angle, so that is the correct measure.

Example 2:

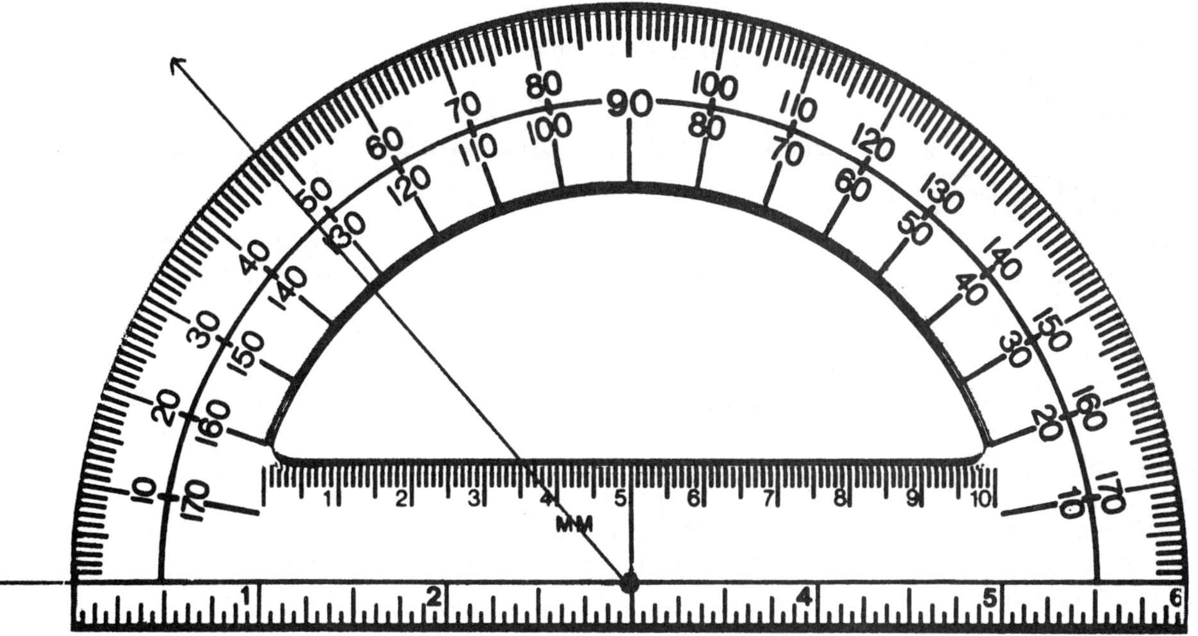

The angle is clearly acute. The two readings on the protractor are about 48° and about 132°. 48° is the measure of an acute angle so that is the correct measure.

Example 3:

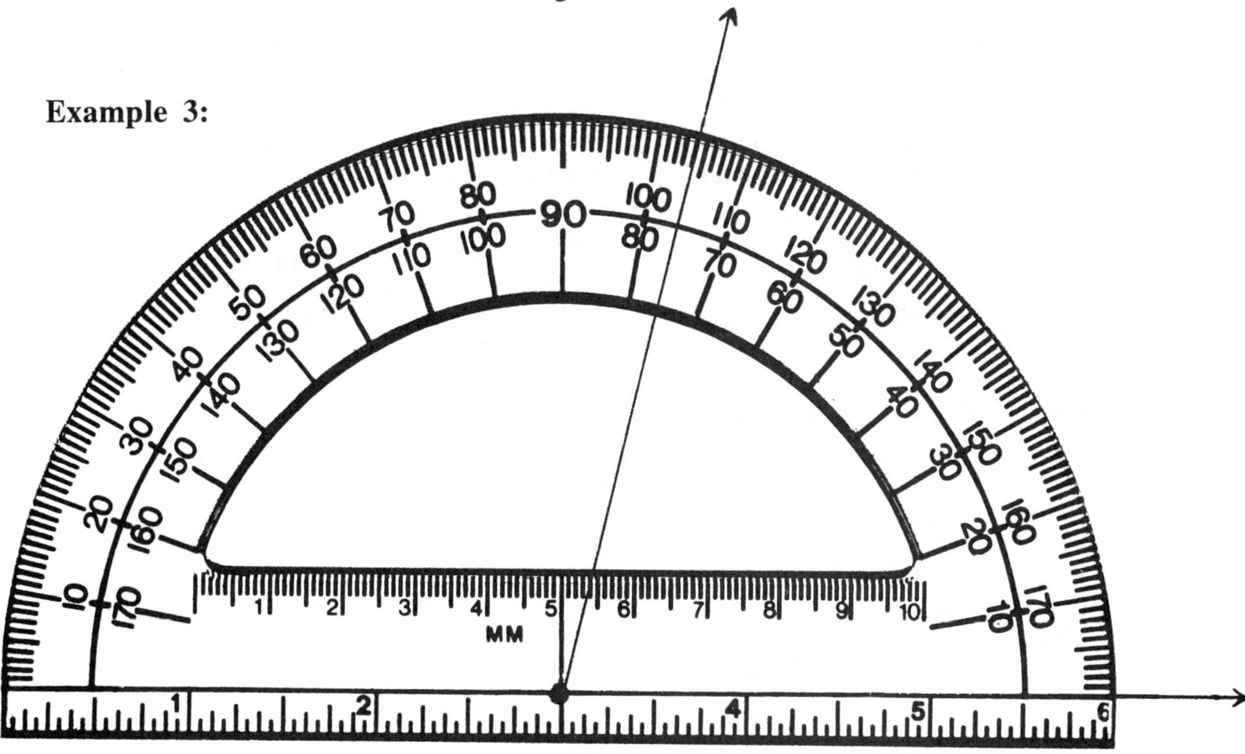

The angle above is clearly acute. The two readings on the protractor are about 76° and about 104°. 76° is the measure of an acute angle so that is the correct measure.

The tangent of an acute angle is the ratio of rise to run when that angle is positioned so that one side is horizontal. To compute the tangent of 74° do the following key strokes.

On a calculator that displays one number at a time on the screen:

[74] [tan]

On a graphics calculator or scientific calculator that displays the entire arithmetic expression:

[tan] [74] [enter]

Try the above on your calculator. Your answer should be approximately 3.49. If you got approximately -5.74 instead, check with your instructor or your manual to put your calculator in degree mode (instead of radian mode).

Exercise 3--The Protractor, Angles, and Slope

Purpose of the Exercise

- To learn how to read the scale on a protractor
- To relate the size of an angle to the ratio of rise to run
- Relate the ratio of rise to run of an acute angle to the tangent of the angle

Materials Needed

- Protractor
- Calculator
- Graph paper

1. In each of the following, lay the protractor on the angle with the vertex of the angle at the dot on the protractor and the side going along the bottom line of the protractor. Extend the lines as needed. Determine whether the angle is an acute, obtuse, or right angle. Record the readings on both scales of the protractor and determine which reading is the appropriate reading for your angle.

a.

Acute or Obtuse?

Reading 1 ______

Reading 2 ______

Correct reading ______

b.

Acute or Obtuse?

Reading 1 _______

Reading 2 _______

Correct reading ____

c.

Acute or Obtuse?

Reading 1 ______

Reading 2 ______

Correct reading ______

d.

Acute or Obtuse?

Reading 1 _______

Reading 2 _______

Correct reading ____

e.

Acute or Obtuse?

Reading 1 ______

Reading 2 ______

Correct reading ______

f.

Acute or Obtuse?

Reading 1 _______

Reading 2 _______

Correct reading ____

g.

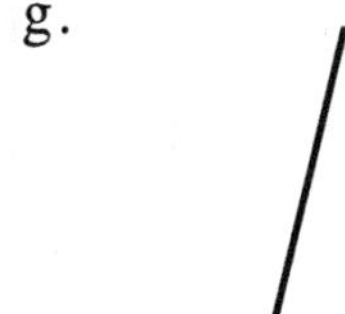

Acute or Obtuse?

Reading 1 ______

Reading 2 ______

Correct reading ______

h.

Acute or Obtuse?

Reading 1 _______

Reading 2 _______

Correct reading ____

i.

Acute or Obtuse?

Reading 1 ______

Reading 2 ______

Correct reading ______

j.

Acute or Obtuse?

Reading 1 _______

Reading 2 _______

Correct reading ____

On a sheet of graph paper, use your protractor and draw the acute angles in each of the following problems. Draw one side of the angle as a horizontal line. Using the length of a square on your graph paper as a unit, estimate the rise and run and compute the ratio of rise to run in decimal form.

Use your calculator to compute the tangent of the angle and compare it to the computed ratio of rise to run.

2. For a 11° angle,
   a. $\frac{\text{rise}}{\text{run}} \approx$ ______ (in decimal) b. the tangent of 11° = __________

3. For a 27° angle,
   a. $\frac{\text{rise}}{\text{run}} \approx$ ______ (in decimal) b. the tangent of 27° = __________

4. For a 45° angle,
   a. $\frac{\text{rise}}{\text{run}} \approx$ ______ (in decimal) b. the tangent of 45° = __________

5. For a 52° angle,
   a. $\frac{\text{rise}}{\text{run}} \approx$ ______ (in decimal) b. the tangent of 52° = __________

6. For a 60° angle,
   a. $\frac{\text{rise}}{\text{run}} \approx$ ______ (in decimal) b. the tangent of 60° = __________

7. For a 76° angle,
   a. $\frac{\text{rise}}{\text{run}} \approx$ ______ (in decimal) b. the tangent of 76° = __________

Vocabulary

A **central angle in a circle** is one that is formed by two of the radii.

A **straight angle** is one that measures exactly 180°. The two sides of a straight angle form a straight line.

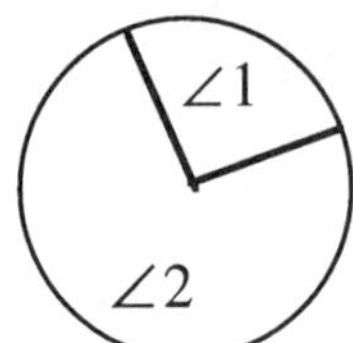

In the circle shown, the two radii form two central angles, the smaller one marked $\angle 1$ and the larger one, marked $\angle 2$. The symbol $\angle$ means "angle."

No matter how you divide up the circle with radii, the sum of the measures of all of the central angles is 360°. But to find that out first hand, do the following exercise.

Exercise 4--Angles in a Geometric Shape

Purpose of the Exercise

- To learn about the central angles of a circle

Materials Needed

- Protractor

1.

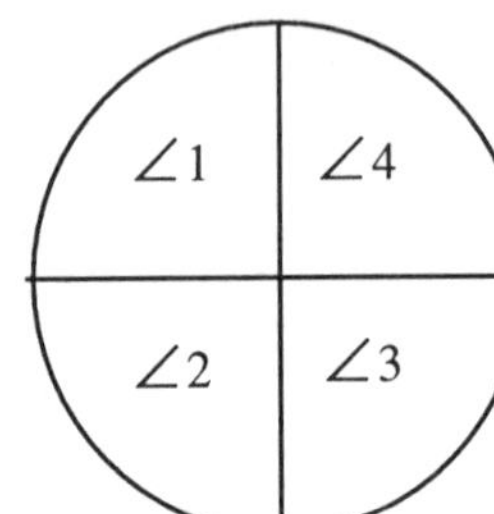

In the circle at left, there are four central angles. Measure each angle with your protractor and record your result. What is the sum of the four angles?

$\angle 1 =$ ______ $\angle 2 =$ ______ $\angle 3 =$ ______ $\angle 4 =$ ______

$\angle 1 + \angle 2 + \angle 3 + \angle 4 =$ ___________

2.

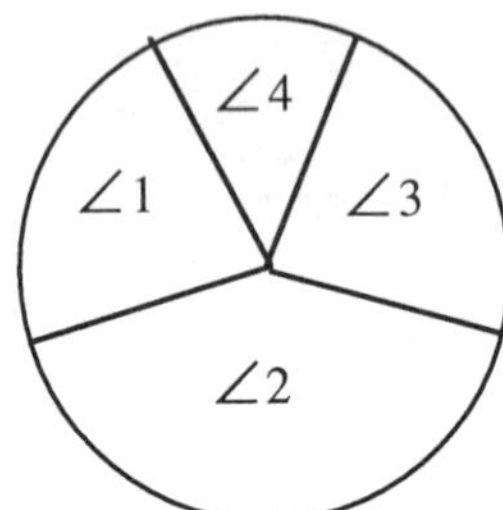

In the circle at left, there are four central angles. Measure each angle with your protractor and record your result. What is the sum of the four angles?

$\angle 1 =$ ______ $\angle 2 =$ ______ $\angle 3 =$ ______ $\angle 4 =$ ______

$\angle 1 + \angle 2 + \angle 3 + \angle 4 =$ ___________

3.

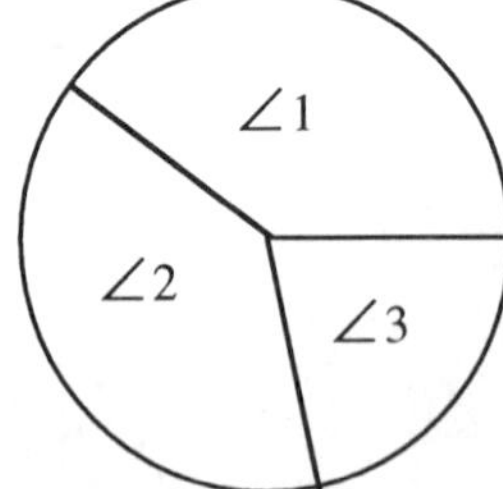

In the circle at left, there are three central angles. Measure each angle with your protractor and record your result. What is the sum of the three angles?

$\angle 1 =$ ______ $\angle 2 =$ ______ $\angle 3 =$ ______

$\angle 1 + \angle 2 + \angle 3 =$ ___________

4.

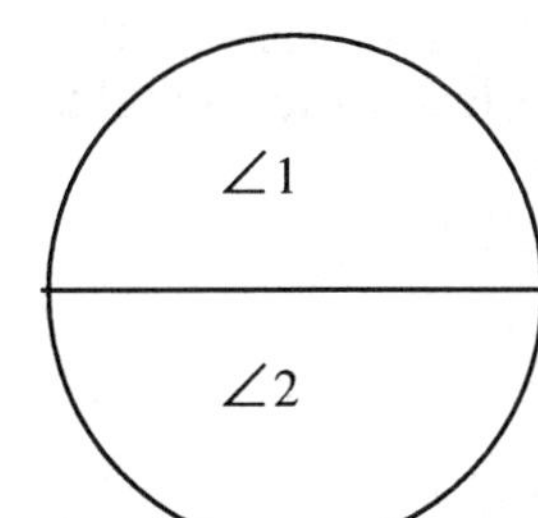

In the circle at left, there are two central angles. Measure each angle with your protractor and record your results. What is the sum of the angles?

$\angle 1 =$ ______ $\angle 2 =$ ______

$\angle 1 + \angle 2 =$ ____________

5.

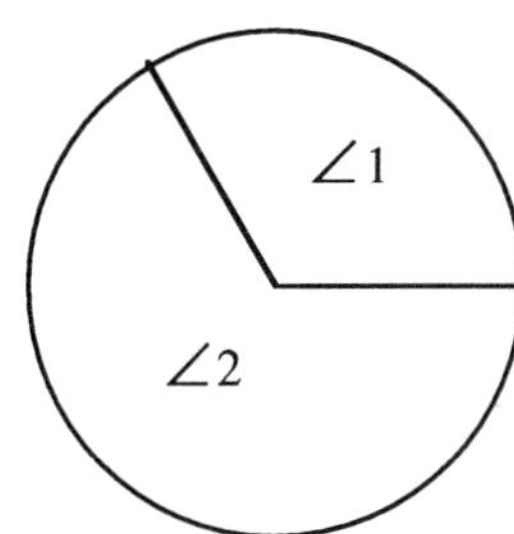

In the circle at left, there are two central angles. Measure each angle with your protractor and record your results. What is the sum of the angles? **Hint: To measure an angle larger than 180°, break it into a straight angle + a smaller angle.**

$\angle 1 =$ ______ $\angle 2 =$ ______

$\angle 1 + \angle 2 =$ ____________

6.

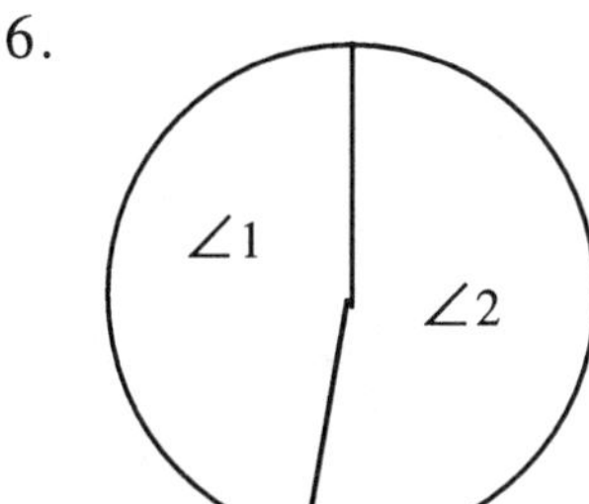

In the circle shown on the left, there are two central angles. Measure each angle with your protractor and record your results. What is the sum of the angles?

$\angle 1 =$ ______ $\angle 2 =$ ______

$\angle 1 + \angle 2 =$ ______________

7.

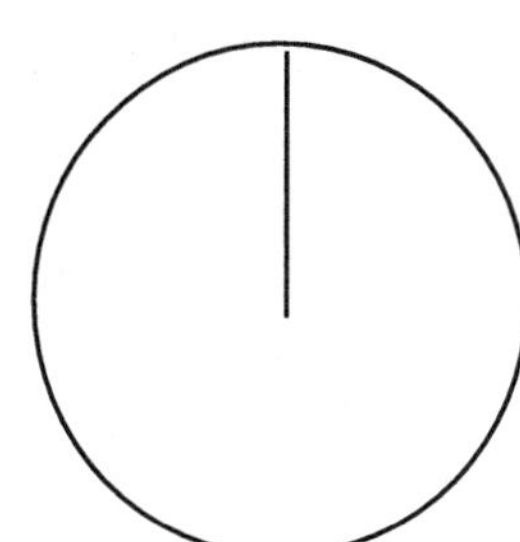

In the circle shown on the left, there is only one central angle. What is the measure of the angle?

$\angle 1 =$ ______

8. **Observations from the previous problems**. In any one circle, the measures of all of the central angles add up to the same number of degrees. Some of your totals may vary slightly because of inaccuracy in reading the protractor; however, if all of your readings were accurate, what is the sum of the degrees in a circle? ___________

   If any of your sum varied greatly from 360°, rework the problem. Check your protractor readings and total.

9. How many degrees are in a straight angle? ________ In a right angle? ________

10. Experiment with triangles. Draw several different shaped triangles and measure the three angles of the triangle. Record each angle and sum of the angles of the triangle.

Triangle 1: $\angle 1 =$ ____ $\angle 2 =$ ____ $\angle 3 =$ ____ $\angle 1 + \angle 2 + \angle 3 =$ ______

Triangle 2: $\angle 1 =$ ____ $\angle 2 =$ ____ $\angle 3 =$ ____ $\angle 1 + \angle 2 + \angle 3 =$ ______

Triangle 3: $\angle 1 =$ ____ $\angle 2 =$ ____ $\angle 3 =$ ____ $\angle 1 + \angle 2 + \angle 3 =$ ______

11. A quadrilateral is a four-sided geometric figure. Draw several differently shaped quadrilaterals. Measure the four angles and find the total of their measures.

What is the sum of the measures of the four angles in a quadrilateral? ____________

Slopes, Ramps, and Trusses--An Out-of-Class Project and Report

Purpose of the Exercise

- To learn how slopes are used in the real world
- To learn about legal and safe accessibility ramps

Materials Needed

- Graph paper
- Calculator

1. Find out about some of the most common slopes of roofs. Visit or call a lumber yard and find out what the pitches are on pre-assembled trusses. Which slopes might represent an A-frame roof, and what is the maximum pitch a roofer is willing to walk on without building a platform? Illustrate each of these slopes by drawing a line on graph paper. If you are given the slopes in angle form, use a protractor to draw the angle and then draw a line perpendicular to one side to get the slope. (Label each picture.)

2. Find out what is a recommended safe slope for an accessibility ramp (that a person with upper body strength can navigate without assistance.) Illustrate this slope by drawing a line on graph paper. Using string, a line level, and measuring tape, calculate the slope of several public sidewalks or wheelchair ramps and see if they meet specifications. If you know someone in a wheelchair, interview them to find out from them which of these are navigable. If you are given the slopes in angle form, use a protractor to draw the angle. From the drawing, determine the rise, run, and slope of the ramp.

3. Represent all slopes as ratios as well as in decimal form.

Turn in

- The answers to the questions and pictures of the slopes, well labeled on graph paper.
- A paragraph describing any difficulties you had in completing this project.

Review of Terms

- A **ratio** of one number to another is a fraction in which the first number is the numerator and the second number is the denominator. A ratio can be written several ways. The ratio of 4 to 5, can be written as 4:5, 4/5, or 4 to 5 or as 0.8 to 1. (page 308)
- A **proportion** is an equation of the following form: ratio 1 = ratio 2. (page 309)
- A **conversion factor** is a ratio of 1 that is used as a multiplier to convert from one unit to another. (page 315)
- A **scale drawing** is a drawing where all of the lengths/distances on the drawing are of the same proportion as the actual object or area. Maps and floor plans are examples of scale drawings. (page 324)
- The **scale** tells us how to convert lengths on the scale drawing to actual lengths (distances). (page 324)
- A **rate** is how fast something is changing or is being consumed. (page 326)
- A **slope** of a line is the ratio of rise to run. Stated differently, it is a measure of the change in vertical position divided by a horizontal distance. (page 335)
- A **right angle** is one that is made with two perpendicular lines. The measure of a right angle is 90°. An **acute angle** is an angle whose measure is less than 90°. An **obtuse angle** is an angle whose measure is greater than 90°. (page 338)
- A **central angle in a circle** is one that is formed by two of the radii. A **straight angle** is one that measures exactly 180°. The two sides of a straight angle form a straight line. (page 342)

Review of Processes

- To **solve a proportion** there are two strategies. One is to make repeated educated guesses and use the calculator to substitute values for the variable until a particular value makes the two ratios (compared in decimal form) equal. The algebraic strategy is to solve the equation for the variable using rules of equations. (pages 310, 311)
- To **convert from one unit to another**, multiply by the appropriate conversion factor(s) until the old unit is canceled and the new unit is left. (pages 315-318)
- **Rate conversions** are very similar to other unit conversions. The difference is there may be units to cancel in both the numerator and the denominator. (page 331)

Formulas (page 327)

d = rt distance = rate times time. (If the rate is in miles per hour, distance is measured in miles and time in hours.)

Total cost = (cost per day) times (the number of days)

Total miles the car will travel on a particular number of gallons of gas = (miles per gallon) times (the number of gallons)

Total price = (price per unit) times (number of units)

Summary Test--Chapter 5

Part 1: No calculators allowed.

1. In each of the following, set up an appropriate proportion, solve the equation, and answer the question. Show your work.

   The ratio of red to white beans is 5 to 9.

   a. How many white beans would be in a pile of 28 beans? ____________

   b. How many red beans would be in a pile of 28 beans? ____________

2. Solve for the variable in the following proportions. You may express the solution in fraction or decimal form.

   a. $\frac{4}{3} = \frac{x}{10}$ x = __________ b. $\frac{2}{7} = \frac{7}{a}$ a = ____________

   c. $\frac{5}{8} = \frac{15}{t}$ t = __________ d. $\frac{6}{11} = \frac{r}{100}$ r = ____________

3. For each of the following ratios, circle the closest equivalent ratio from the three choices given.

a. 4789 to 2068	0.4 to 1	2.3 to 1	3.2 to 1
b. 18,967 to 20,102	0.5 to 1	1 to 1	2 to 1
c. 89 to 15	4 to 1	5 to 1	6 to 1

4. In the following angles, circle the most likely ratio of rise to run from the three choices given.

   a.

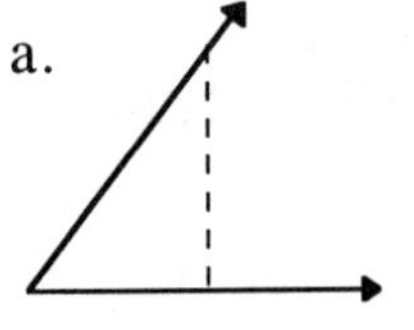

   5 to 1 0.2 to 1 1.4 to 1

   b. 3 to 1 1 to 3 1 to 5

5. For each of the following angles, circle the most likely angle measurement.

   a. 38° 15° 99°

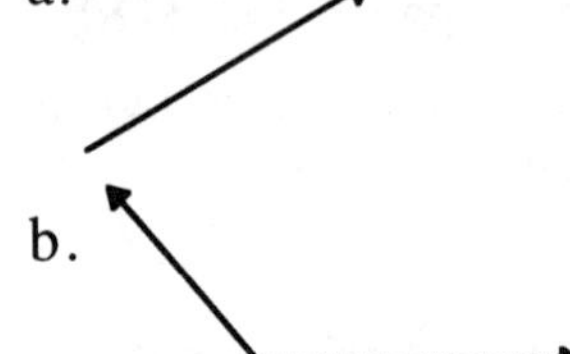

   b. 75° 170° 128°

Part 2: Calculator recommended.

6. In a school district of 22,312 total students, there are 4568 Hispanic students,

   a. What is the ratio of Hispanic to total students? ____________________

   b. What is the ratio of Hispanic to non-Hispanic students? ______________

   If the total number of students grew to 35,290 students and the ratio of Hispanic to non-Hispanic students remained the same,

   c. About how many students would be Hispanic? ____________________

   d. About how many students would be non-Hispanic? ______________

7. Compute the following rates.
   a. An airplane travels 1,870 miles in 4.5 hours. What is the speed of the plane in miles per hour?
   b. Jesse typed 986 words in 19 minutes. What is Jesse's typing speed (in words per minute?)

8. Make the following unit conversions. Write in the unit that is given, the unit that is wanted, and the path to make the conversion. Show the work.

   a. Convert 79 inches to yards (1 yard = 36 inches).
   Given: ____________ Wanted: ___________ Path: ________________

   Answer: 79 inches = ____________________ yards

   b. Convert 587 grams to pounds (1 kilogram = 1000 grams and 1 pound = 0.45 kilograms). There are two conversions.
   Given: ____________ Wanted: ___________ Path: ________________

   Given: ____________ Wanted: ___________ Path: ________________

   Answer: 587 grams = ____________________ pounds

   c. Convert 87 liters to quarts (1 quart = 0.95 liters, 1 liter = 1.06 quarts).
   Given: ____________ Wanted: ___________ Path: ________________

   Answer: 79 liters = ____________________ quarts

   d. Convert 79 square inches to square feet.
   Given: ____________ Wanted: ___________ Path: ________________

   Answer: 79 inches = ____________________ yards

   e. Convert 3 hours and 43 minutes to hours.
   Given: ____________ Wanted: ___________ Path: ________________

   Answer: 3 hours and 43 minutes = ____________________ hours

9. □ □ □ ■ ■ ■ ■ ■

   From the above drawings, write the following ratios. Write your ratios as fractions.
   a. The ratio of unshaded rectangles to shaded rectangles: ____________________
   b. The ratio of shaded rectangles to unshaded rectangles: ____________________
   c. The ratio of shaded rectangles to total rectangles: ________________________
   d. The ratio of unshaded rectangles to total rectangles: ______________________
   e. If there were 4182 total rectangles with about the same ratio of shaded to unshaded rectangles, which of the following is the most likely combination of shaded and unshaded rectangles?

1568 unshaded 2614 shaded	924 unshaded 3258 shaded	3596 unshaded 586 shaded

   Explain why the following ratios are obviously wrong.
   f. The ratio of unshaded to the total number of rectangles is 3025/4182.
   Explanation: ____
   g. The ratio of shaded to the total number of rectangles is 4003/4182.
   Explanation: ____

10. Make the following rate conversion (1 mile = 1.61 kilometers and 1 kilometer = 0.62 miles). 55 miles per hour is how many kilometers per hour? ______________________

Unit Conversion Chart

English Measurements

Length

12 inches =	1 foot
3 feet =	1 yard
5280 feet =	1 mile

Time

60 seconds =	1 minute
60 minutes =	1 hour
24 hours =	1 day
7 days =	1 week
365 days =	1 year

Liquid

3 teaspoons =	1 tablespoon
2 cups =	1 pint
2 pints =	1 quart
4 quarts =	1 gallon

Weight

16 ounces =	1 pound
2000 pounds =	1 ton

1 pint (liquid) = 1 pound (liquid)

Metric Measurements

Length

1 millimeter =	0.001 meter	
1 centimeter =	10 millimeters	= 0.01 meter
1 decimeter =	10 centimeters	= 0.1 meter
1 **METER** =	10 decimeters	= 1 meter
1 dekameter =	10 meters	= 10 meters
1 hectometer =	10 dekameters	= 100 meters
1 kilometer =	10 hectometers	= 1000 meters

Weight

1 milligram =	0.001 gram	
1 centigram =	10 milligrams	= 0.01 gram
1 decigram =	10 centigrams	= 0.1 gram
1 **GRAM** =	10 decigrams	= 1 gram
1 dekagram =	10 grams	= 10 grams
1 hectogram =	10 dekagrams	= 100 grams
1 kilogram =	10 hectograms	=1000 grams
1 metric ton =	1000 kilograms	

Liquid and Dry Measure

1 milliliter =		= 0.001 liter
1 centiliter =	10 milliliters	= 0.01 liter
1 deciliter =	10 centiliters	= 0.1 liter
1 **LITER** =	10 deciliters	= 1 liter
1 dekaliter =	10 liters	= 10 liters
1 hectoliter =	10 dekaliters	= 100 liters
1 kiloliter =	10 hectoliter	= 1000 liters

English to Metric Conversions

1 inch	≈	2.54 centimeters
1 foot	≈	0.30 meter
1 yard	≈	0.91 meter
1 mile	≈	1.61 kilometers
1 quart	≈	0.95 liter
1 pound	≈	0.45 kilograms

Metric to English Conversions

1 centimeter	≈	0.39 inches
1 meter	≈	3.28 feet
1 meter	≈	1.09 yards
1 kilometer	≈	0.62 mile
1 liter	≈	1.06 quarts
1 kilogram	≈	2.20 pounds

Chapter 6 Percent and Percentage

Key Topics

- The connection between percents and cents
- Estimating percents
- Computing percents
- Using percents in making predictions
- Circle graphs and bar charts
- Percent change

Materials Needed for the Exercises

- Calculator
- Tracing paper
- Ruler

Section 6-1 *Cents, Percent, and Percent Estimates*

We seldom read a magazine or newspaper or watch a news show without seeing some statistic presented in terms of percent. Percent is a way to understand very large numbers that are otherwise difficult to comprehend. If we try to understand the federal budget, for example, we find that the numbers are very large and it is difficult to determine whether a one-million-dollar expenditure is a lot of money or a little. To judge that, we have to put it in perspective. We need to know how the million dollars relates to the greater picture of federal spending. To compare the very large numbers involved in the federal budget (billions and trillions of dollars), we use ratios or percents.

Percent means "**per hundred**," 49% means 49 per hundred. If 49% of babies born are male, this means that for every 100 babies born, 49 are male. Using our knowledge of ratios and fractions, 49 per hundred is written as the ratio of 49 to 100, or in fraction form as $\frac{49}{100}$, or as a decimal, 0.49. In other words, $49\% = \frac{49}{100} = 0.49$.

The word "**percentage**" which means "**by the hundred**," is closely related to percent. The percentage is the number you get when you multiply a percent (which is a rate) times a number. For example, suppose 35% of the cats in a cat show are Siamese and there are 400 cats in the cat show. Then the percentage of Siamese cats in the cat show is 140 Siamese cats because 35% of 400 is 0.35 times 400 which equals 140.

Let's begin our work with percent and percentage by learning how they relate to fractions and decimals. We will start by connecting cents and percents. Writing percent in decimal form is mathematically the same as writing cents in dollar form.

In the following examples, we will demonstrate what is meant by a percent and how to write percent in decimal form.

Example 1:

100 cents is $1.00. 100 percent of $1.00 is 100 cents ($1.00)

100% (100 percent), written in decimal form, is 1 (or 1.00).

Written as a ratio, 100% is 100 out of $100 = \frac{100}{100} = 1$

$100\% = 1$

Example 2:

75 cents is $0.75 75 percent of $1.00 is 75 cents ($0.75)

75% (75 percent), written in decimal form, is 0.75.

As a ratio, $75\% = \frac{75}{100} = 0.75$

Example 3:

7 cents is $0.07 7 cents is also 7 percent of $1.00

7% (7 percent), written in decimal form, is 0.07.

As a ratio, $7\% = \frac{7}{100} = 0.07$

Example 4:

192 cents is $1.92 192 cents is also 192 percent of $1.00

192% (192 percent), written in decimal form, is 1.92.

As a ratio, $192\% = \frac{192}{100} = 1.92$

Now we will expand the concept of calculating percentage using money amounts other than $1.00.

Example 5: Find 75% of $3.00.

As shown above, 75% of one dollar is $0.75, so 75% of $3.00 can be computed by taking **75% of $1.00 three times.**

75% of $3.00 = $0.75(3)= $2.25

Example 6: Find 28% of $5.00.

As shown above, 28% of one dollar is $0.28, so 28% of $5.00 can be computed by taking **28% of $1.00 five times.**

28% of $5.00 = $0.28(5)= $1.40

Example 7: Find 16% of $7.00.

As shown above, 16% of one dollar is $0.16, so 16% of $7.00 can be computed by taking **16% of $1.00 seven times.**

16% of $7.00 = $0.16(7)= $1.12

Exercise 1--Pennies and Dollars

Purpose of the Exercise

- To practice connections between cents and percents
- To practice writing a percent as a decimal and a decimal as a percent

1. In each of the following, put the cents in dollar notation and convert the percents to decimal form.

a. 89 ¢ = ____________________ b. 65 ¢ = ____________________

89% = ____________________ 65% = ____________________

c. 17 ¢ = ____________________ d. 145 ¢ = ____________________

17% = ____________________ 145% = ____________________

e. 9 ¢ = ____________________ f. 1¢ = ____________________

9% = ____________________ 1% = ____________________

g. 456 ¢ = ____________________ h. 83 ¢ = ____________________

456% = ____________________ 83% = ____________________

i. 10 ¢ = ____________________ j. 52 ¢ = ____________________

10% = ____________________ 52% = ____________________

2. Without a calculator, compute the following percents.

a. 50% of $1.00 = _______________ 50% of $3.00 = _______________

Show how you set up the multiplication to compute 50% of $3.00. ______

b. 80% of $1.00 = ______________ 80% of $7.00 = ______________

Show how you set up the multiplication to compute 80% of $7.00. ______

c. 20% of $1.00 = ______________ 20% of $4.50 = ______________

Show how you set up the multiplication to compute 20% of $4.50. ______

d. 32% of $1.00 = ______________ 32% of $5.00 = ______________

Show how you set up the multiplication to compute 32% of $5.00. ______

e. 9% of $1.00 = ______________ 9% of $2.00 = ______________

Show how you set up the multiplication to compute 9% of $2.00. ________

f. 1% of $1.00 = ______________ 1% of $52.00 = ______________

Show how you set up the multiplication to compute 1% of $52.00. ______

g. Write a sentence or two explaining how you would get 83% of some number.

____

____

h. Write a sentence or two explaining how you would get 147% of some number.

____

____

Let's summarize our findings by writing the process to convert a percent to decimal and to compute a particular percent of a number.

Convert a Percent to Decimal Form

Write the percent as a ratio (the number per 100) and then change the fraction to decimal form (divide by one hundred). A shortcut is to move the decimal point two places to the left in the number.

For example, 187% = 187/100 = 1.87 written in decimal form.

Convert Decimal Form to Percent

Reverse the above process. The shortcut is to move the decimal point two places to the right in the number.

For example, 0.04 written as a percent is 4%.

Compute a Particular Percent of a Number

First, convert the percent to decimal form and
Second, multiply the decimal form by the number.

Percent Thermometers: Estimating Percentages

As we learn to do calculations involving percents, we will work with a "percent thermometer" to help us visually understand percents and to estimate answers. We will use the estimates to determine whether our calculations are reasonable. There are several types of arithmetic problems involving percents, and it is sometime difficult to remember whether we multiply or divide and if we divide, which number goes on top. Knowing what a reasonable answer would be can help us to determine whether our process is correct. We will also learn algebraic processes for determining answers to percent problems.

A **percent thermometer** is just a scale going from 0% to 100% that looks like a thermometer.

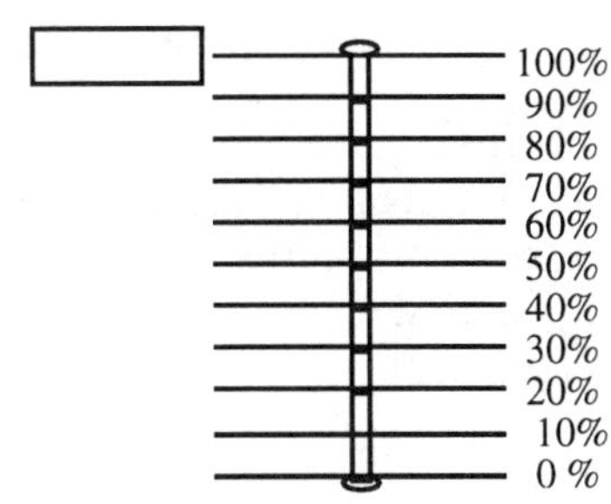

Example 8: According to an article, *Understanding Brain Injury*, at the Neurotrauma ♦ Law Nexus site on the World Wide Web, there are over 2 million traumatic brain injuries a year, and 76% of those injuries occur among children under 15 years of age. To use a percent thermometer to estimate 76% of 2 million, we first put the number 2 million in the box at the 100% mark on the left side of the thermometer and put a 0 at the bottom. We then treat the thermometer as a vertical scale and estimate what number would fit on the scale at the 76% mark. The 50% mark would be at 1 million, and the 76% (which is about three-quarters) mark would be a little over half way in between 50% and 100%. Let's guess 1.6 million.

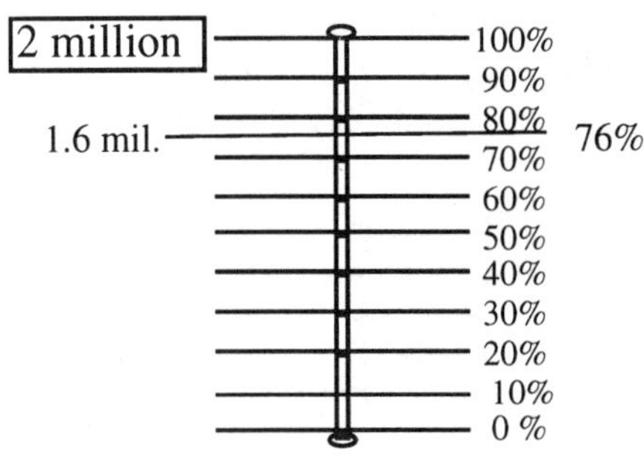

76% of 2 million = (0.76)(2) = 1.52 million. Our guess was close.

Example 9: When computing more than 100% of a number, we go off the scale, above the 100% mark. Suppose the new price of a suit is 115% of the original price of $250.00. To compute the new price, we take 115% of 250. We know that the answer will be more than 250 because 115% is above 100%. In fact, it is 15% higher than 100%. You know that 10% of 250 is 25, so let's guess that 15% of 250 is about 30, so 115% of 250 is apprximately 280.

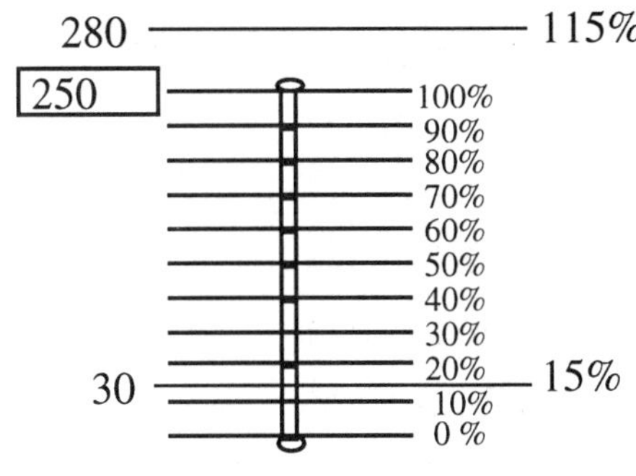

Estimate: 30 15% mark115% of 250 = (1.15)(250) = 287.5. Our guess was close. The price of the suit is now $287.50

Exercise 2--Percent Thermometer

Purpose of the Exercise

- To get a visual sense how percentages of numbers fit on a scale
- To solve problems involving percent

Materials Needed

- Calculator
- Tracing paper

For each of the following problems, determine a reasonable answer to the question by using the percent thermometer. To help guess the answers you may want to fill in percentages for some of the easy to compute percents such as 10%, 25% (1/4), 50% (1/2), and 75% (3/4).

Use your calculator to check your estimate by converting the percent to decimal and performing the multiplication. In order to use the same drawing many times, use tracing paper to lay over the drawing to make the estimates.

1.

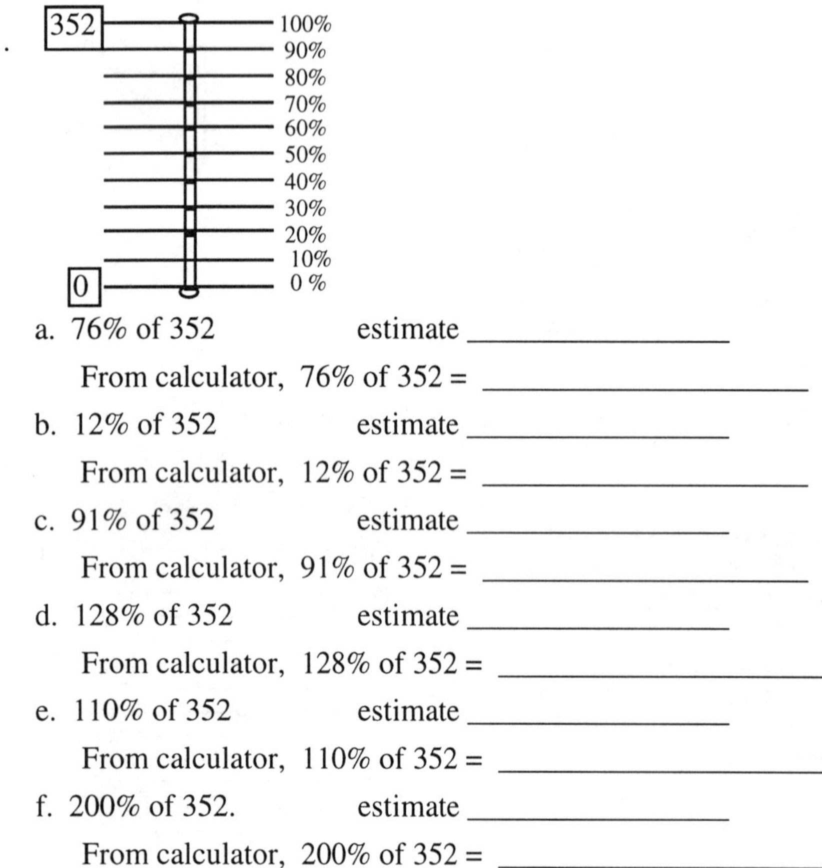

a. 76% of 352 estimate ________________

From calculator, 76% of 352 = ____________________

b. 12% of 352 estimate ________________

From calculator, 12% of 352 = ____________________

c. 91% of 352 estimate ________________

From calculator, 91% of 352 = ____________________

d. 128% of 352 estimate ________________

From calculator, 128% of 352 = ____________________

e. 110% of 352 estimate ________________

From calculator, 110% of 352 = ____________________

f. 200% of 352. estimate ________________

From calculator, 200% of 352 = ____________________

2. 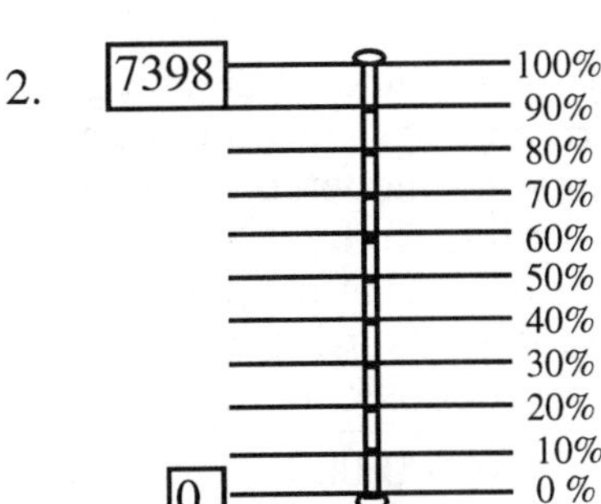

a. 81.5% of 7398 estimate ________________

From calculator, 81.5% of 7398 = ____________________

b. 29.6% of 7398 estimate ________________

From calculator, 29.6% of 7398 = ____________________

c. 8% of 7398 estimate ________________

From calculator, 8% of 7398 = ____________________

d. 2% of 7398 estimate ________________

From calculator, 2% of 7398 = ____________________

e. 51% Of 7398 estimate ________________

From calculator, 51% of 7398 = ____________________

f. 125% of 7398 estimate ________________

From calculator, 17% of 7398 = ____________________

3. Try your luck at making reasonable guesses. **Without your calculator**, pick the best percentage from the four choices. Use tracing paper and the percent thermometer as needed to make the estimates. After you finish, compute the percentages with your calculator to see how well you did.

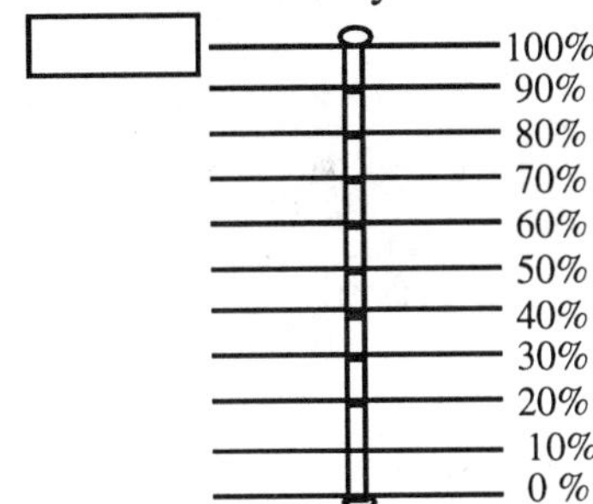

a. 83% of 3578 ≈	900	3000	4560	2140
b. 14.2% of 876 ≈	420	730	900	120
c. 252% of 375 ≈	600	950	1500	300
d. 132% of 2120 ≈	2800	4000	2000	1600
e. 17% of 11 ≈	7	21	2	6
f. 52% of 20 ≈	17	25	7	11
g. 42% of 999 ≈	723	41	410	211
h. 6.9% of 250 ≈	19	5	35	70

Now that we know how to find a reasonable estimate and how to compute a percentage (that is, a particular percent of a number), we now need to learn how to determine what percent one number is of another. For example, in 1997 the amount budgeted for Social Security in the federal budget was about \$364.2 billion while the total federal budget was about \$1.630 trillion. To compare the numbers, we might determine what percent 364.2 billion is of 1.630 trillion. We already know how to make comparisons using ratios. The ratio of Social Security spending to the total budget was $\frac{364,200,000,000}{1,630,000,000,000} \approx 0.22 =$ 22%. 22% is slightly less than 25% or 1/4, so the amount budgeted for Social Security spending in 1997 was about 22% (almost one-fourth) of the federal budget.

Since determining what goes in the numerator and denominator, or whether we multiply or divide can be confusing in problems involving percents, we will use the percent thermometer to help determine a reasonable answer as a double-check that our process is correct.

Example 10: 78 is what percent of 265?

We will use the percent thermometer to estimate the percent. First fill in 10% and 50% of 265. We notice that the answer falls between the two and make a guess. Let's guess 30%. Now let's write the ratio and compute the percent to see how close we came. $\frac{78}{265} \approx 0.294 = 29.4\%$, which is close to our estimate. (Had we made the common mistake of reversing the numerator and denominator in our computations, we would have gotten $265/78 \approx 3.40 = 340\%$. Notice that 340% is not reasonable, so we would check the process.)

265	100%
	90%
	80%
	70%
	60%
132.5	50%
	40%
78	30%
	20%
26.5	10%
0	0 %

Exercise 3-- Computing Percent

Purpose of the Exercise

- To practice computing percent

Materials Needed

- Tracing paper

In the following problems, use the percent thermometer to guess at a reasonable answer. Then write the appropriate ratio to determine the percent.

☐	100%
	90%
	80%
	70%
	60%
	50%
	40%
	30%
	20%
	10%
	0 %

1. 83 is what percent of 279? Guess ________ Ratio ________ = ________%

2. 54 is what percent of 279? Guess ________ Ratio ________ = ________%

3. 599 is what percent of 279? Guess ________ Ratio ________ = ________%

4. 187 is what percent of 279? Guess ________ Ratio ________ = ________%

5. 876 is what percent of 1000? Guess ________ Ratio ________ = ________%

6. 1000 is what percent of 876? Guess ________ Ratio ________ = ________%

7. 38 is what percent of 280? Guess ________ Ratio ________ = ________%

8. 63 is what percent of 70? Guess ________ Ratio ________ = ________%

9. 678 is what percent of 2789? Guess ________ Ratio ________ = ________%

10. 678 is what percent of 600? Guess ________ Ratio ________ = ________%

11. 49 is what percent of 78? Guess ________ Ratio ________ = ________%

12. 78 is what percent of 49? Guess ________ Ratio ________ = ________%

Example 11: Another category of percent problem comes when we know, for example that 52% of a particular community is female and that there are 4798 females in the community. From that information, we can determine how many people, total, there are in the community. On the percent thermometer, the total population (the unknown in this case) goes in the box at the 100% mark. 4798 goes at the 52% mark.

52% of some number is 4798. What is the number?

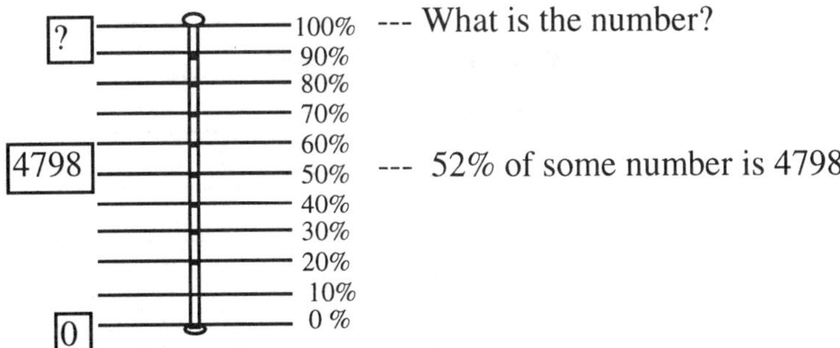

First: Place 4798 on the percent thermometer at the 52% mark. The number we are seeking is the one that will go in the box.

Second: Guess at the number that will go in the box.

First guess: 9500. Notice that the number that goes in the box will be almost double 4798.

52% of 9500 is 4940 which is too high. Our next guess will be less than 9500.

Second guess: 9000

52% of 9000 is 4680 which is too low.

Using our previous guesses to help us make better educated guesses and using the calculator to compute 52% of various numbers until we find the answer, we find that 52% of <u>9227</u> is 4798.04 which is as close as we will get with whole numbers. To answer the question, the number is about 9227.

In the next section, we will learn to solve algebraically the three basic types of problems involving percent. But for now, when we want the find the total number of items, given a percentage, we will use the percent thermometer to determine a reasonable answer and a calculator to find an answer through repeated guesses.

Exercise 4--Percent Thermometer

Purpose of the Exercise

- To visually estimate the total number of items, when a percentage is known

Materials Needed

- Tracing paper
- Calculator

In each of the following, compute the total number of items from the given percentage. First determine a reasonable answer by using the percent thermometer and then make repeated guesses, checking each guess with a calculator until you find the answer. Record the final answer and how you check to see that it is right.

100%
90%
80%
70%
60%
50%
40%
30%
20%
10%
0 %

1. 10% of a number is 4759. What is the number? __________

   Check ______________________________

2. 83% of a number is 4759. What is the number? __________

   Check ______________________________

3. 25% of a number is 37. What is the number? __________

   Check ______________________________

4. 120% of a number is 37. What is the number? __________

   Check ______________________________

5. 245% of a number is 37. What is the number? __________

   Check ______________________________

6. 92% of a number is 87. What is the number? __________

   Check ______________________________

7. 92% of a number is 563. What is the number? __________

   Check ______________________________

8. 92% of a number is 22. What is the number? __________

   Check ______________________________

9. 57% of a number is 7845. What is the number? __________

   Check ______________________________

10. 108% of a number is 93. What is the number? __________

    Check ______________________________

Section 6-2 *An Algebraic Look at Percent Problems*

When working with problems involving percent, if we just memorize a process, it is easy to get confused whether we multiply the numbers involved or divide the numbers involved and, if we divide, which number is the numerator and which is the denominator. There are two things we do to help us remember what the process is. One is to use the percent thermometer to see when the answer is reasonable. Another systematic way is to set up a calculation involving percents as an algebraic equation and then solve the equation to get the answer. As we learned from taking 73% of 183, for example, the answer is (0.73)(183) = 133.59. To take a percent of a number, we get the following generic formula.

(the percent, written as a decimal) times (the number) = percentage

Notice that the percent is multiplied by the number that follows the phrase "percent of" or "% of" in the problem.

Now let's set up some percent problems as equations using the above statement. In any of the percent problems, we know two of the values in the above statement and we are to solve for the third, unknown value. As before when we worked with algebra, we will assign a variable to the unknown.

Review of the Rules of Equations

1. We can add the same number to both sides of the equation.
2. We can subtract the same number from both sides of the equation.
3. We can multiply both sides of the equation by the same number.
4. We can divide both sides of the equation by the same nonzero number.

In the following examples, the rounding is done differently in different problems. In real life, the choice of where to round would be determined by the practical situation. For example, when working with people, the percentage would almost always be expressed as a whole number since people are generally counted in whole numbers, not in parts. Whether to write 87.937% as 87.94%, 87.9% or 88% depends on the precision needed in the particular situation.

Example 1: Compute 23% of 14.

In this case the unknown is the "answer" in the statement above, so the equation becomes

$$x = (0.23)(14)$$
$$x = 3.22$$

Since 1.4 is at the 10% mark and 23% would be more than double that, the answer looks reasonable.

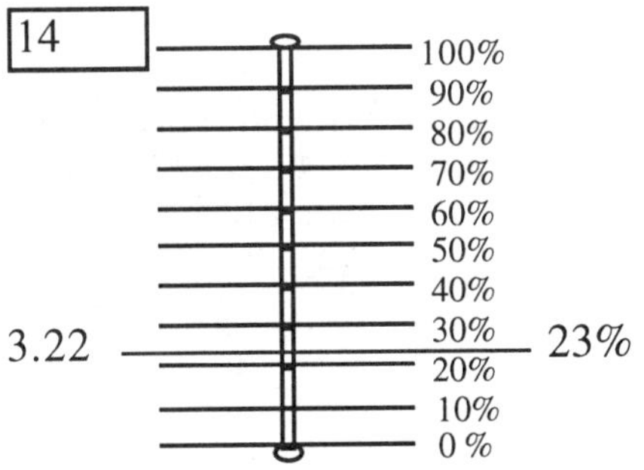

14
100%
90%
80%
70%
60%
50%
40%
30%
3.22
23%
20%
10%
0 %

The problem in Example 1 is the most straightforward type of the percent problems because all we have to do is to multiply the numbers on the right side of the equation to get the answer.

Example 2: What is 7% of 389?

$t = (0.07)(389\ )$

$t = 27.23$

Since 10% of 389 is 38.9 and 7% is slightly below the 10% mark, the answer looks reasonable.

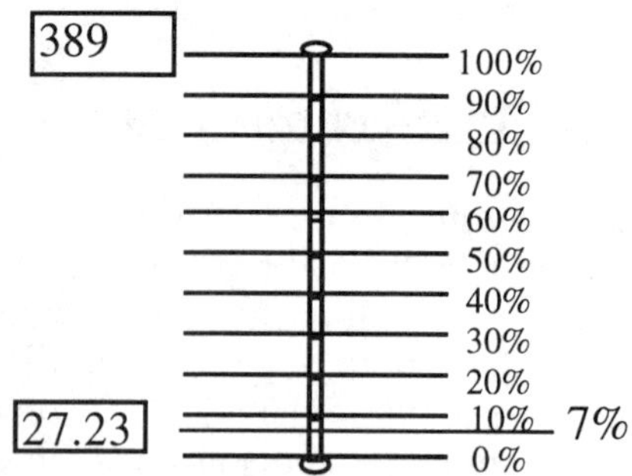

Example 3: 389 is 7% of what number?

In this example, the unknown quantity follows the pharse "% of." This means we multiply the percent times the variable. That leaves 389 on the other side of the equation.

$0.07n\ = 389$

$\frac{0.07n}{0.07} = \frac{389}{0.07}$ Divide both sides of the equation by 0.07.

≈ 55.57

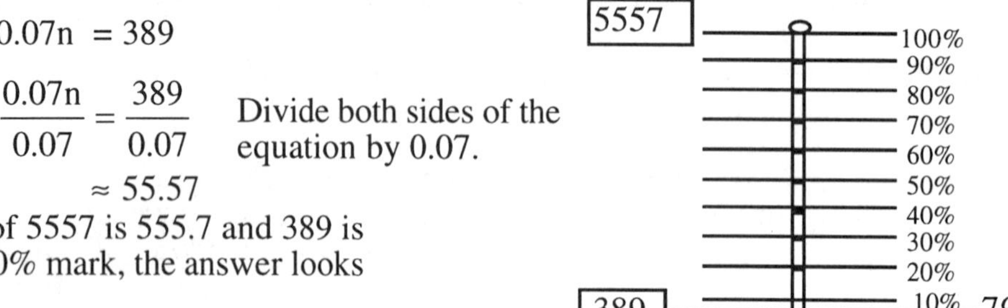

Since 10% of 5557 is 555.7 and 389 is below the 10% mark, the answer looks reasonable.

Example 4: 85 is what percent of 99?

Written in algebra, 85 = (a)(99). Or more concisely,

$99a = 85$

$\frac{99a}{99} = \frac{85}{99}$ Divide both sides of the equation by 99.

$a \approx 0.859$

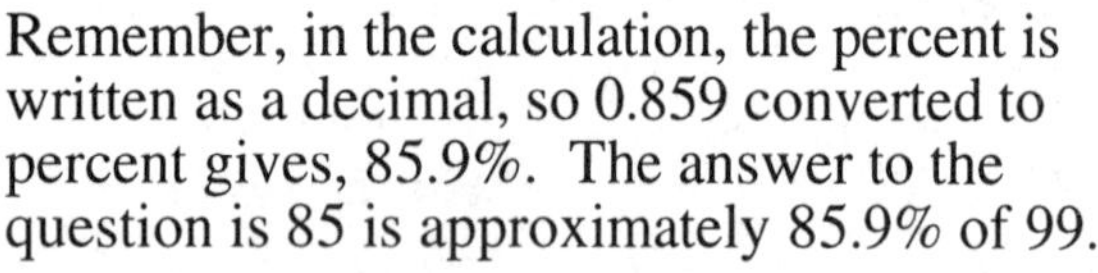

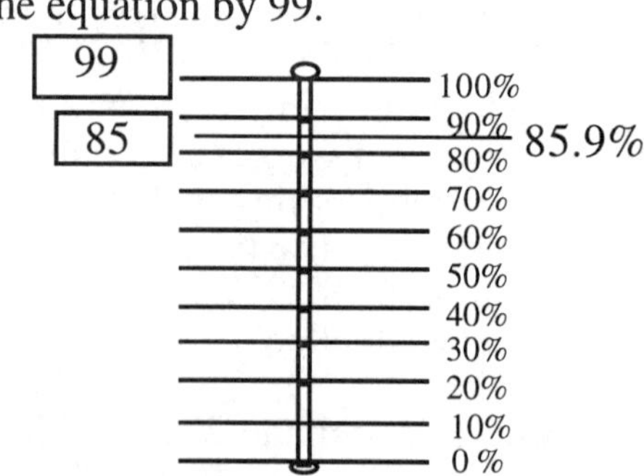

Remember, in the calculation, the percent is written as a decimal, so 0.859 converted to percent gives, 85.9%. The answer to the question is 85 is approximately 85.9% of 99.

99 is very close to 100, and 85 is 85% of 100, so the answer is reasonable.

Example 5: 857 is what percent of 99?

Written algebraically, 857 = (a)(99. Or more concisely,

$99a = 857$

$\frac{99a}{99} = \frac{857}{99}$ Divide both sides by 99.

$a \approx 8.657$

857

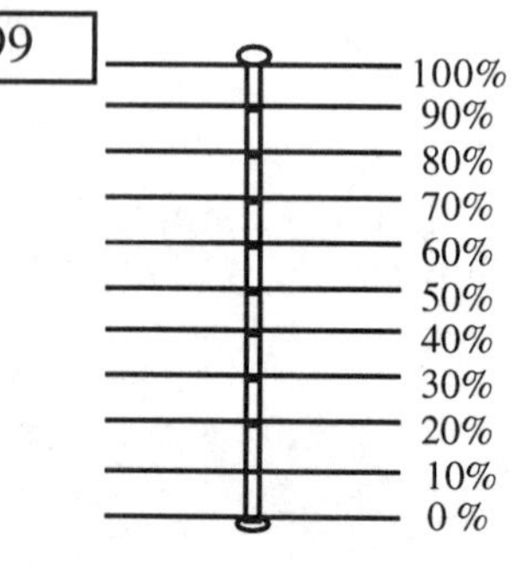

To answer the question, 857 is approximately 865.7% of 99. At first this might seem like a strange answer, but if we check visually, we see that 857 is much larger than the number, 99, in the box. In fact it is more than 8 times as large as 99, so the percent of 99 associated with 857 would be way above the scale. The entire scale would have to be repeated more than 8 times above 100%.

In each of the following examples,

First: Determine how to place the information on the percent thermometer.
Second: Get a rough estimate by guessing.
Third: Set up an algebraic equation to solve the problem.
Fourth: Answer the question in a full English sentence.
Fifth: Check the reasonableness of the answer and compare it with your estimate.

Example 6: 352 is 74% of what number?

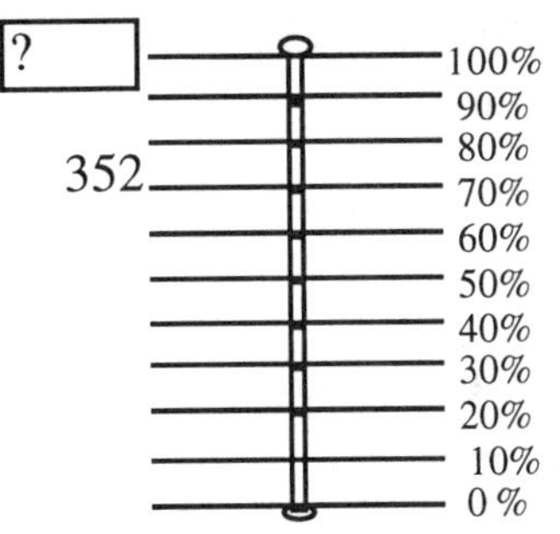

The unknown is the total number of items.
Estimated answer: 460 (higher than 352)
Algebra equation: 352 = 0.74x
Solution: x = 352/0.74 ≈ 475.7
Answer: 352 is 74% of 475.7 (approximately)
Check: 0.74(475.7) = 352.018 ≈ 352

Example 7: 74% of 352 is what number?

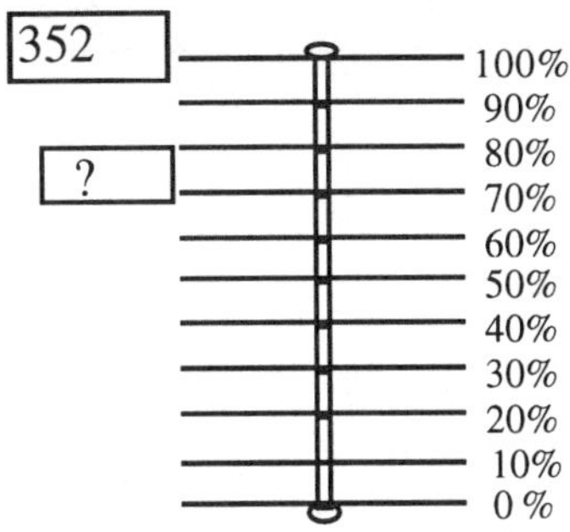

The unknown is a number on the scale.
Estimated answer: 300 (less than 352)
Algebraic equation: 0.74(352) = x
Solution: x = 260
Answer: 74% of 352 is 260
Check: 0.74(352) = 260

Example 8: 352 is what percent of 74?

74

100%
90%
80%
70%
60%
50%
40%
30%
20%
10%
0 %

The unknown is the percent.
Estimated answer: 400% (352 is more than 4 times as large as 74)
Algebra equation: 352 = 74t
Solution: t = 352/74 ≈ 4.758
4.758 written as a percent is 475.8%
Answer: 352 is approximately 475.8% of 74.
Check: 4.758(74) = 354.09 (close)

Example 9: 74 is what percent of 352?

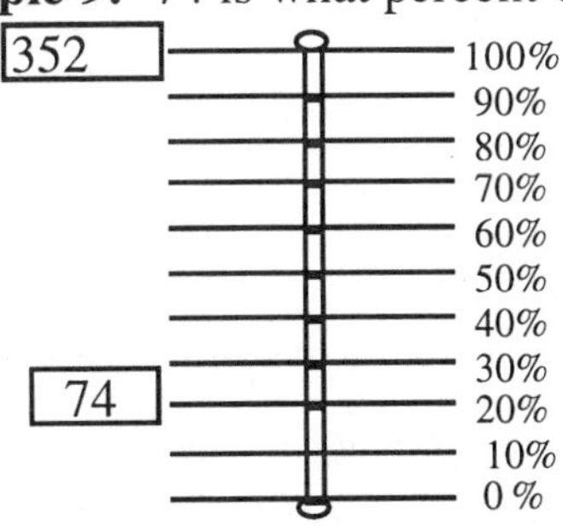

The unknown is the percent.
Estimated answer: 25%

Algebraic equation: 74 = 352s
Solution: s = 74/352 ≈ 0.21
0.21 written as a percent is 21%
Answer: 74 is approximately 21% of 352.
Check: 0.21(352) = 73.92 (close)

When checking the work, the answer may not be exact because of rounding.

Exercise 1--All Three Kinds of Percentage Problems

Purpose of the Exercise

- To practice estimating solutions to percent problems
- To practice computing solutions to percent problems using an algebraic equation

Materials Needed

- Calculator
- Tracing paper

In each of the following, write down the information that is given and the unknown information into the algebraic equation and solve the equation. Answer the question in a complete sentence. As in the examples above, state what is unknown, the percent, the total number of items, or a number on the scale. Show your work.

1. 352 is what percent of 74?
   Unknown: ______________________
   Alg. eqn: ______________________
   Solution: ______________________
   Answer: ______________________
   Check: ______________________

2. What is 220% of 87?
   Unknown: ______________________
   Alg. eqn. ______________________
   Solution: ______________________
   Answer: ______________________
   Check: ______________________

3. 6% of a number is 879?
   Unknown: ______________________
   Alg. eqn: ______________________
   Solution: ______________________
   Answer: ______________________
   Check: ______________________

4. 88 is what percent of 17?
   Unknown: ______________________
   Alg. eqn. ______________________
   Solution: ______________________
   Answer: ______________________
   Check: ______________________

5. 17 is what percent of 65?
   Unknown: ______________________
   Alg. eqn: ______________________
   Solution: ______________________
   Answer: ______________________
   Check: ______________________

6. What is 130% of 369?
   Unknown: ______________________
   Alg. eqn. ______________________
   Solution: ______________________
   Answer: ______________________
   Check: ______________________

7. 893 is what percent of 2056?
   Unknown: ______________________
   Alg. eqn: ______________________
   Solution: ______________________
   Answer: ______________________
   Check: ______________________

8. What is 7.3% of 215?
   Unknown: ______________________
   Alg. eqn. ______________________
   Solution: ______________________
   Answer: ______________________
   Check: ______________________

9. What percent of 600 is 700?
   Unknown: ______________________
   Alg. eqn: ______________________
   Solution: ______________________
   Answer: ______________________
   Check: ______________________

10. 7800 is what percent of 7000?
    Unknown: ______________________
    Alg. eqn. ______________________
    Solution: ______________________
    Answer: ______________________
    Check: ______________________

11. What is 29.3% of 900?

Unknown: ______________________

Alg. eqn: ______________________

Solution: ______________________

Answer: ______________________

Check: ______________________

12. 8.2% of a number is 88. What is the number?

Unknown: ______________________

Alg. eqn. ______________________

Solution: ______________________

Answer: ______________________

Check: ______________________

We see percents and quantitative data many places--in newspapers, magazines, on television, and in sales in stores. Taxes are frequently determined as a percent of purchase prices or property values. We even see percents on cans of food. To understand the world around us, we need to understand percents and how to interpret information involving percents. The following examples will show how we can interpret data given to us in different forms--in tables, charts, and in words.

Example 10: A parade float is decorated with 7293 red roses, plus white roses, yellow roses, and pink roses. The percents of red, white and yellow roses are as given in the chart. On the chart, fill in the percentage of yellow, pink and white roses and the total number of roses on the float and the percent of pink roses.

Color of Roses	Percentage of (Number of) that Color of Roses	Percent of Total Number of Roses
Red	7293	53.9%
White		17.2%
Yellow		11.3%
Pink		
Total		100%

To fill in the chart, we will first fill in the missing percent of pink roses. Since the percent of roses must add up to 100%, the percent of pink roses is equal to 100 minus the total of the percents of the other roses. We can set this up as an algebra equation.

Let the unknown percent of pink roses be the variable x.
Adding the percents of colored roses to get the 100% total, we get

$53.9 + 17.2 + 11.3 + x = 100$ Add the numbers on the left.

$82.4 + x = 100$ Subtract 82.4 from both sides.

$82.4 + x - 82.4 = 100 - 82.4$

$x = 17.6$

17.6% of the roses are pink roses.

To determine the number of each type of rose, find the total number of roses on the float. We know that 7293 red roses make up 53.9% (a little over one-half) of the total number of roses. Let's set up the equation where 53.9% is written in decimal form as 0.539 and T is the total number of roses.

$0.539T = 7293$

$\frac{0.539T}{0.539} = \frac{7293}{0.539}$ Divide both sides by 0.539.

$T \approx 13{,}530$ Round the answer to the nearest whole rose.

Check: 53.9% of 13,530 is $(0.539)(13{,}530) \approx 7293$

Since we now know the total number of roses, finding the number (percentage) of the other colors of roses involves simply multiplying each percent by the total number of roses.

White roses:	17.2% of 13,530 is (0.172)(13,530) ≈ 2327
Yellow roses:	11.3% of 13,530 is (0.113)(13,530) ≈ 1529
Pink roses:	17.6% of 13,530 is (0.176)(13,530) ≈ 2381

Now we will check to see if all of the colors of roses add up to 13,530.

7293 + 2327 + 1529 + 2381 = 13,530.

In this case, our total came out to be exactly 13,530. It is possible for the work to be correct but the totals to be very slightly off because of rounding in the computations of the percentages.

Example 11: Suppose the 75,000 square miles of land in a state is proportioned as shown in the following pie chart, determine how much land is in each category.

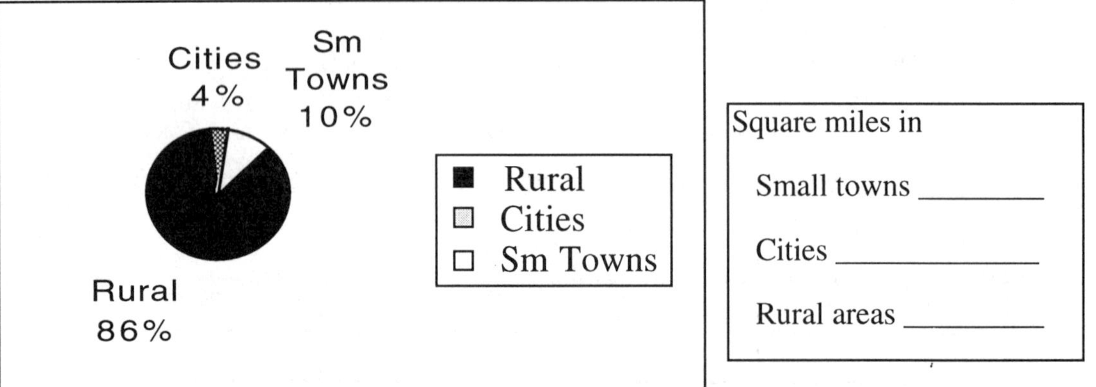

The percents are given in the chart. Computations of the square feet in the different types of areas in the state are straightforward.

Small towns:	10% of 75,000 is (0.10)(75,000) = 7500 square miles
Cities:	4% of 75,000 is (0.04)(75,000) = 3000 square miles
Rural areas:	86% of 75,000 is (0.86)(75,000) = 64500 square miles

Check: The total number of square miles in the state is 75,000 = 7500 + 3000 + 64500.

Example 12: Reading and interpreting

The *Labor Force Statistics from the Current Population Survey, Displaced Workers Summary* states that the employment status in February 1996 is as follows: "Of the 4.2 million workers with 3 or more years of tenure who had been displaced over the period from January 1993 to December 1995, 74 percent were reemployed and 13 percent were unemployed when surveyed in February 1996."

a. According to the statement, how many workers with three or more years of tenure had been displaced during the period from January 1993 to December 1995?
   Answer: 4.2 million workers

b. How many of those workers were reemployed when surveyed in February 1996?
   Answer: 74% of 4.2 million were reemployed. (0.74)(4.2) 3.1 million

c. How many of those workers were unemployed when surveyed in February 1996?
   Answer: 13% of 4.2 million were unemployed. (0.13)(4.2) 0.55 million

d. What percent of the workers were neither reemployed nor unemployed according to the February 1996 survey? (There may be several reasons why people do not fit into either category. For example, a person who chooses not to seek work is not classified either as employed or unemployed.)

Answer: 74% + 13% = 87% is the percent reemployed or unemployed.
This leaves 100% - 87% = 13% neither reemployed or unemployed.

e. How many of the displaced workers were neither reemployed nor unemployed according to the February 1996 survey?

Answer: 13% of 4.2 million. $(0.13)(4.2) \approx 0.55$ million

Check: The number reemployed + number unemployed + everyone else is
3.1 + 0.55 + 0.55 = 4.2 million total displaced workers.

Example 13:

According to information on the pint carton, *Haagen-Daz* vanilla swiss almond ice cream has 4 servings per carton and each serving is $\left(\frac{1}{2}\right)$ cup. Each serving has 310 calories, with190 of the calories per serving come from fat.
According to the above information,

a. How many cups of ice cream are in the carton?

Ans: There are four, one-half-cup servings. $(4)\left(\frac{1}{2}\right) = 2$ cups.

b. How many calories are in the whole pint?

Ans: Each of the four servings has 310 calories. 4(310) = 1240 calories.

c. How many calories in the carton come from fat?

Ans: Each of the four servings has 190 fat calories. 4(190) = 760 fat calories.

d. What percent of the calories come from fat?

We could work this two ways. In one serving, 190 of the 310 calories come from fat, so $190 = 310x$ where x is the percent (written as a decimal) of the 310 calories that come from fat. Solving the equation, $x \approx 0.6129$, so approximately 61% of the calories come from fat.

If we used the total 760 fat calories out of the total 1240 calories, we get $760 = 1240x$. Solving the equation, $x \approx 0.6129$, so we still get approximately 61% of the calories come from fat.

Exercise 2--Percent Applications and Word Problems

Purpose of the Exercise

- To work with percents and data in written form, chart form, and from tables
- Practice solving equations

Materials Needed

- Calculator

Solve the following word problems. Use additional paper to show your work.

1. The sales tax rate in one town is 5.93%. Dustin buys a car. The agreed upon price between Dustin and the sales person is $8975.

   a. How much is the sales tax on the car?

   b. What is the price of the car after taxes?

2. Sarah sells real estate and works on commission. She gets 9% of her sales. She sells a house for $215,970. How much money does she make on the sale?

3. In a partnership agreement, Christopher is a 20% partner, Brittany is a 49% partner, and Jesse is a 21% partner.

   a. How much is each partner's share of a $17,950 profit? ____________________

   b. How much is each partner's share of a $47,899 building repair? ____________

4. The price marked on a television set is $359.99. Chad paid $398.73 for the television after taxes were added on. What was the tax rate?

5. A discount co-op sells all items at a 8.7% markup over the cost of the item. The items on the shelf are marked at cost. Terry buys items marked at $7.89, $0.45, $0.98, $4.37, $5.91, $1.88. The sales tax rate is 5.5%.

   a. What is the total price (before taxes) of the items after the markup? _________

   b. What is the tax amount on the purchases? ______________________________

   c. What is the total price Terry pays for the items? ______________________

6. A store is going out of business and advertises that everything in the store is 60% off the marked price. The sales tax rate is 6.75%. Jill buys a dress marked $82.57, shoes marked $56.98, jeans at $42.76, and a T-shirt marked $13.75.

   a. What is the total discount price of Jill's purchases after the 60% discount? _____

   b. How much does Jill have to pay including taxes? ______________________

7. According to the Nabisco, reduced fat Nilla Wafers package, one serving is 8 wafers. There are about 11 servings per container, and each serving contains 120 calories with 20 of the calories coming from fat.

   a. About how many wafers are in the box?

   b. About how many calories are in the wafers in the box?

   c. What percent of the calories come from fat?

   d. About how many calories are in one wafer?

   e. About how many fat calories are in one wafer?

8. Puzzle Problem--How Rich Have You Become?

Suppose you hear on the news that two wealthy people, Joseph Moreno (who left an estate worth $40,343,000) and Natalie Harris (who left an estate worth $124,492,000) have died. When Mr. Moreno and Ms. Harris were very young they had a child who was put up for adoption. They lost contact with each other as well as the baby. Mr. Moreno's estate is to be divided evenly among four children: the three children he raised as well as the one he lost track of as a baby. Ms. Harris's situation is similar. Her stated wishes are to divide her money evenly among her 6 children: the 5 children she raised and one who was adopted at birth. Your parents (who adopted you at birth) call you and tell you that these were your biological parents and of course you are the lost child of Mr. Moreno and Ms. Harris. Suppose this is verified.

a. What is the value of your total inheritance?

b. What would your inheritance be if the lawyers and court fees for Mr. Moreno's estate took 40% of the estate in fees and the lawyers and court fees for Ms. Harris was 28.3% of the estate?

c. Write a report clearly showing your computations and answers.

9. In 1996, people graduating from law school had an average debt of $40,300, not including any unpaid undergraduate loans. If the loan is a 10-year, 8% loan, the average monthly payment is about $489. The percent of their total income that goes toward paying the monthly loan payment is as follows for the typical attorney in the following types of practices.

Public Defender	Public Service	Legal Services	Government	Small-Firm Associate	Large-Firm Associate
25%	23%	27%	20%	15%	12%

Source: Education Resources Institute and *USA Today*, January 10, 1997.

a. What is the average salary of a public defender? (Hint: 25 percent of the salary makes up the $489 monthly payment.)

b. What is the average salary of an attorney in public service?

c. What is the average salary of an attorney in legal services?

d. What is the average salary of an attorney in the government?

e. What is the average salary of a small-firm associate?

f. What is the average salary of a large-firm associate?

10. In 1995, there were 196,116 women serving in the military. Their distribution according to service branch and enlisted vs. officer is shown in the chart below.

	Officers	Enlisted
Army	10,786	57,260
Navy	7,899	47,931
Marine Corps	690	7403
Air Force	12,068	52,079

Source: Defense Department and *USA Today*, November 18, 1996.

a. How many women officers were there in 1996?

b. What percent of the women were officers in 1996?

c. What percent of the women were enlisted in 1996?

d. What percent of the women were officers in the Marine Corps in 1996?

e. What percent of the women were in the Marine Corps in 1996?

f. Which branch has the most women?

g. Which branch has the highest percent of women officers out of the total women in their branch?

11. There were 4,572 students enrolled at Maple Woods Community College, Kansas City, Missouri in the fall of 1995. Those students fell into the categories shown in the pie chart below.

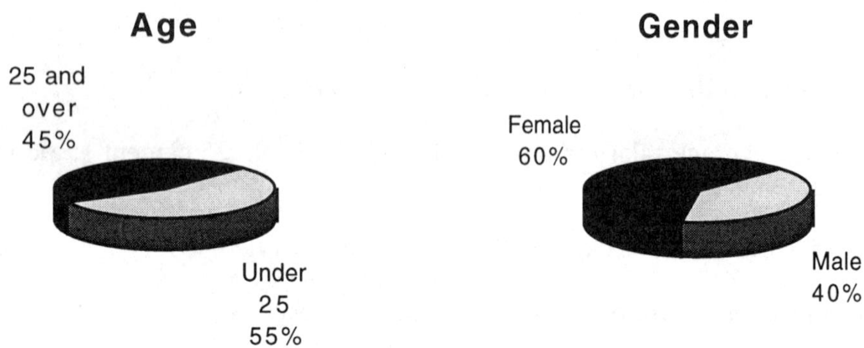

a. How many female students were enrolled at Maple Woods?

b. How many male students were enrolled at Maple Woods?

c. How many Maple Woods students were 25 and older?

d. How many Maple Woods students were under 25?

12. Solve each of the following algebra equations for the varible.

a. $0.69\text{ w} = 3245$ w = ________

b. $(34.5)(289) = t$ t = ________

c. $K(7895) = 2678$ K = ________

d. $\frac{7}{9}x = 3789$ x = ________

e. $\frac{9}{7}x = 3789$ x = ________

f. $\left(\frac{22}{79}\right)(567) = v$ v = ________

g. $85m = 23$ m = ________

h. $\frac{x}{11} = \frac{14}{19}$ x = ________

I. $\frac{5}{9} = \frac{3278}{x}$ x = ________

Projects to be done outside of class. Pick one.

1. If your city/state has a sales tax, find out the sales tax rate. What would be the price of a $32,870 car after taxes were added on?
2. Some schools have a tuition refund policy if you drop a class within a certain time period. See if your school has such a policy. If it does, determine how much it would cost you to enroll in a particular class and drop it during the first week of school.
3. Check your records for your housing expenses and the money you earned for the last two months
   How much money did you spend on rent/mortgage? ________
   How much money did you spend on paying utilities? ________
   How much money did you make? ________
   What percent of the money you made was spent on rent/mortgage? ________
   What percent of the money you made was spent on utilities? ________
4. Check your monthly gas bill (or some other utility) for a 12-month period. For each month, compute that month's percent of the bills for the entire year.
5. Pose a question about a current political issue or candidate. You might want to put the answers in multiple choice format. Ask the question to 50 people you do not know. Record their answers. Report on the percent of people giving each response.
6. For this project, you will need a food scale that measures either in ounces or grams. Check and record the net weight of a can of vegetables (peas, corn, etc.). Drain the liquid off of the vegetables and weigh the drained vegetables. What percent of the net contents (by weight) is liquid? What percent is the vegetable? How much did the can of vegetables cost? What is the price per ounce (or gram) of the drained vegetables? Repeat this experiment with two or three different brands of the same type of product and compare the cost per ounce (or gram) before and after draining the liquid.
7. Buy two different grades of ground beef. Record the weight and price of each. Brown the two grades of ground beef in separate skillets and use a strainer to drain off the excess liquid and fat. Weigh the meat left after draining off the fat and liquid. What is the price per pound of each grade of meat before cooking? What is the price per pound of each grade of meat after cooking and draining off the liquid and fat? Which grade of meat is cheaper? In each grade of meat, what is the percent of liquid and fat? Write a report of your findings.

Section 6-3 *Percent to Fraction, Fraction to Percent*

Percents and fractions are closely related--both represent parts of a whole. As we already know, 25%, written as a ratio is $\frac{25}{100}$ which reduces to $\frac{1}{4}$. To further recognize the connection between percents and fractions, notice that 25% of 160 is (0.25)(160) = 40. One-fourth of 160 is also 40.

Any percent can be written as a fraction and any fraction can be written as a percent. The arithmetic process is very simple.

Convert a Percent to Fraction

Write the percent as a ratio (per one hundred) and then reduce the fraction if needed.

Examples: $75\% = \frac{75}{100}$ which reduces to $\frac{3}{4}$.

$17\% = \frac{17}{100}$ which does not reduce.

$280\% = \frac{280}{100}$ which reduces to $\frac{14}{5}$.

Convert a Fraction to Percent

Divide the fraction to put it in decimal form and then convert the decimal to percent. Since some fractions have decimal representations that do not terminate, we round the percent to a particular place or express it as a mixed number.

Examples: $\frac{2}{3} = 0.66\overline{6} \approx 0.667$ $\quad \frac{2}{3} = 66\frac{2}{3}\% \approx 66.7\%$ rounded to one decimal place.

$\frac{1}{13} \approx 0.077$ $\quad \frac{1}{13} \approx 7.7\%$

$\frac{5}{4} = 1.25$ $\quad \frac{5}{4} = 125\%$

To check our conversions and see whether our answer is reasonable, we can use a thermometer with a percent scale on one side and a scale from 0 to 1 on the fraction side. The scale works like the linear units we have used before, except it is vertical instead of horizontal.

Example 1: Convert $\frac{2}{7}$ to a percent.

$\frac{2}{7} \approx 0.29 = 29\%$

On the fraction thermometer, estimate $\frac{2}{7}$ of the way from 0 to 1. 29% looks reasonable.

Exercise 1--The Fraction Thermometer

Purpose of the Exercise
- To practice converting fractions to percents and percents to fractions

Materials Needed
- Calculator

In problems 1 - 14, convert each fraction to a decimal then to a percent. Do Problems 1-14 without a calculator. Use a calculator as needed on problems 15 - 22 to convert the fraction to percent.

1. $\frac{1}{4}$ = ____________ 2. $\frac{1}{4}$ = ____________ %

3. $\frac{1}{2}$ = ____________ 4. $\frac{1}{2}$ = ____________ %

5. $\frac{3}{4}$ = ____________ 6. $\frac{3}{4}$ = ____________ %

7. $\frac{1}{20}$ = ____________ 8. $\frac{1}{20}$ = ____________ %

9. $\frac{3}{20}$ = ____________ 10. $\frac{3}{20}$ = ____________ %

11. $\frac{7}{20}$ = ____________ 12. $\frac{7}{20}$ = ____________ %

13. $\frac{1}{3}$ = ____________ 14. $\frac{1}{3}$ = ____________ %

15. $\frac{55}{97} \approx$ ____________ % 16. $\frac{29}{78} \approx$ ____________ %

17. $\frac{97}{55} \approx$ ____________ % 18. $\frac{78}{29} \approx$ ____________ %

19. $\frac{22}{19} \approx$ ____________ % 20. $\frac{19}{22} \approx$ ____________ %

21. $\frac{2}{53} \approx$ ____________ % 22. $\frac{3}{21} \approx$ ____________ %

In each of the following convert the percent to a fraction and then to decimal form. **Do not use a calculator.**

	fraction		decimal		fraction		decimal
23. 20% =	_______	=	_________	24. 90% =	_______	=	_________
25. 70% =	_______	=	_________	26. 75% =	_______	=	_________
27. 100% =	_______	=	_________	28. 120% =	_______	=	_________
29. 83% =	_______	=	_________	30. 17% =	_______	=	_________
31. 207% =	_______	=	_________	32. 43% =	_______	=	_________

In each of the following, convert percent to decimal. Use a calculator as needed.

33. 80.5% = ________________ 34. 108.97% = ________________

35. 3% = ________________ 36. 3.7% = ________________

37. 14.9% = ________________ 38. 7.896% = ________________

39. 267.987% = ________________ 40. 0.5% = ________________

41. $17\frac{4}{5}\%$ = ________________ 42. $\frac{1}{3}\% \approx$ ________________

43. 12.975 % = ________________ 44. 252% = ________________

45. 300% = ________________ 46. 97.25% = ________________

Section 6-4 *Sampling and Approximations in 2-Dimensional Space*

Many times percents or ratios are used to estimate crowd sizes or room capacity or to predict trends. For example, to estimate crowd size, you would take a small portion of the area, count the number of people in the sample area, estimate the total area, and then compute the percent of the total area that the sample area represents. When someone reports that 100,000 people attended a march on Washington, they certainly did not count that many people. They more likely counted the number of people in a small area, and estimated how long the parade was at any point in time (maybe from a helicopter). With this information, they used ratios or percents to estimate the crowd size. When there are such marches, it is difficult, if not impossible, to count the number of people accurately. The estimates vary, according to where the sample is taken and how long the march is estimated to be. For this reason, one person's crowd estimate may vary greatly from another person's estimate.

In the exercises in this section, you will work with geometric shapes and ratios or percents within those shapes to estimate the capacity of the entire area. The problems are much more predictable than crowd estimates; however you should get the idea of how to use ratios or percents to make predictions from a sample. To set up the ratios or percents, you will need to compute or estimate areas. You may need to review formulas from Chapter 3.

Exercise 1--Approximation of Area Using Percent or Proportions

Purpose of the Exercise

- To approximate percents of areas in various geometric shapes

Materials Needed

- Ruler
- Calculator

1. Suppose the floor of a room is shown as below.

Picture 1

The floor space represented by vertical lines is where chairs will be placed. The right area is a raised platform for displays.

a. Without measuring, estimate the percent of the platform to the total floor space and the percent of the seating area to total floor space.

Estimated percents: ________________

b. Now measure the entire rectangle, the platform and the seating area to the nearest 1/8 inch. Include appropriate units with your answers.

Entire rectangle: Length = ______ Width = _______ Area = ________

Platform: Length = ______ Width = _______ Area = ________

Seating Area: Length = ______ Width = _______ Area = ________

c. Compute the following percents.

The percent of the area of the platform to the total floor space = ___________

The percent of the area of the seating area to the total floor space = __________

2. In picture 2, the width of the room from picture 1 is decreased but the length is the same.

Picture 2

a. Without measuring, visually estimate the percent of the platform to the total floor space and the percent of the seating area to total floor space.

Estimated percent: ___________________

b. Now measure the entire rectangle, the platform, and the seating area. Measure to the nearest 1/8 inch.

Entire rectangle:	Length = _______	Width = ________	Area = _________
Platform:	Length = _______	Width = ________	Area = _________
Seating Area:	Length = _______	Width = ________	Area = _________

c. Compute the following percents.

The percent of the area of the platform to the total floor space = ____________

The percent of the area of the seating area to the total floor space = __________

3. In pictures 1 and 2 above, suppose 1 inch represents 8 feet.

a. What are the actual dimensions and the areas (in feet and square feet) of each of the regions represented in the pictures? Include appropriate units with your answers.

Picture 1:

Entire rectangle:	Length = _______	Width = ________	Area = _________
Platform:	Length = _______	Width = ________	Area = _________
Seating Area:	Length = _______	Width = ________	Area = _________

Picture 2:

Entire rectangle:	Length = _______	Width = ________	Area = _________
Platform:	Length = _______	Width = ________	Area = _________
Seating Area:	Length = _______	Width = ________	Area = _________

b. The percent of the area of the platform to the total floor space in picture 1: ______

c. The percent of the area of the seating area to the total floor space in picture 1: ____

d. The percent of the area of the platform to the total floor space in picture 2: ______

e. The percent of the area of the seating area to the total floor space in picture 2: ____

Generalizations: Notice that the percent of stage area to total floor space is the same whether the measurements are in feet or in inches. (There may be slight variation due to round-off errors in unit conversions.)

4. Suppose we decide to use the theater space shown in picture 1 to set up exhibits for a job fair. Each booth or table will take up the same amount of space. We set up the exhibits in the stage area and find that there is room for 12 booths. Devise a plan to estimate how many booths you would expect to be able to put in the entire theater if all of the chairs are removed.

Chairs	12 booths

Picture 1

   a. About how many booths would be in the larger (seating area?) _______________

   b. About how many booths would be in the entire theater? ____________________

   c. Show how you determined the number of booths.

   d. Explain why the number of booths in the seating area and the entire theater is an estimate and in reality the number might be a little off.

5. Devise a method of approximating the number of squares in the following drawing. **Do not count the squares**. Write a paragraph explaining how you made the estimate. What formulas and concepts did you have to use to make the approximation?

Estimated number of squares ______

6. Devise a method of approximating the number of circles or squares in the following drawings. **Do not count the squares or circles**.

Estimated number of white circles ______________

Estimated number of squares ____________

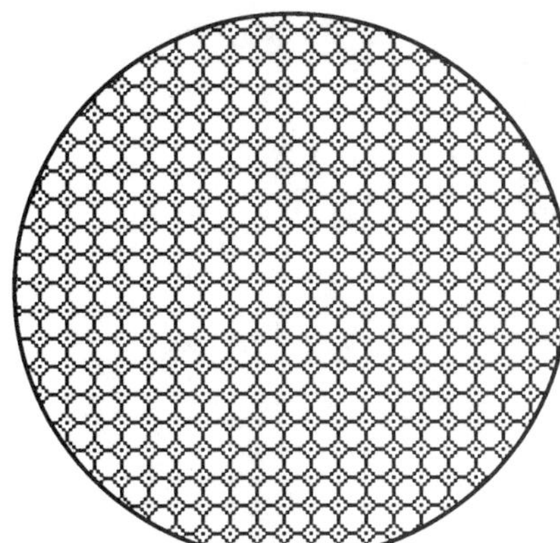

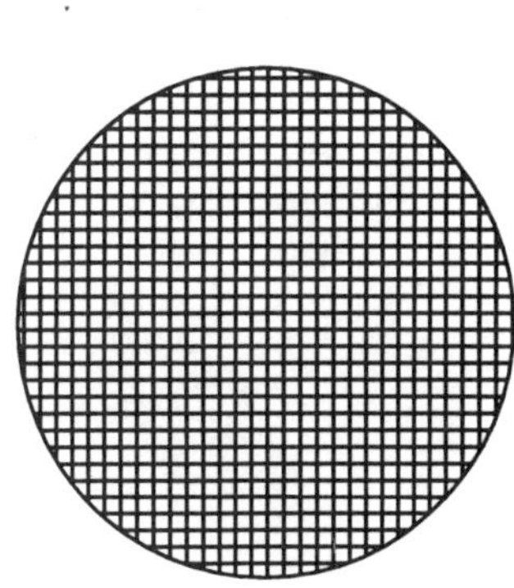

7. Estimate the number of squares in the following picture.

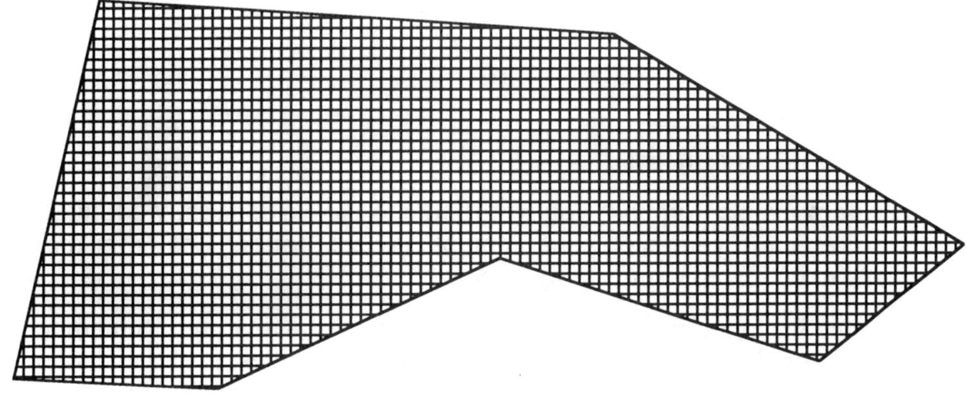

8. If the same squares used to fill the rectangle in Problem 5 were used to fill each of the following shapes, how many squares would you expect to be in the shape?

   a. A triangle is 10 inches high and has a base of 15 inches. No. of squares _________

   b. A trapezoid has bases of 12 inches and 14.2 inches and a height of 9.5 inches. No. of squares _________

   c. A circle with diameter of 19 inches. No. of squares ____________

9. Estimate what percent of the following circle is shaded. How many squares would you expect if the entire circle were shaded? Explain how you arrived at your results.

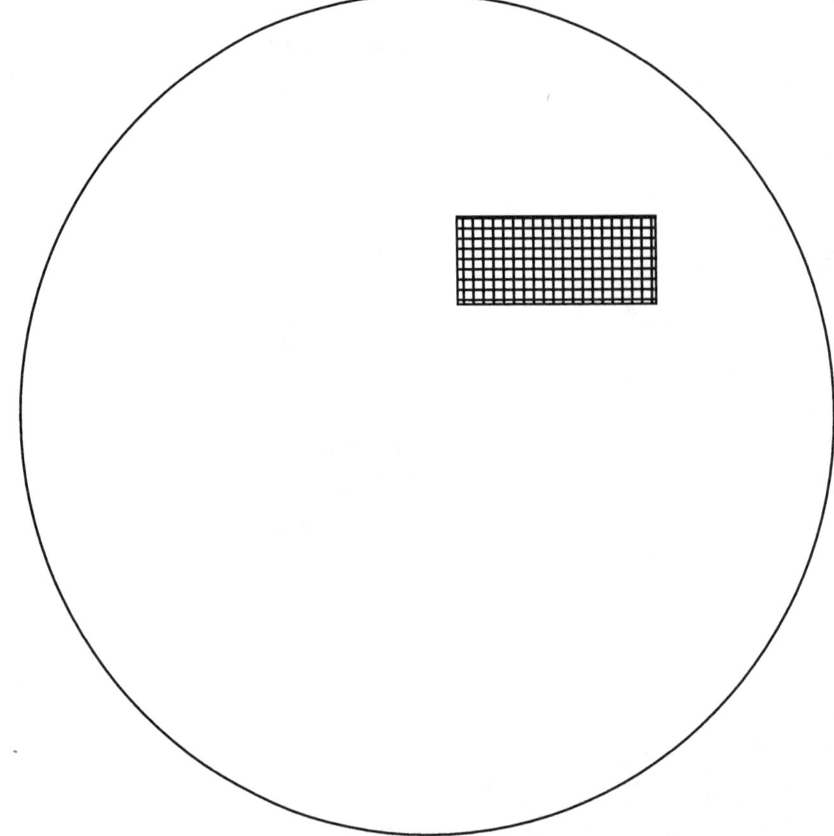

Projects using percent to estimate quantities.

1. **Backyard/house/apartment event:** Suppose you would like to have a concert, wedding, sales party, or some other event in your back yard. You plan to rent folding chairs for the event so that each guest can be seated comfortably.

    1. Decide how much of the backyard/house/apartment you can use for seating. Remember to save room for aisles and the performers/ the happy couple/ the salesperson, and so on.
    2. Set up 4 to 6 folding chairs (or your kitchen chairs) to estimate the amount of space required for the chairs. Be sure to consider the comfort of your guests in spacing the chairs.
    3. Estimate how many guests you can invite to your event. (Remember you want to be able to seat all guests.)
    4 Find a party store or company that will rent folding chairs. How much will it cost to rent the chairs?

    Write a report about your experiment describing your experiment and conclusions. Clearly show the computations you made. Explain why the results you got are estimations.

2. **Size of auditorium estimate:** If you are in a large public area such as a concert hall, a large auditorium, or a convention center, devise a way (without counting every chair if there are chairs) to estimate how many people will fit in the area when it is full.

    Write a complete report describing what kind of area you chose--a large room with benches, chairs, no chairs--and how you did the estimations. Show your computations. Why are your answers approximate instead of exact?

Section 6-5 *Percent Change*

In studying populations, budgets, profits, or trends in energy consumption, we frequently measure and study changes in the quantities over periods of time. Many times the actual increase/decrease is not very useful information without the perspective of the percent the quantity changes.

Example 1: If profits increased by $1,025, we cannot determine whether the increase is substantial or insignificant unless we know how much the profits were before the increase. If a company's profits started at $2,548,000 and increased by $1025, the percent increase, $\frac{1025}{2548000} \text{x} 100 \approx 0.04\%$ is not even close to a 1% change, so it is very insignificant. However, if the company's profits started at $765 and increased by $1025, the percent change is $\frac{1025}{765} \text{x} 100 \approx 134\%$. This is a significant increase in profits.

Vocabulary

Mathematically speaking, **percent change** from one number to a another number is

$$\frac{\text{second number} - \text{first number}}{\text{first number}} \text{ x } 100\ \%$$

If the second number is larger than the first number, there is a **percent increase** and the computation above turns out to be **positive**.

If the second number is smaller than the original number, there is a **percent decrease** and the computation above turns out to be **negative**.

Have you ever wondered what news reports mean when they say that the growth in the federal debt has slowed? Does this mean that the government owes less money than before or that there is not as much new debt as in previous years?

In the exercise below, you will look at some data about the economics of the nation as reported by *USA Today*, September 30, 1996. The sources were Economic Report of the President and the Commerce Department. *USA Today* compared the nation's economic status under the last four years of the Reagan administration, the Bush administration and the Clinton administration--in progress.

Exercise 1--The Federal Budget and Percent Change

Purpose of the Exercise

- To gain an understanding of percent change by comparing growth in the federal debt during the Reagan, Bush, and Clinton administrations
- To interpret other real data involving percent change

Materials Needed

- Calculator

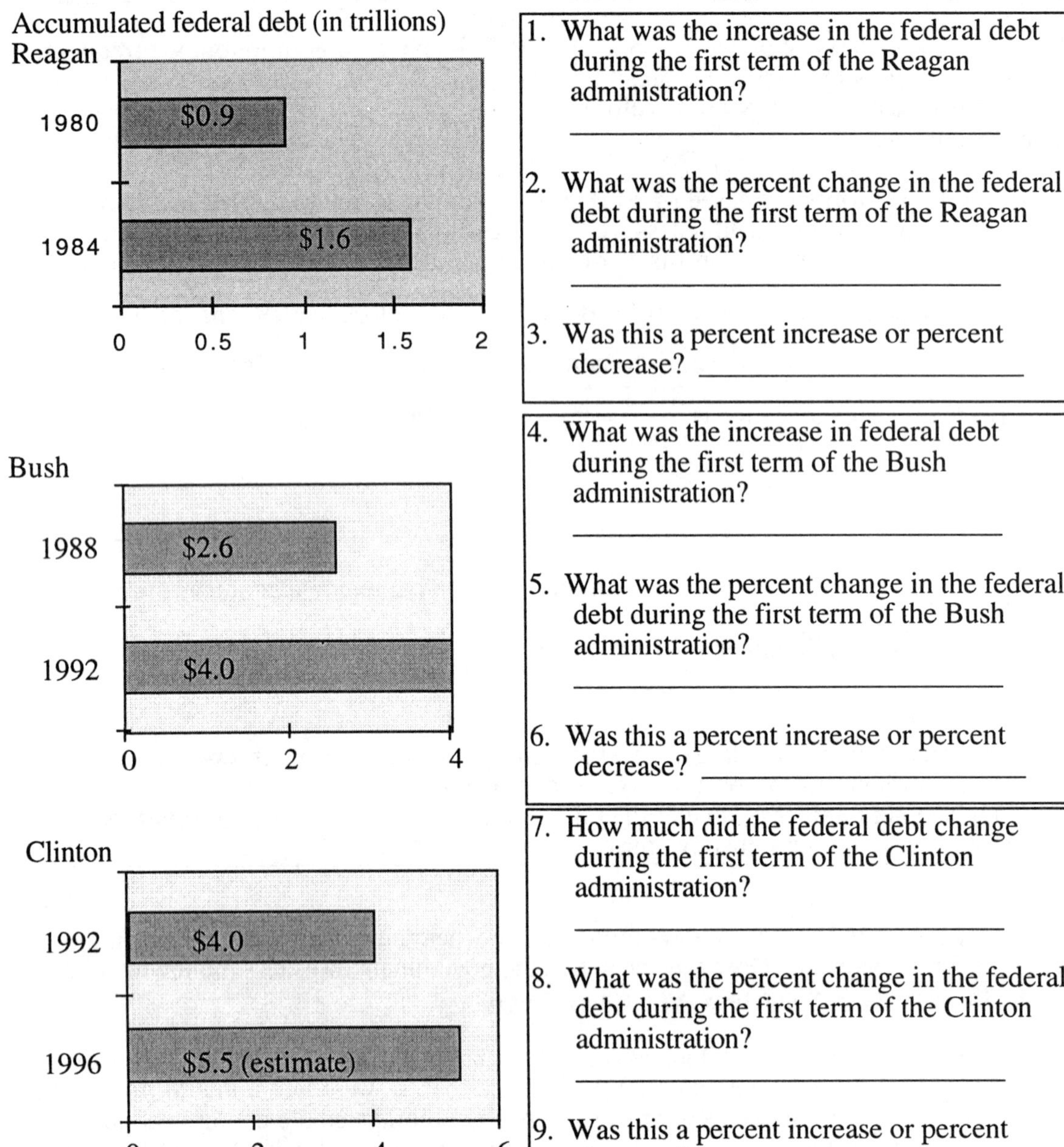

1. What was the increase in the federal debt during the first term of the Reagan administration?

   ____

2. What was the percent change in the federal debt during the first term of the Reagan administration?

   ____

3. Was this a percent increase or percent decrease? ______________________

4. What was the increase in federal debt during the first term of the Bush administration?

   ____

5. What was the percent change in the federal debt during the first term of the Bush administration?

   ____

6. Was this a percent increase or percent decrease? ______________________

7. How much did the federal debt change during the first term of the Clinton administration?

   ____

8. What was the percent change in the federal debt during the first term of the Clinton administration?

   ____

9. Was this a percent increase or percent decrease? ______________________

10. During which administration did the federal debt increase the most (in dollars)? Write a few sentences to explain your decision.

11. Which administration had the highest **rate** of growth (percent increase) in the federal debt? Write a sentence or two to explain your choice.

12. Has the federal debt increased or decreased from one administration to another?

13. Has the rate of growth in the federal debt increased or decreased from one administration to another?

Review of Concepts

- **Percent** means "per hundred." Percent is a rate. (page 350)

  49% means 49 per hundred or 49/100. 49% = 0.49. All of something consists of 100% of the items, objects and so forth. Less than 100% represents part of a whole. More than 100% represents more than one whole.

- **Percentage** means "by the hundred." The percentage is the number you get when you multiply a percent times a number. (page 350)

Review of Processes

- **To write a percent as a decimal**, write the percent as a ratio (per one hundred) and then divide (move the decimal point 2 places left). (page 352)
- To **convert a decimal to a percent**, reverse the above process (move the decimal point 2 places right). (page 352)
- To **compute a particular percent of a number**, (page 352)

  First, convert the percent to decimal form and

  Second, multiply the decimal form by the number.
- To write a **percent as a fraction**, write the percent as a ratio (per one hundred), and then reduce the fraction if needed. (page 370)
- To write a **fraction as a percent.** (page 370)
  1. Divide the fraction to change it to decimal form.
  2. Convert the decimal form to percent.
- **Percent change** = $\frac{\text{second number} - \text{first number}}{\text{first number}}$ x 100%. (page 377)

  Percent increase is percent change that is **positive**.
  Percent decrease is percent change that is **negative**.

Summary Test--Chapter 6

Part 1: Do not use a calculator.

1. Solve the following percent problems.

   a. 18 is what percent of 200? __________ b. What number is 12% of 80? _______

   c. 30% of a number is 60. d. What is a $\frac{1}{2}$% of 20? ______________

   What is the number? ______________

2. In each of the following, pick the best answer from the choices given.

a. 29.23% of 3567 ≈	150	500	1034	3025
b. 92.7% of a number is 763 The number ≈	830	701	9	1500
c. About what percent of 87 is 103.59?	85%	118%	89%	230%
d. 215% of 89.9873 ≈	45	300	195	19.5

3. In each of the following, convert the percent to decimal.

a. $\frac{1}{2}\% =$ ________________ b. $4\frac{1}{2}\% =$ ________________

c. 4.5 % = ________________ d. 17.8% = ________________

4. In each of the following, convert the fraction to a percent.

a. 4/5 = ____________________ b. 29/100 = ____________________

c. 9/4 = ____________________ d. 3/10 = ____________________

Part 2: Calculator recommended.

5. Solve the following percent problems.

a. 875 is what percent of 739? ______ b. What is 4.6% of 278? __________

c. 67.3% percent of a number is 287 What is the number? __________

d. 127.3% of a number is 287 What is the number? __________

Solve the following word problems.

6. a. An item in the store costs $8.95. The local sales tax is 4.75%.

How much is the tax on the item? ______________________________

How much does the item cost including the tax? ________________

b. A dress regularly priced at $92.98 is on sale for 15% off.

What is the sale price of the item? ______________________________

7. Sarah bought a computer priced at \$2350.78. The sales tax rate is 6.8%.

a. How much was the tax amount? ____________________

b. How did the computer cost including tax? ____________________

8. Before receiving a 7.5% raise, Jerry's annual salary was \$32,957. What is Jerry's salary after the raise?

9. Michelle invested \$5780 and earned 205% over a five-year period. How much did Michelle make from her investment?

10. Perri bought a house for \$45,700 and 20 years later sold the house for \$178,500.

a. How much did the house increase in value?

b. What was the percent change in the value of the house?

11. Pat bought a house for \$215,700 and five years later sold the house for \$178,500.

a. How much did the house decrease in value?

b. What was the percent change in the value of the house?

Challenge Questions

12. Mrs. Carey's estate of $350,790 is to be divided in the following way: the attorney's fee is 31% of the entire estate; the "Francis Center for Battered Women" receives $7500; $3400 goes to the church's building fund; $6000 goes to a scholarship fund; 11% of the amount remaining is to be paid to each of Mrs. Carey's sisters, Ruth and Gretchen; the money left over is to be divided evenly among Mrs. Carey's three children.

    a. How much money is Ruth's share of the estate?

    b. What percent of the entire estate does Ruth receive?

    c. What is the attorney's fee?

    d. What percent of the money goes to the church?

    e. How much money does each of Mrs. Carey's children receive?

    f. What percent of the entire estate does each of Mrs. Carey's children receive?

Chapter 7 Statistical Information, Pie and Bar Charts and Other Graphical Data

Key Topics

- Mean, median, and range
- Intrpretating graphical data
- Statistics of world awareness

Materials Needed for the Exercises

- Calculator
- Measuring tape, yardstick or meter stick
- Lid or compass and a protractor, or a computer spreadsheet program (to draw pie charts)

Section 7-1 *Mean, Median, and Range*

We use the word **average** to mean several different things in English. When we say someone is average for some trait, we generally mean that most people have that trait. In mathematics, there are several ways to determine what might be "typical." Although there is no sure formula for determining which measure might be the best indicator of what typical is, it is good to be familiar with all measures and understand what the measure means. The range and the three measures of "central tendency" are defined below.

Vocabulary

The numeric or arithmetic **average** of several numbers is just the sum of all of the numbers divided by how many numbers there are. (This is the average we worked with in Chapter 1.) The arithmetic average is also called the **mean**.

When the numbers are put in order, the **median** is the middle number. If there are two middle numbers, the median is the numeric average of the two middle numbers.

The **mode**, simply stated, is the number that occurs most often. The mode is the most informative when there are many data points where we group the data before computing the mean, median, and mode. Since working with grouped data is beyond the scope of this book, we will not work with the mode.

The **range** of values of several numbers is
(the largest number) - (the smallest number).

To determine the range and the median, it is helpful to put the data points in numeric order (either descending or ascending). For computing the average, the order does not matter. In the examples below, the numbers are already listed in order.

Example 1: The average of the numbers 45, 49, and 53.

$(45 + 49 + 53) \div 3 = 147/3 = 49$

The median is 49.

The range is 53 - 45 = 8.

Example 2: The average of the numbers 14, 97, 103 and 172.

$(14 + 97 + 103 + 172) \div 4 = 386/4 = 96.5$

The median is $(97 + 103) \div 2 = 200 \div 2 = 100$.

The range is 172 - 14 = 158.

Example 3: The average of the numbers 16.3, 17.4, 17.9, 18.2, 18.2.

$(16.3 + 17.4 + 17.9 + 18.2 + 18.2) \div 5 = 88/5 = 17.6$

The median is 17.9.

The range is 18.2 - 16.3 = 1.9.

Exercise 1--Finding Arithmetic Averages (Group)

Purpose of the Exercise

- To better understand the arithmetic average (mean)

Materials Needed

- Calculator

1. Discuss the question: If a town's working population makes an average salary of $45,000 a year and the unemployment rate is very low, would we assume the community has a strong economy?

2. Read the following situations for Smalltown, USA that has 15 employed adults. The unemployment rate is very low. Compute the sum, average, median salaries, and range of salaries. Verify that the average salary in each situation is $45,000.

   a. The 15 adults make annual salaries as follows:

$66,800	$63,200	$53,004	$51,240	$50,003
$47,305	$46,783	$45,207	$44,272	$40,100
$39,205	$36,300	$34,450	$29,200	$27,883

   The sum of all the salaries is __________

   The average of all of the salaries is (sum of salaries/15) ≈ ______________

   The median salary is ____________

   The range of salaries is ___________

   b. The 15 adults make annual salaries as follows:

$427,260	$26,780	$24,007	$23,900	$22,945
$22,056	$19,203	$16,006	$15,087	$14,990
$14,204	$13,990	$13,201	$12,360	$11,025

   The sum of all the salaries is __________

   The average of all of the salaries is (sum of salaries/15) ≈ ______________

   The median salary is ____________

   The range of salaries is ___________

Questions for Discussion

3. If the average income in a community is $45,000, is this community necessarily affluent?

4. Under what circumstances does a numerical average indicate a "typical" value?

5. Do the median salary and the range of salaries help indicate what is typical? Explain.

Exercise 2--Computing Averages with Real Data (Two of the Questions Are Group Activities)

Purpose of the Exercise

- To apply arithmetic concepts to real-life situations

Materials Needed

- Calculator
- Measuring tape, yardstick or meter stick

1. According to the Bureau of Labor Statistics, the number of people unemployed each month in 1995 are as follows:

Jan	Feb	Mar	Apr	May	Jun	Jul	Aug	Sep	Oct	Nov	Dec
7498	7183	7237	7665	7492	7384	7559	7431	7451	7249	7432	7380

   (Source: Bureau of Labor Statistics.)

   where the numbers are in thousands (7498 for January means 7498 thousand, or 7,498,000 people were unemployed).

Find or compute the following numbers showing the calculations when appropriate.

Largest number of unemployed people in any one month ________________

Smallest number of unemployed people in any one month ________________

Range ____

Average number of unemployed people per month ______________________

Median number unemployed ____________________________________

2. The number of cases of cholera and deaths from cholera in 1991 in Africa, Europe, the Americas, and Asia are as follows: (Source: Peter H. Gleick*Water in Crisis, A Guide to the World's Fresh Water Resources.*)

	Africa	Europe	N. America	S. America	Asia
Cases	153,367	316	2,718	388,502	49,791
Deaths	13,998	9	34	3968	1,286

a. Compute the average number of cholera cases per continent. ________________

b. Compute the average number of deaths due to cholera per continent. ___________

c. Does the average number of cases of cholera represent the "typical" number of cases per continent? Explain.

d. Does the average number of cholera deaths represent the "typical" number of deaths per continent? Explain.

3. According to the Heath Services Cost Review Commission, Spinal Cord Injuries (SCI) and Traumatic Brain Injuries (TBI) occurred in Maryland in the frequencies shown in the chart below over the years 1990 through 1993.

1990		1991		1992		1993	
SCI	TBI	SCI	TBI	SCI	TBI	SCI	TBI
1608	4114	1521	4032	1404	4195	1250	3652

a. What was the average number per year of spinal cord injuries in Maryland during the years 1990-1993? ______________________________

b. Does the average represent "typical" number of spinal cord injuries? Explain.

c. What was the average number per year of traumatic brain injuries in Maryland during the years 1990 - 1993? ______________________________

b. Does the average represent the "typical" number of traumatic brain injuries? Explain.

Section 7-2 *Circle Charts and Bar Graphs*

Many times charts and graphs based on percentages make it is easier to comprehend numbers and the relative size of numbers. This is especially true when the numbers are very large, as in population demographics and federal budget numbers. First let's see how to put information into a table and interpret the data with circle charts and bar graphs.

Example 1:

A large container holds 9278 red beans and 17,963 white beans. We can compute the percentage of red beans to the total number of beans by first computing the total number of beans.

There are 27,241 beans in the container.

$\frac{9278}{27241} \approx 0.340589$, so the percentage of red beans to total beans is about 34%.

$\frac{17963}{27241} \approx 0.65941$, so the percentage of white beans to total beans is about 66%.

Since the red beans and the white beans make up all of the beans in the container, the sum of the two percents should add up to 100%.

$$34\% + 66\% = 100\%$$

Here is a table giving the percentage (the number) and the percent of each color of bean in the container.

Color of Beans	**Percentage of Beans (The Number of Beans)**	**Percent of Beans**
Red beans	9,278	34%
White beans	17,963	66%
Total	**27,241**	**100%**

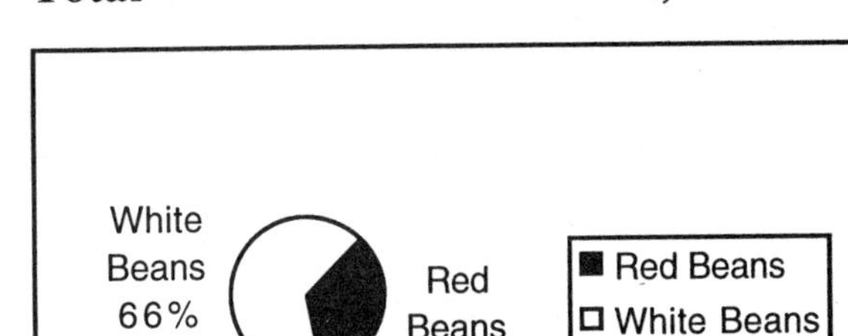

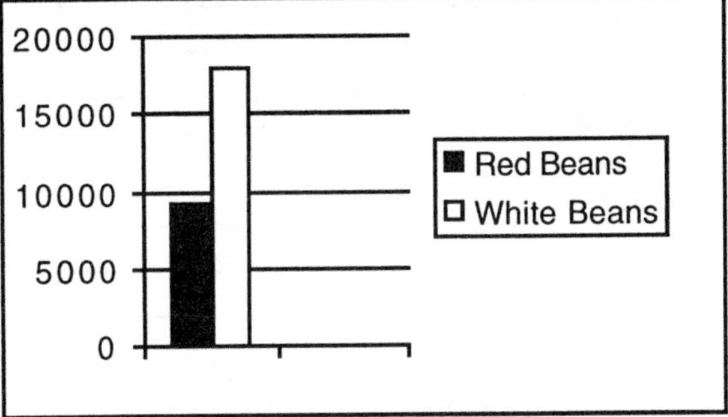

The chart above represents the data in a pie chart. Notice how the picture represents clearly that there are almost twice as many white beans as red beans.

The above bar graph shows a comparison of the numbers of the red beans and white beans. Visually we see that the bar representing white beans is about twice as high as the bar for red beans. Reading the scale, we also see that there are between 5000 and 10,000 red beans (closer to 10,000) and that there are between 15,000 and 20,000 white beans.

Exercise 1--Computing Percents, Percentages, and Making Circle Graphs

Purpose of the Exercise

- To practice computing percents from data in a table
- To practice computing percentages (number of items) when the total and the percents are given
- To practice representing data in pie charts
- To practice recognizing reasonable numbers

Materials Needed

- Calculator
- Lid or compass and a protractor, or a computer spreadsheet program (to draw pie charts)

In each of the following, fill in the blanks in the tables. On a separate sheet of paper, make a pie chart to represent the data. To determine the angle, remember there are 360° in a circle. Take the appropriate percent of 360° to get the angle size. Use the protractor to draw the angles.

If you have a computer and spreadsheet program, enter the data in the spreadsheet to help produce the chart.

1.

Type of Tree	Percentage of Trees	Percent of Total
Oak	297	
Maple	567	
Elm	82	
Sycamore	123	
Total		

2.

Type of Lunch	Percentage of Lunches	Percent of Total
Roast Beef	7892	
Chicken	276	
Fish	88	
Vegetarian	415	
Total		

3.

Breed of Cat	Percentage of Cats	Percent of Total
A		39.2%
B		2.5%
C		17.9%
D		13.9%
E		4.7%
F		21.8%
Total	**87,291**	

4.

Hair Color	Percentage of People	Percent of Total
Brown		51%
Red		13%
Blond		12%
Black		
Total	**3,598**	

In each of the following, find the mistakes in the table or chart.

5\.

Item	Percentage of Items	Percent of Total
A	875	42%
B	801	38.4%
C	276	17.3%
D	132	6.3%

What is wrong? ______________________________

6\.

Item	Percentage of Items	Percent of Total
A	510	17%
B	780	29%
C	1050	35%
D	1260	42%

What is wrong? ______________________________

7. Explain what is wrong with each of the circle charts for the following data.

Item	Percentage of Items	Percent of Total
A	987	44.5%
B	245	11%
C	789	35.5%
D	201	9%

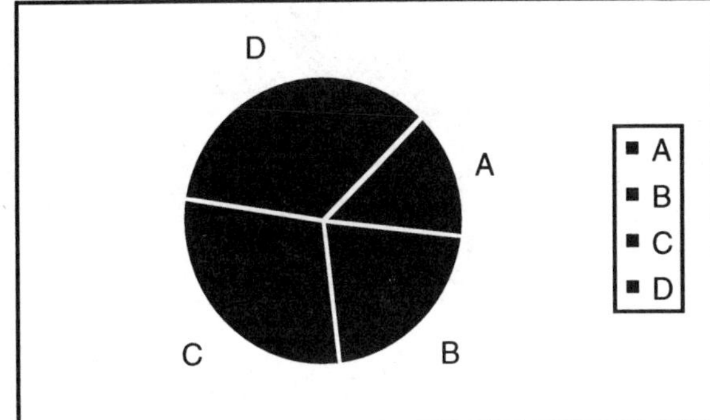

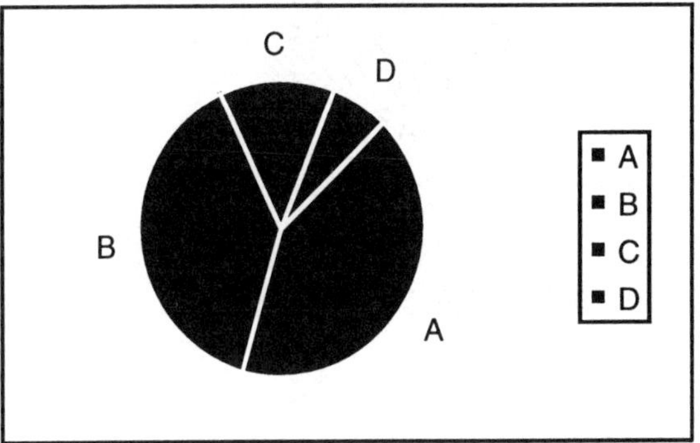

What is wrong? ____________________ What is wrong? ____________________

8. **Without using a calculator**, determine what is wrong with each of the bar graphs for the following data.

Item	Percentage of Items	Percent of Total
A	66	
B	231	
C	657	
D	906	

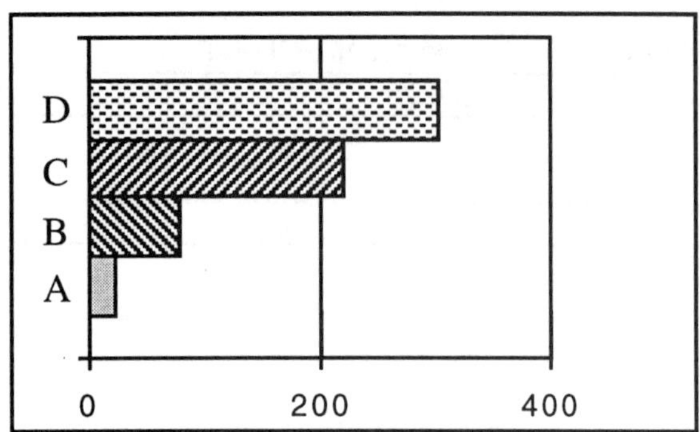

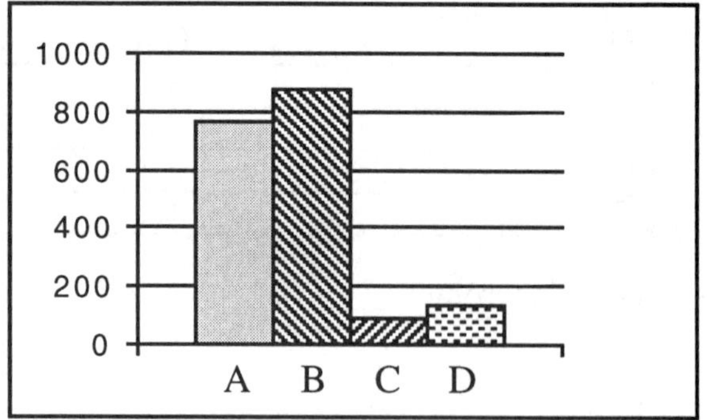

What is wrong? ____________________ What is wrong? ____________________

9. **Without using a calculator**, determine what is wrong with each of the graphs for the following data.

Item	Percentage of Items	Percent of Total
A	21,027	
B	21,876	
C	34,578	
D	79,876	

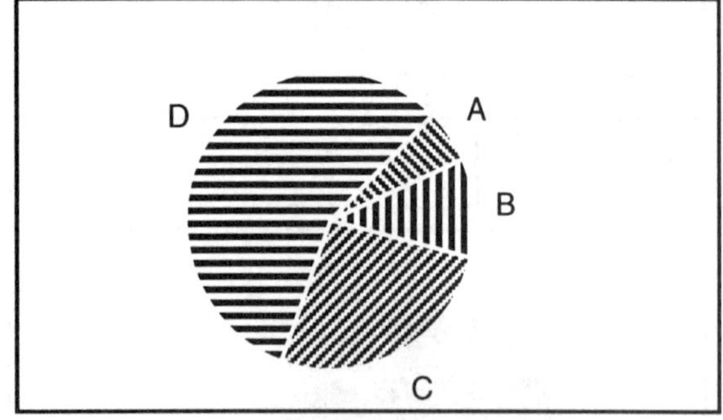

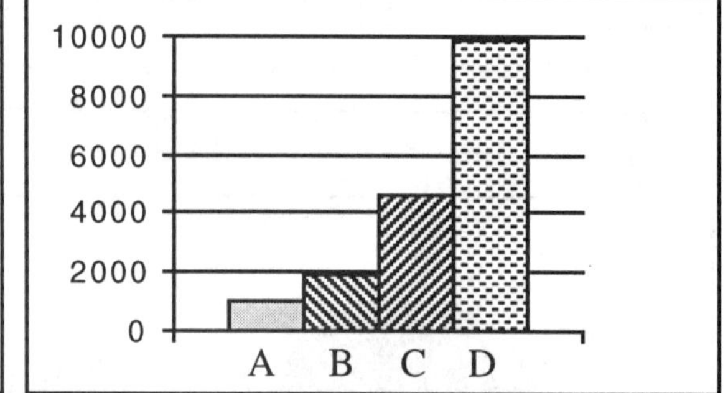

What is wrong? ____________________ What is wrong? ____________________

10.

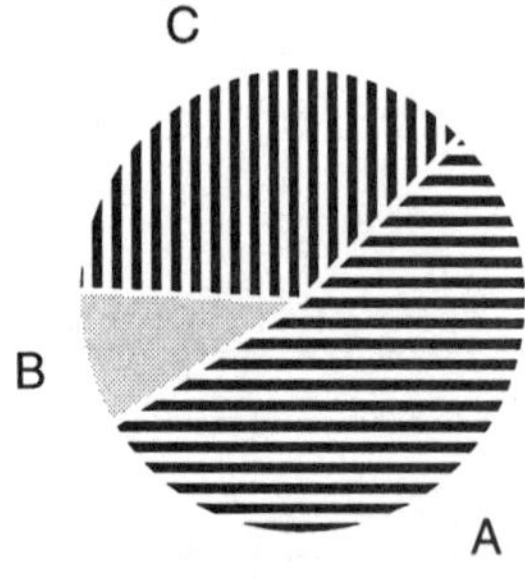

Which of the following combination of number of items could fit the data in the chart?

a. 125 of A 99 of B 105 of C

b. 97 of A 17 of B 67 of C

c. 789 of A 400 of B 500 of C

11.

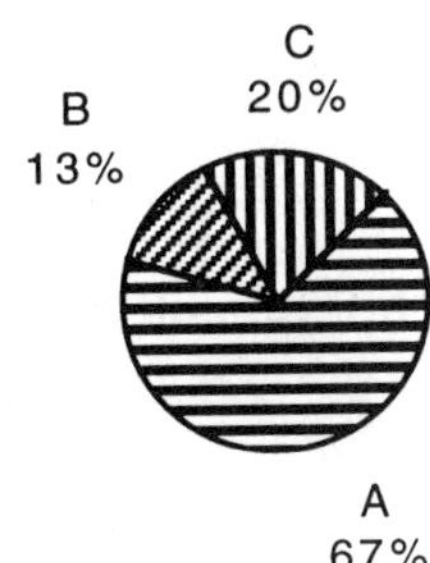

Compute the percentage of each of the items if the total number of items is 7865.
A ____________
B ____________
C ____________

In Section 6.5, we defined and worked with percent change. In the following exercises, we will work more with percent change.

Percent Change from One Number to Another Number

$$\frac{\text{second number} - \text{first number}}{\text{first number}} \times 100\ \%$$

Percent **decrease** is negative and percent **increase** is **positive**.

Exercise 2--Percent Change

Purpose of the Exercise

- To practice computing percent change using real data

Materials Needed

- Calculator

The following chart shows the total number of female ASE-certified technicians (for automotive repair) from 1988 through 1995. Answer the questions using the information in the chart. The first one is done as an example. (Source: *USA Today*, November 18, 1996.)

Year	Number of ASE Certified Technicians
1988	556
1989	614
1990	654
1991	737
1992	849
1993	1086
1994	1329
1995	1592

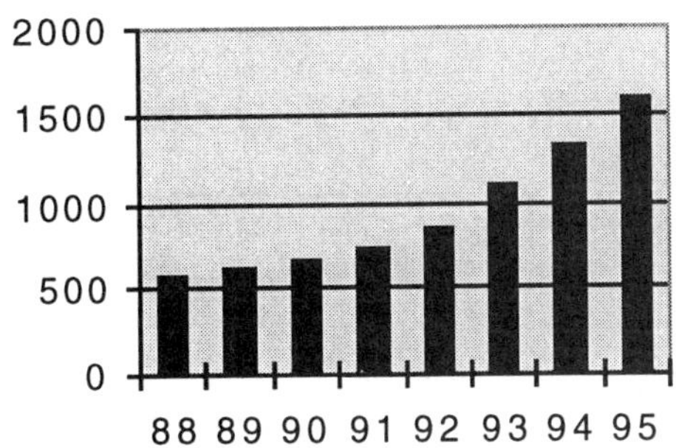

1. Compute the increase and percentage increase in number of female certified technicians from 1989 to 1993.

   Increase = 1086 - 614 = 472

   Percent increase = $\frac{472}{1086} \times 100 \approx 43\%$

2. Compute the increase and percentage increase in number of female certified technicians from 1988 to 1995.

   Increase = ________________

   Percent increase = ________________

3. Between which two years was the there the smallest increase in number of female certified technicians? ____________

What is the increase? _______ What is the percent increase? _______________

4. Between which two years was there the largest increase in number of female certified technicians? _______________

What is the increase? _______ What is the percentage increase? _______________

In each of the following, read the paragraph and answer the questions.

5. According to the National Insurance Crime Bureau (as reported by *USA Today*, September 30, 1996) there was a 45% increase in the number of motorcycles, scooters, and all-terrain vehicles stolen from 1992 to 1995. There were 6362 stolen in 1992.

   a. According to these numbers, how many motorcycles, scooters, and all-terrain vehicles were stolen in 1995?

   b. According to these numbers, how many more motorcycles, scooters, and all-terrain vehicles were stolen in 1995 than in 1995?

   c. What was the percent increase in thefts of motorcycles, scooters and all-terrain vehicles from 1992 to 1995?

6. According to *USA Today* (November 18, 1996), the average price of a new car today is about $20,000 which accounts for one-half of the median family income. In 1996, it costs on average about 42.5 cents per mile, or $532 a month, to own and operate a new car.

   Many people do not drive new cars. Thirty-five percent or 67 million cars on the road (in 1996) were more than 10 years old. More than 45 million used vehicles were sold in America in 1995.

   a. Determine from the paragraph above what is the median family income was for 1996.

   b. If you own a new car, estimate the cost of driving the car for a 245-mile trip. Explain why this amount is not necessarily accurate for your situation (even if you own a new car.)

   c. From the data in the article, determine approximately how many vehicles were on the road in 1996.

   d. Discussion question: If your company pays you 32 cents a mile for a business trip, discuss under what circumstances the 32 cents a mile would cover the cost of the trip and when it would not.

Section 7-3 *Demographics and World Awareness*

As the world is ever changing and "getting smaller" according to the expression, people of one culture, religion, and so on become more and more in contact with people of different cultures and religions. Because we are more integrated locally and globally than we used to be, we sometimes study our differences and similarities statistically to help us gain a broader understanding of ourselves and where we fit in a larger world. There are many ways we approach studying cultures and peoples, but understanding demographics--the statistics of populations--plays an important part in the study of cultures and populations. This section focuses on analyzing real data involving people and populations.

Exercise 1--Demographics, Economics, and other Population Studies

1. The data in the chart below reflect results of a 1988 survey of eighth graders about their school activities outside of class.

Race/Ethnicity	School Varsity Sports	Intra-mural Sports	Music	Dance/ Drama	Science Fairs	Student Paper/ Yearbook
Total	47.9%	42.5%	39.8%	31.4%	28.3%	21.5%
American Indian and Native Alaskan	46.6%	44.2%	31.4%	28.9%	31.5%	21.0%
Asian and Pacific Islander	43.1%	47.3%	36.5%	32.2%	29.4%	24.7%
Hispanic	44.4%	39.5%	31.1%	30.7%	22.9%	20.5%
Black	48.3%	45.0%	42.2%	30.9%	33.8%	27.5%
White	48.4%	42.2%	40.9%	31.5%	27.9%	20.5%

The data in the chart come from *Statistical Record of Native North Americans*, ©1993, Marlita A. Reddy, editor; published by ITP.

In interpreting the chart, the number, 31.1%, under Music on the row of Hispanic means that 31.1% of Hispanic eighth graders were involved in school music activities that were not part of a class.

a. Add the percents across the row describing Asian and Pacific Islander student participation in extra school activities. Total = ____________________

b. Explain why the percents add up to more than 100%. ____________________

____

c. Is there enough information to tell how many students are involved in at least one extracurricular school activity? ________ Explain ____________________

____

d. i. Which group has the largest **percent** participation in science fairs?

____

ii. Does the chart give enough information to determine which group has the greatest number of students participating in science fairs? ________ Explain

____

2. The following chart shows the average number of hours spent per week by eighth graders on outside reading, homework, and television watching, by race/ethnicity.

	Outside Reading	Homework	TV Total
American Indian and Native Alaskan	1.7	4.7	23.3
Asian and Pacific Islander	1.9	6.7	21.4
Hispanic	1.6	4.7	22.6
Black	1.6	5.2	27.6
White	1.9	5.7	20.8

The data in the chart come from *Statistical Record of Native North Americans*, ©1993, Marlita A. Reddy, editor; published by ITP.

a. For each of the blocks, compute the percent of time spent in a week on the different activities. Fill the percent in on the chart next to the number of hours spent.

b. Which ethnic group spends the most time per week doing homework? __________

c. Which ethnic group spends the most time per week watching TV? ____________

d. Which ethnic group spends the least time per week doing homework? __________

3. In 1980, there were 23,357 Navajo students living on reservations in Arizona, New Mexico, and Utah. Of these students the types of schools they attended were as follows:

Tribal schools	134
Bureau of Indian Affairs day school	1,861
Bureau of Indian Affairs boarding school	5,514
Other public schools	12,997
Other private schools	725
Not reported	2,126

The data taken from *Statistical Record of Native North Americans*, ©1993, Marlita A. Reddy, editor; published by ITP,

a. Confirm that the number of students in the schools and the number not reported add up to 23,357.

b. What percent of students attended "other public schools"--public schools other than the Bureau of Indian Affairs schools? ________________________________

c. What percent of students attended a Bureau of Indian Affairs day school or boarding school? ____

d. What percent of students were reported as attending some kind of school? _______

4. According to the U.S. Bureau of the Census, as stated in *Statistical Abstract of the United States 1995* (115th Edition, Bernan Press).
   The population of the United States in 1990 was 248,718,291.
   The population of the United States in 1980 was 226,542,199.
   The total area of the United States is 3,717,796 square miles.
   The total land area of the United States is 3,536,278 square miles.
   The total water area of the United States is 181,518 square miles.

   a. How much did the population increase from 1980 to 1990? ________________

   b. What was the **percent increase** in population from 1980 to 1990? ____________

   c. What is the average square mile of land area per person? ________________

   d. Geographically, what percent of the United States is water? ______________

5. Parents who received child support in 1991 had average incomes and support payments as follows:

	Mothers and Fathers	Mothers	Fathers
Average total money income	$19,217	$18,144	$33,579
Average Child Support received	$2,961	$3,011	$2,292

   The data come from the *Statistical Abstract of the United States 1995*, 115th edition, Bernan Press.

   a. What percent of the total income of mothers and fathers came from child support?
   ____

   b. What percent of the total income of mothers came from child support? ___________

   c. What percent of the total income of fathers came from child support? ___________

   d. Discussion question: Compute the average of $18,144 and $33,579. Discuss why the average of both groups together is much closer to the average income of the mothers than to the average income of the fathers and is not the average of $18,144 and $33,579.

6. The racial/ethnic distribution of the U.S. workforce is as follows:

Hispanic	8.1%	White	77.9%
Black	10.4%	Native American	0.6%
Asian	2.8%		

   The total number of people 25 years and older in the U.S. workforce in 1989 was 154,144,000 people. (The data come from *Statistical Record of Hispanic Americans*, Marlita A. Reddy, editor, ©1993, ITP Publishing Company, Source: U.S. Census Bureau, Bureau of Labor Statistics.)

   Based on this workforce and the percents above, about how many people were in the workforce from each of the racial/ethnic categories?

   Hispanic __________ White ___________ Black __________

   Asian ___________ Native American ____________

Exercise 2--*The World Almanac* (Library Projects or Internet Projects)

1. World Populations and Religions

Using the latest information from *the World Almanac* or from Internet searches,

a. Find the world's population.
b. Based on demographic studies, determine the major religions of the world. After you make this determination, be sure to include an "other" category to cover people who do not fit in the religions you chose. You may break the "Christian" religion into several categories or treat it as one religion.
c. How many people fit into each of the above categories? What percent is each of the above of the total world's population.
d. Repeat steps 1-3 for the U. S. population and its religious demographics.

Turn in a written report.

Write a report explaining what you found and any computations you may have made. Explain why the figures are estimates. Illustrate the results with a pie chart. You may use a computer spreadsheet if you like.

Be prepared to present a report to the class.

If you are called on to give a report, have your charts ready to explain the percentages and percents.

2. World's Largest Continents and Nations by Population and Land Area

Using the latest information from *the World Almanac* or from Internet searches,

a. Find the world's population and the total land area (in square miles).
b. Based on population, list the five largest continents, their populations, and the percents of the world's total population they represent.
c. How many people populate the rest of the world and what percent do they represent?
d. Based on land area, list the five largest continents, their land area, and the percent of the world's total land area.
e. Pick a nation of your ancestral descent. How many people are in that country and what percent of the world's total population does that nation represent? What percent of the total land area does that nation occupy?

Turn in a written report.

Write a report explaining what you found and any computations you may have made. Explain why the figures are estimates. Illustrate the results with a pie chart. You may use a computer spreadsheet if you like.

Be prepared to present a report to the class.

If you are called on to give a report, have your charts ready to explain the percentages and percents.

Vocabulary

- The **arithmetic average** or **mean** is the sum of the numbers divided by how many numbers there are. (page 384)
- The **median** is the middle number when the numbers are put in order. If there are two middle numbers, the median is the average of the two middle numbers. (page 384)
- The **range** of values of several numbers is (the largest number) - (the smallest number). (page 384)

Summary Test--Chapter 7

Part 1: Do not use a calculator.

1. Explain what is wrong with each of the charts for the data given.

Color of Paper	Percentage of Reams in Inventory
Blue	97
Red	119
Green	87
White	145

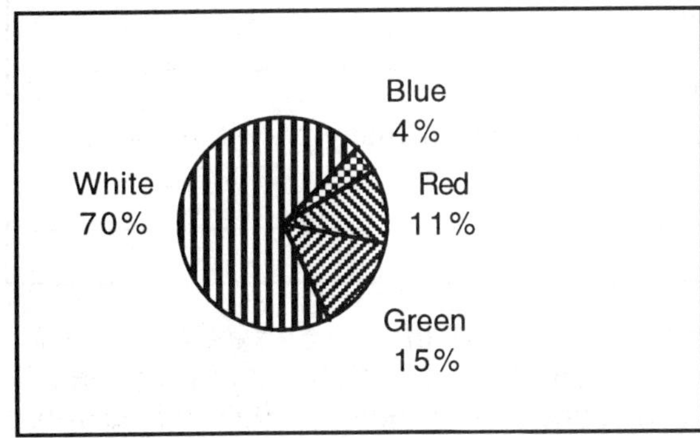

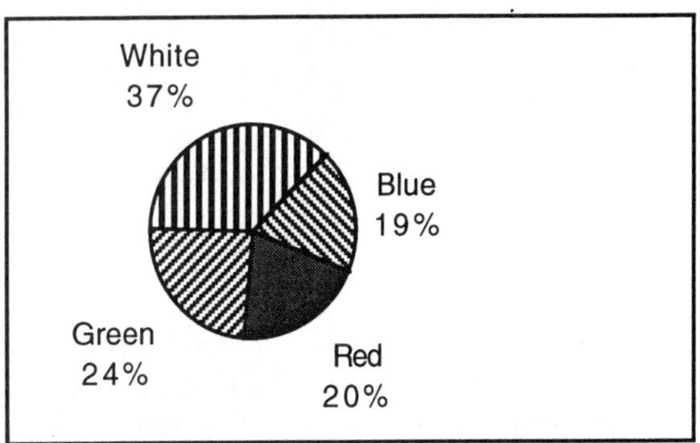

a. Explain what is wrong with the above picture.

b. Explain what is wrong with the above picture.

2. Find the average, median, and range of the numbers, 3, 9, 11, 5, 4, 2, and 8.

Part 2: Calculator recommended.

3. a. Complete the following table.

Name of Item	Percentage of Items	Percent in Category
A	289	
B	765	
C	130	
D	52	
E	23	

Total ________________ ________________

b. Find the average number of items in the five categories. ____________

a. Explain how it is possible for the percents to add to approximately 100% but not exactly.

4. If there is a total of 162 items in all six categories, answer the questions about the following pie chart.

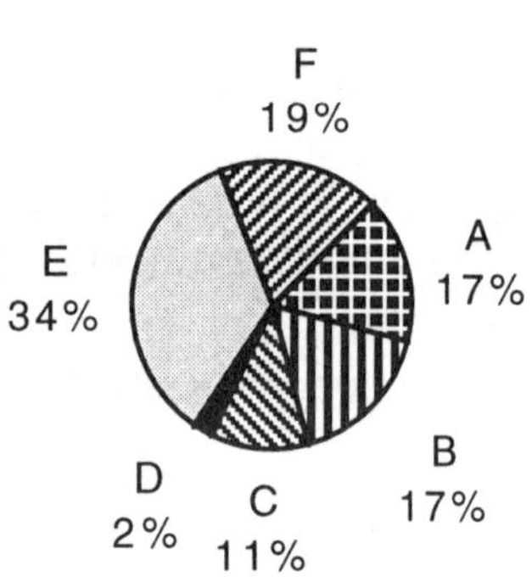

Find the percentage (number of items) in each category?

A ________________

B ________________

C ________________

D ________________

E ________________

F ________________

5. Answer the questions by interpreting the following bar chart.

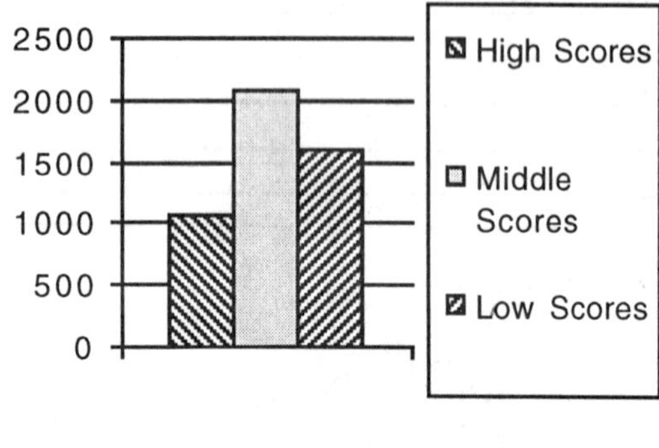

Approximately how many people made high scores?

____

Approximately how many people made middle scores?

____

Approximately how many people made low scores?

____

6. According to the *Labor Force Statistics from the Current Population Survey, Displaced Workers Summary*, as of February, 1996 4.2 million workers with three or more years of tenure were displaced from their jobs over the period from January 1993 to December 1995. Forty-four percent of the displaced workers lost their jobs due to plant closings or moves. How many workers lost their jobs due to plant closings or moves?

7. According to the *1997 Britannica Book of the Year*, Hong Kong's population of about 6.3 million falls into the following age categories.

Ages		Population in Age Group (in millions)
0-14	19%	
15 - 29	23%	
30 - 44	30%	
45 - 59	15%	
60 - 74	10%	
75 and older	3%	

a. Complete the chart to show approximately how much of the population lies in each age group.

b. Draw a bar graph to show the distribution of population.

Chapter 8 Floor Plan Exercises

Key Topics

- To use and test the skills learned throughout the book

Floor Plan 1 (Group)
Percents, Area, Unit Pricing, and Unit Conversion

Purpose of the Exercise
- To practice skills and concepts learned throughout the book

Materials Needed
- Calculator

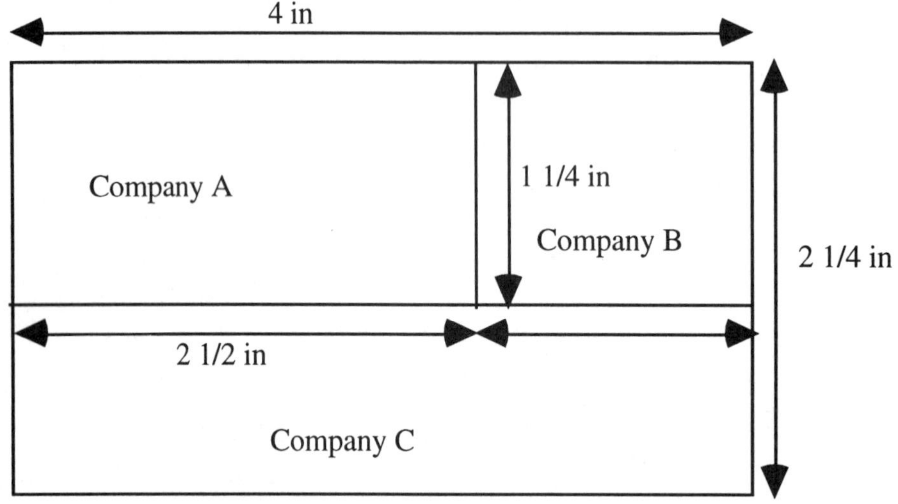

Answers the following questions. Show your computations.

1. What are the dimensions of the space that each company occupies on the drawing?

2. If 1 inch represents 15 feet, what are the dimensions of the actual space in the building that each company occupies?

3. How many square feet does each company occupy?

4. How many square feet does the floor contain?

5. What percentage of total space does each company occupy?

6. If the building's heating bill is $753, how much is each company's share if the bill is prorated according to floor space?

7. Rent for the entire building is $2430 per month. Prorate this bill among the companies according to floor space occupied.

8. What is the <u>annual</u> rent per square foot of this building?

Floor Plan 2 (Group)
Percents, Areas, Unit Pricing, and Unit Conversion

Purpose of the Exercise
- To practice skills and concepts learned throughout the book

Materials Needed
- Calculator

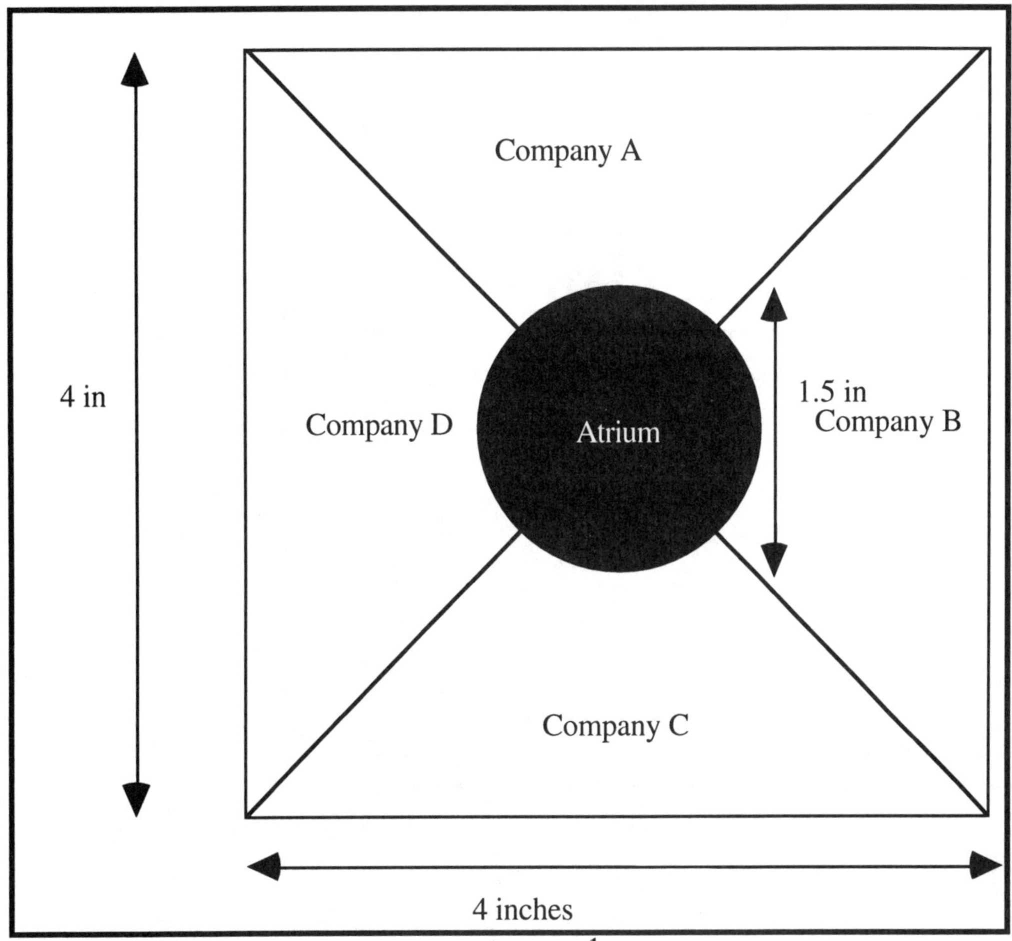

The above square is 4" X 4" and the circle is $1\frac{1}{2}$ inches in diameter. 1" on the drawing represents 14' in the building.

Answer the following questions. Show your computations.

1. Find the total space in square feet for each company and for the atrium.
2. If space in the building rents for $13.00 annually per square foot, how much is each company's monthly rent?
3. Suppose the atrium is to be recarpeted and the carpet selected costs $14.27 per square yard, installed. How much will it cost to install the carpet? Explain why is this an estimated cost, not necessarily representing the exact cost.
4. Suppose the building's electric bill of $1329.32 for July is prorated among the companies according to their floor space as a percentage of the entire building's space. Compute each company's share. How does this change if they also prorate the cost for the atrium?

Floor Plan 3 (Group)
Percents, Areas, Unit Pricing, Unit Conversion

Purpose of the Exercise

- To practice skills and concepts learned throughout the book

Materials Needed

- Calculator

Front of Building

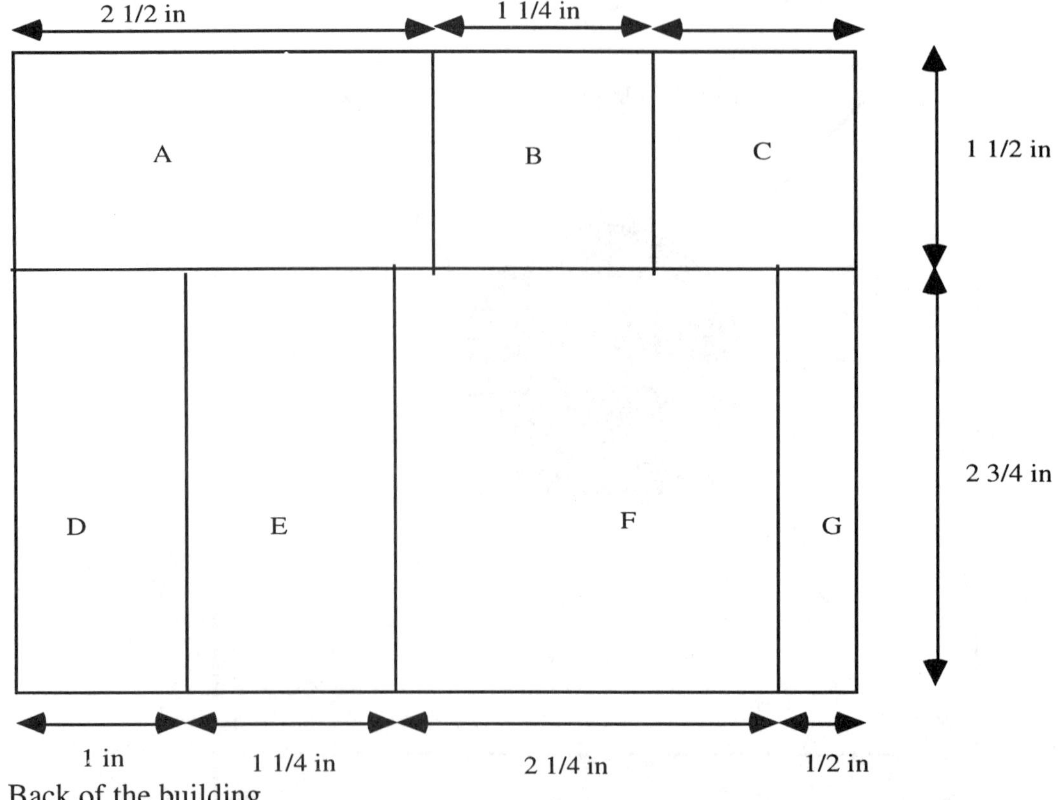

Back of the building

The annual rent for stores in the front of the building is $16.50 per square foot and the annual rent for the back spaces rent is $13.75 per square foot. 1 inch on the drawing represents 20 feet in the building. Answer the following questions. Show your computations.

1. How much space in square feet does the company in store C occupy?
2. When all of the spaces are rented, what is the total monthly rent for the building?
3. If store E is vacant for a month, how much rent for the building can the landlord collect for that month?
4. Suppose the landlord decides to modernize the building the front of the building. The estimated cost is $15.35 per linear foot. How much will the remodeling project cost?
5. If the landlord raises the rent by $1.20 per square foot for the stores in the front of the building and by $0.95 for the stores in the back spaces, how long will it take to recuperate the renovation costs providing all units remain rented?

Floor Plan 4 (Group)
Percents, Areas, Unit Pricing, and Unit Conversion

Purpose of the Exercise

- To practice skills and concepts learned throughout the book

Materials Needed

- Calculator
- Ruler

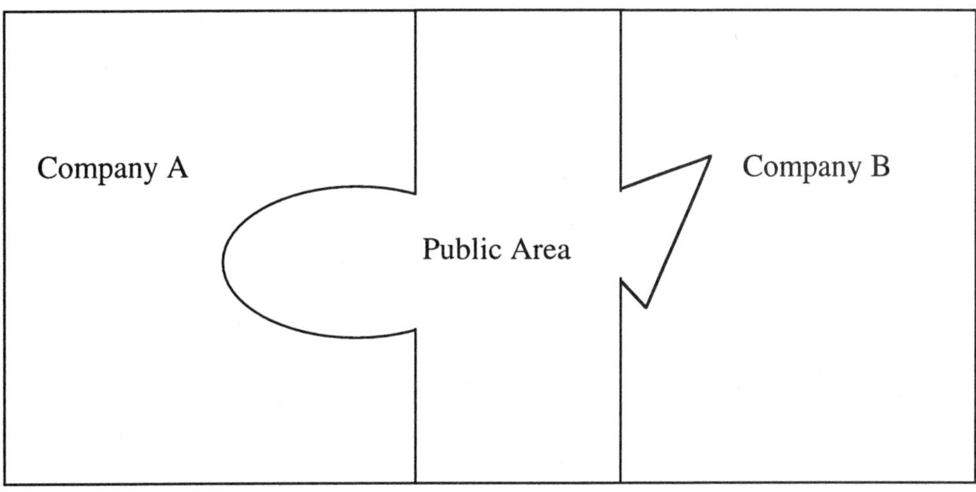

If 1 cm in the above diagram represents 33 meters in a building. Using your ruler, measure the distances and approximate each of the following. If necessary, draw a grid over the areas to approximate the odd shaped areas.

1. The floor space in the building (in square meters).

2. The floor space in the building (in square feet).

3. The floor space of each company (in square meters).

4. The floor space in the public area (in square meters).

5. About what percentage of the total floor space does each company occupy?

6. Prorate a utility bill of $5283.97 according to percentage of floor space. (Assume the utilities for the shared public areas are paid by the building management company and are computed included in the rent).

7. How much is each company's monthly rent if the annual rent is $2.95 per square meter?

8-2 *Reasonableness Exercises*

Answer each of the following questions on a separate sheet of paper. Do not use a calculator.

1. The following drawing represents a floor plan for a building. The total rent for the building is \$2798.27 a month.

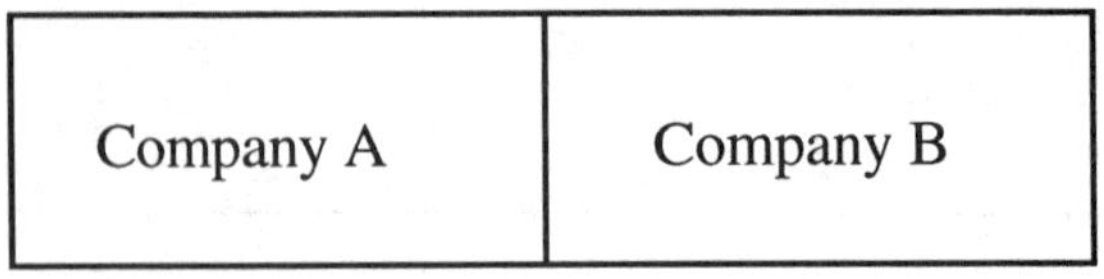

Company A	Company B

   Which of the following gives the best estimate of the prorated rent bill? Explain how you decided.

   a. Company A pays \$1399 and Company B pays \$1399.
   b. Company A pays \$2000 and Company B pays \$798.
   c. Company A pays \$1600 and Company B pays \$1600.
   d. Company A pays \$1500 and Company B pays \$1300.

2. The following drawing represents a floor plan for a building and the total rent for the building is \$5295 a month.

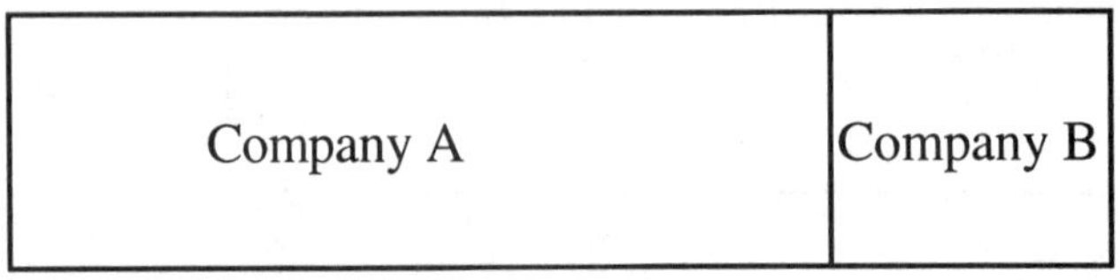

Company A	Company B

   Explain why none of the prorated bills below are correct and provide a reasonable estimate of the rents paid by Company A and Company B.

   a. Company A pays \$2647.50 and Company B pays \$2647.50.
   b. Company A pays \$1620 and Company B pays \$2340.
   c. Company A pays \$3105 and Company B pays \$2195.
   d. Company A pays \$2421 and Company B pays \$3632.

   Reasonable estimate ______________________________

3. The following drawing represents a floor plan for a building. The entire building is 170 ft. by 90 ft.

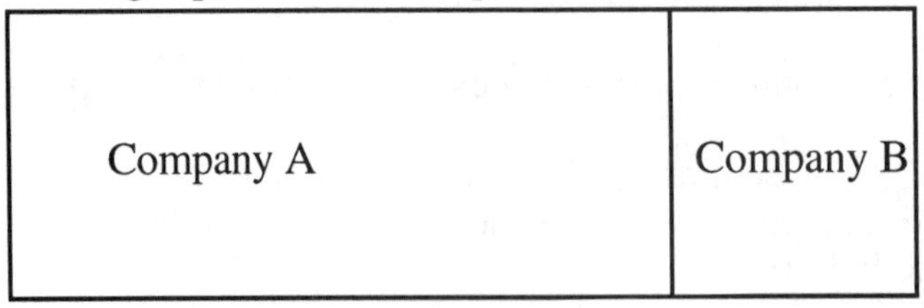

Company A	Company B

   Explain what is wrong with each of the following estimates of floor space for Company A and Company B and provide a reasonable estimate of the floor space occupied by each company.

   a. Company A has 8000 ft^2 and Company B has 7300 ft^2 of floor space.
   b. Company A has 14,286 ft^2 and Company B has 5714 ft^2 of floor space.
   c. Company A has 4371 ft^2 and Company B has 10929 ft^2 of floor space.
   d. Company A has 71,430 ft^2 and Company B has 28,575 ft^2 of floor space.

   Reasonable estimate ______________________________

4. In the following diagram, approximate the amount of floor space in Company A and Company B if the total floor space in the building is 191,000 square feet. Describe how you made your estimation.

Company A	Co. B

5. The heights of the 38 patients who visited Dr. Reinhardt on Monday ranged from 4'2" to 6'3". Explain why it is unlikely that the average height of her patients for Monday was 4'3".

6. Choose is the best approximation for the sum, product, or quotient.

   27.932 + 41.03578 + 87.923 + 94.99   53   137   251   352

   (5,295.87)(789.67)   41,819,890   4,181,198   418,198   41,819

   795.87÷18   4   44   444   4444

7. Some estimates of the national debt put it at about $5 trillion. The population of the United States is about 276 million. Which of the following best estimates each person's share of the national debt?
   $200 $2000 $20,000 $200,000

8. Mary has a bank balance of $239.52 before making the following deposits and writing the following checks. Which of the choices best estimate her bank balance after the transactions.
   Deposit: 34.75   Check: $147.83   Check: $47.74   Deposit: $22.51   Check $145.83

   Bank balance ≈ $45.00   - $45.00   - $90.00 $90.00

9. Which is bigger, 9/8 U or 1.5 U?

10. Which is bigger, $\frac{x}{2}$ or $\frac{x}{1/2}$?

11. Which is the best estimate of the ratio of rise to run?

   1 to 4.1   1 to 1.4   1.4 to 1

12. Choose the best estimated answer for the following problems involving percents.

a. 135.3% of 462.8 ≈	60,000	6000	600
b. 1.9% of 9087.35 ≈	170	1700	17,000
c. 82.73 is what percent of 76.6?	108%	93%	63%
d. 205.9% of a x is 17.6. x ≈	12%	9	36

13. If the ratio of coffee drinkers to noncoffee drinkers is 6:5, which of the following correctly sets up a proportion to estimate how many of the 7895 attendees of a conference are coffee drinkers?

   a. $\frac{x}{7895} = \frac{6}{5}$ b. $\frac{x}{7895} = \frac{6}{11}$ c. $\frac{x}{7895} = \frac{5}{6}$

Summary Exam

Work the problems in the exam on a separate sheet of paper. Write the question and show your work.

Part 1: Do not use a calculator.

1. Do the following arithmetic problems. Show your work on a separate sheet of paper.

a. 347.2 + 18 = ______ b. 87 - 0.956 = ______

c. (-93)(4) = ______ d. -93 ÷ -4 = ______

e. (4,765.987)(100) = ______ f. (7,765.987) ÷ (100) = ______

g. $\frac{3}{5} - \frac{8}{5} =$ ______ h. $\frac{3}{5} - \frac{2}{3} =$ ______

i. $(-7)\left(\frac{4}{5}\right) =$ ______ j. $7 + \frac{4}{5} =$ ______

k. $\left(\frac{10}{51}\right) \div 5 =$ ______ l. $(7)\left(\frac{-1}{5}\right) =$ ______

m. $\frac{12}{11} \div \frac{-2}{5} =$ ______ n. $\frac{-789}{11} + \frac{800}{11} =$ ______

2. Choose the best estimate for the quotient, product or sum.

a. $\frac{4295.38}{2150.993} \approx$	1.3	2	2.6	3
b. $(34.987)(11.9913) \approx$	4100	39	3900	420
c. $890 + 811 + 219 + 560 \approx$	2500	3500	1500	3000
d. $\frac{789}{2098} \approx$	1.2	0.9	0.4	0.1
e. $\frac{8.9}{4.51} \approx$	0.5	1.2	1.9	3.2

3. For each of the following, decide whether the answer will be positive or negative or whether there is not enough information to tell. Circle your choice.

a. a positive number times a negative number	positive	negative	not enough info
b. a positive number + a positive number	positive	negative	not enough info
c. a positive number - a positive number	positive	negative	not enough info
d. a negative number + a negative number	positive	negative	not enough info
e. a negative number ÷ a negative number	positive	negative	not enough info
f. a negative number + a positive number	positive	negative	not enough info

4. For each of the following, round as indicated.

a. 789.36457 to three significant digits ______
b. 789.36547 to three decimal places ______
c. 789.36547 to the nearest tenth ______
d. 789.36547 to the nearest ten ______

5. Circle the best estimated answer for each of the following.

a. 81% of 5789 ≈	3000	4700	6900	5600
b. 124% of 936 ≈	700	825	950	1200
c. 29% of 3478 ≈	1200	1700	2100	2400

6. Solve the following equations for the variable.

a. $5x - 7 = 4$ b. $\frac{x}{5} = \frac{3}{4}$

c. $\frac{3}{5} = t - 5$ d. $3(2 - x) = x + 9$

7. Use the appropriate distributive, commutative, and associative laws to multiply and simplify each of the following.

a. $5(3 - x) =$ b. $5(3)(-x) =$

c. $\frac{3}{5}\left(\frac{15x}{2}\right)\left(\frac{4}{3}\right) =$ d. $7\left(\frac{1}{14} + \frac{x}{49}\right) =$

7. Evaluate the following for a = 3, b = -1, and c = -2

a. $-2abc =$ b. $4c^2 =$

8. The Eagles' ratio of wins to losses is 3 to 2. If they played 25 games, how many games did they win?

Part 2: Calculator recommended.

9. Compute each of the following. Write your answer rounded to two decimal places.

a. $(-3.567)(84) - (-83.6)\div(42.1) - 5.6^2 =$ ____________________

b. $(-29)^2 - 14.5(-10.3) + 5.6(12.3 - 18.9) =$ ____________________

10. ○ ○ ○ ○ ○ ○ ○ ○ ○

a. In the above diagram, what are the following ratios?

The ratio of shaded to unshaded circles is ________________________

The ratio of unshaded circles to the total number is ________________________

b. If the ratio stayed approximately the same in a larger collection of circles and there were 7395 circles, how many circles would be shaded and how many would be unshaded?

The number of unshaded circles would be ________________________

The number of shaded circles would be ________________________

11. Compute the total population, percents, and percentages. Put the information into the chart as shown.

Type of Dog	Percentage in each Breed	Percent of Total Population
German Shepherd	52	
Poodle	29	
Yorkie	17	
Collie	52	
Golden Retriever	45	
Total population		100%

12. Suppose the population in a particular community had religious affiliations as shown on the pie chart below. The total population is 17,392 people. Determine the percentage of (number of) members of each religion, and fill the columns appropriately.

Religion	Percentage of Members	Percent of Total Population
Christian		
Hindu		
Buddhist		
Moslem		
Other		
Total population	17,392	100%

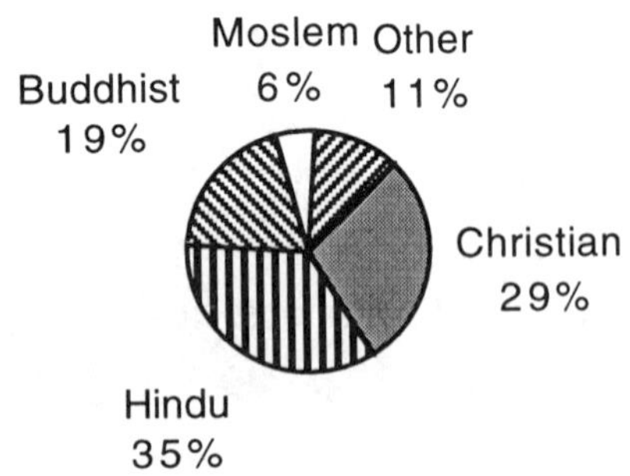

13. Calculate each of the following when T = -13.2 and L = 145.6.
a. $T^2 - 3L =$ ________________ b. $11TL =$ ________________

14. Answer the following questions about the floor plan shown.

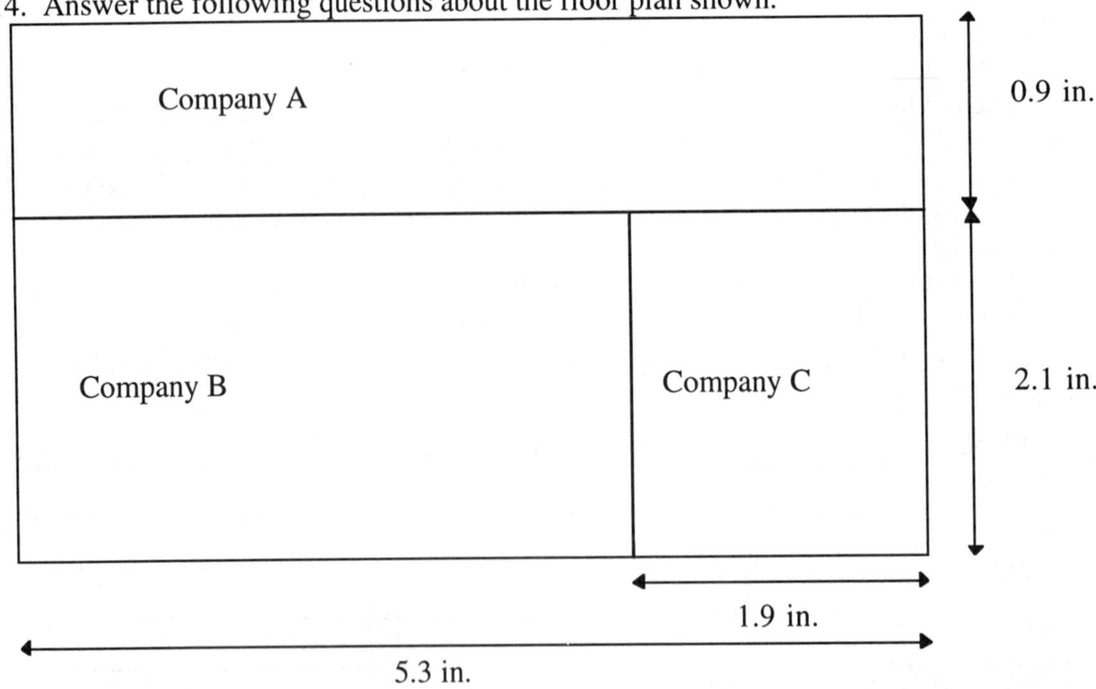

Measuring the dimensions in feet, find each of the following. Include appropriate units with your answers.

a. The dimensions of the building
b. The floor area of the building
c. The dimensions of Company B's space
d. The area of Company B's space
e. The ratio of Company B's space to the entire building
e. The percentage of total building space that Company B occupies
f. Company B's prorated share of a $1,592.46 utility bill.

15. If 1 T = 8 M, make the following conversions.

1 M = ____________ T $1\ T^2$ = ____________ M^2
$1\ M^2$ = ____________ T^2 $5\ M^2$ = ____________ T^2

16. Use the conversion chart at the end of Chapter 5 to make the following unit conversions.

a. 372 ft = ____________ m   b. 372 ft = ____________ mi
c. 62 ft per sec = __________ mi per hr   d. 46 gal = ____________ liters

17. Kevin bought a suit on sale for $82.98. The sales tax rate is 7.25%. How much did Kevin pay for the suit?

18. One car travels North at 37 mph while another car starting at the same place and time travels South at 58 mph. How far are the cars apart after driving $1\frac{1}{2}$ hours?

19. Out of 37 times at bat, Kylin had 19 hits. What is Kylin's batting average?

20. Mark's July electric bill (covering a 31-day period) was $138.93. He used 1374 kilowatt-hours of electricity.
a. What was the average cost of electricity per day?
b. What is the cost of electricity per kilowatt-hour?

21. Kendall's gross pay for August was $3789.65. His pay check was for $2152.87.
a. How much was deducted for taxes, social Security, insurance, and other deductions?
b. What percent of his pay check was deducted for taxes, social security, insurance and other deductions?

22. Challenge Question: Kevin's monthly take-home pay at his current job is $1088. He wants to rent an apartment that costs $630 a month with utilities that run about $80 a month. Kevin pays $320.00 a month for car payment and insurance, and estimates food will cost him about $235 a month. Kevin has found a second part-time job that pays $7.00 an hour and estimates that his take-home pay will be about $5.90 an hour.

a. How many hours will Kevin have to work a month to meet his basic expenses?

b. How many hours a month will Kevin have to work at his part-time job in order to have $150.00 a month for clothes, entertainment and unexpected expenses?

23. Find the radius, area, and circumference of a circle that is 21 feet in diameter.

24. The four sides of a trapezoid are 29 meters, 17 meters, 21 meters, and 36 meters. What is the perimeter of the trapezoid?

Appendix 1 *Names of the Numbers*

This section might be particularly useful if English is not your first language. Even if English is your first language, it gives some insights to why numbers have the names they have.

The digits are:

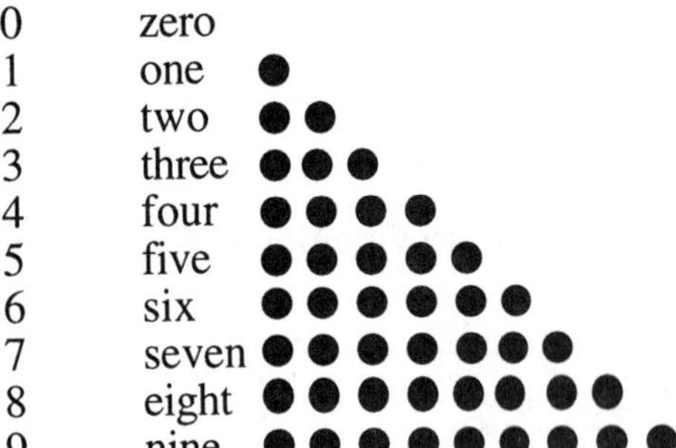

The next pattern of numbers:

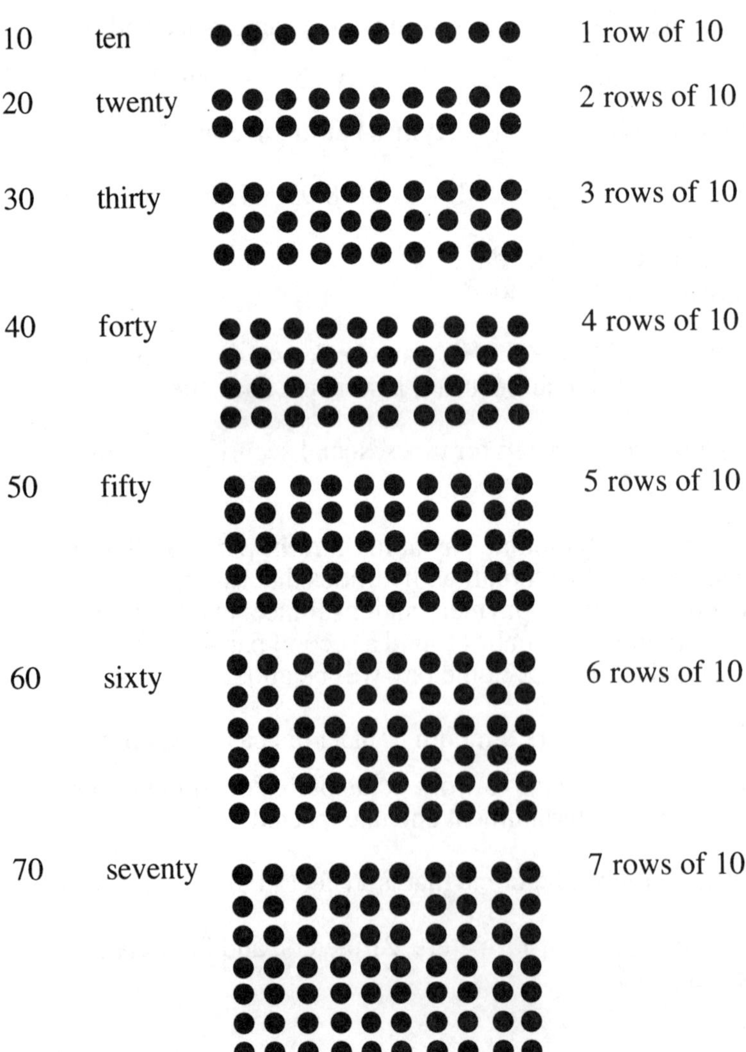

80 eighty 8 rows of 10

90 ninety 9 rows of 10

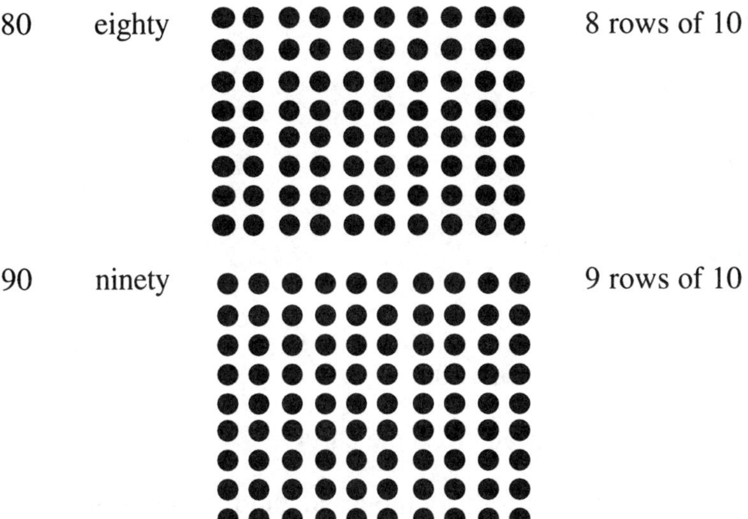

Numbers like 27 mean 20 + 7. Instead of saying "twenty plus seven", we say "twenty-seven." 27 is illustrated below.

27 2 rows of 10 and 1 row of 7.

38 is thirty plus eight, or "thirty-eight"

The numbers between 20 and 100 are all named similarly as the 27 and 38 above. The numbers between 10 and 20 have special names.

11	eleven	10 + 1
12	twelve	10 + 2
13	thirteen	10 + 3
14	fourteen	10 + 4
15	fifteen	10 + 5
16	sixteen	10 + 6
17	seventeen	10 + 7
18	eighteen	10 + 8
19	nineteen	10 + 9

The next groupings of numbers are:

100	one hundred	10 rows of 10 items in each row.
200	two hundred	2 groups of 100
300	three hundred	3 groups of 100
400	four hundred	4 groups of 100
500	five hundred	5 groups of 100
600	six hundred	6 groups of 100
700	seven hundred	7 groups of 100
800	eight hundred	8 groups of 100
900	nine hundred	9 groups of 100

1000	one thousand	10 groups of 100 items
2000	two thousand	2 groups of 1000
3000	three thousand	3 groups of 1000

4000	four thousand	4 groups of 1000
5000	five thousand	5 groups of 1000
6000	six thousand	6 groups of 1000
7000	seven thousand	7 groups of 1000
8000	eight thousand	8 groups of 1000
9000	nine thousand	9 groups of 1000

10,000	ten thousand	10 groups of 1000
20,000	twenty thousand	20 groups of 1000
etc.		

200,000 two hundred thousand 200 groups of 1000

To read a number, break the number into its parts and read the parts.

Example 1: The number 35,000 is a composite of thirty thousand and five thousand so we read it as "thirty-five thousand."

Example 2: 89 is eighty plus 9 so is "eighty-nine."

Example 3: 789 is a composite of seven hundred plus eighty nine, so is read "seven hundred and eighty-nine."

Example 4: The number 35,789 is a composite of thirty-five thousand plus seven hundred plus eighty-nine. 35,789 is read as "Thirty-five thousand, seven hundred and eight-nine."

Example 5: 6,803 is 6 thousand plus 8 hundred plus 3 so is read, "six thousand eight hundred and three."

Example 6: 84,009 is 84 thousand plus 9, so is read' "Eighty-four thousand and nine."

Exercise 1--Naming the Numbers

Purpose of the Exercise

- To practice saying the names of numbers

Write each of the following numbers in words.

1. 59 ____________________
2. 207 ____________________
3. 5207 ____________________
4. 307 ____________________
5. 307,307 ____________________
6. 95,278 ____________________
7. 17,278 ____________________
8. 5,016 ____________________
9. 679 ____________________

Write the following words as numbers.

1. Seven hundred and ninety-three ____________________
2. Eight thousand three hundred fifteen ____________________
3. Eight thousand three hundred fifty ____________________
4. Two hundred and thirteen ____________________
5. Two hundred and thirty ____________________
6. Sixty-two thousand, five hundred and thirty-three ____________________
7. Seven hundred and twenty-five thousand ____________________
8. Seven hundred and twenty-five ____________________
9. Nine hundred and ninety-nine thousand, four hundred forty-four ____________________
10. Five hundred and seven ____________________
11. Five hundred and seventy ____________________
12. Three hundred and sixteen ____________________
13. Three hundred and sixty ____________________

Appendix 2 *Math Facts--Addition and Times Tables*

To understand arithmetic, certain facts are basic enough to memorize. First we need to know the names of the numbers. Times tables and sums of single digits should be memorized, however, they are best remembered if they are understood. The exercises in this appendix are hands-on models for learning and remembering addition and times tables.

Adding Single Digit Numbers

To add two numbers, represent the addition with beans as in the examples below. If the total number of beans adds up to more than 10, regroup the beans as 10 plus the leftovers.

2 + 1 = 3 + =

7 + 5 = 12 +

Regroup these as 10 on the first line, then the 2 left over--this is 10 + 2 = 12

Exercise 1--Adding Single Digit Numbers

Purpose of the Exercise

- To learn or review sums of single digit numbers.

Materials Needed

- 20 beans or other objects.

In each of the following, represent the problem with beans then write the sum of the numbers. If the sum is more than 10, regroup the beans as 10 plus the left over beans to write the answer.

1. 1 + 1 = ____________
2. 1 + 2 = ____________
3. 1 + 3 = ____________
4. 1 + 4 = ____________
5. 1 + 5 = ____________
6. 1 + 6 = ____________
7. 1 + 7 = ____________
8. 1 + 8 = ____________
9. 1 + 9 = ____________
10. 2 + 1 = ____________
11. 2 + 2 = ____________
12. 2 + 3 = ____________
13. 2 + 4 = ____________
14. 2 + 5 = ____________
15. 2 + 6 = ____________
16. 2 + 7 = ____________
17. 2 + 8 = ____________
18. 2 + 9 = ____________
19. 3 + 1 = ____________
20. 3 + 2 = ____________
21. 3 + 3 = ____________
22. 3 + 4 = ____________
23. 3 + 5 = ____________
24. 3 + 6 = ____________
25. 3 + 7 = ____________
26. 3 + 8 = ____________
27. 3 + 9 = ____________
28. 4 + 1 = ____________
29. 4 + 2 = ____________
30. 4 + 3 = ____________
31. 4 + 4 = ____________
32. 4 + 5 = ____________
33. 4 + 6 = ____________
34. 4 + 7 = ____________
35. 4 + 8 = ____________
36. 4 + 9 = ____________

37. 5 + 1 = __________ 38. 5 + 2 = __________ 39. 5 + 3 = __________
40. 5 + 4 = __________ 41. 5 + 5 = __________ 42. 5 + 6 = __________
43. 5 + 7 = __________ 44. 5 + 8 = __________ 45. 5 + 9 = __________
46. 6 + 1 = __________ 47. 6 + 2 = __________ 48. 6 + 3 = __________
49. 6 + 4 = __________ 50. 6 + 5 = __________ 51. 6 + 6 = __________
52. 6 + 7 = __________ 53. 6 + 8 = __________ 54. 6 + 9 = __________
55. 7 + 1 = __________ 56. 7 + 2 = __________ 57. 7 + 3 = __________
58. 7 + 4 = __________ 59. 7 + 5 = __________ 60. 7 + 6 = __________
61. 7 + 7 = __________ 62. 7 + 8 = __________ 63. 7 + 9 = __________
64. 8 + 1 = __________ 65. 8 + 2 = __________ 66. 8 + 3 = __________
67. 8 + 4 = __________ 68. 8 + 5 = __________ 69. 8 + 6 = __________
70. 8 + 7 = __________ 71. 8 + 8 = __________ 72. 8 + 9 = __________
73. 9 + 1 = __________ 74. 9 + 2 = __________ 75. 9 + 3 = __________
76. 9 + 4 = __________ 77. 9 + 5 = __________ 78. 9 + 6 = __________
79. 9 + 7 = __________ 80. 9 + 8 = __________ 81. 9 + 9 = __________

None of the above problems involved adding zero. If you start with a certain number of objects and add zero (or no) objects, you end with the same number you started with.

0 + 0 = 0 1 + 0 = 1 2 + 0 = 2 3 + 0 = 3 4 + 0 = 4 5 + 0 = 5
6 + 0 = 6 7 + 0 = 7 8 + 0 = 8 9 + 0 = 9

Putting together the information from exercise 1 and the sums involving 0, we get the following table of sums.

1+0=1	2+0=2	3+0=3	4+0=4	5+0=5	6+0=6	7+0=7	8+0=8	9+0=9
1+1=2	2+1=3	3+1=4	4+1=5	5+1=6	6+1=7	7+1=8	8+1=9	9+1=10
1+2=3	2+2=4	3+2=5	4+2=6	5+2=7	6+2=8	7+2=9	8+2=10	9+2=11
1+3=4	2+3=5	3+3=6	4+3=7	5+3=8	6+3=9	7+3=10	8+3=11	9+3=12
1+4=5	2+4=6	3+4=7	4+4=8	5+4=9	6+4=10	7+4=11	8+4=12	9+4=13
1+5=6	2+5=7	3+5=8	4+5=9	5+5=10	6+5=11	7+5=12	8+5=13	9+5=14
1+6=7	2+6=8	3+6=9	4+6=10	5+6=11	6+6=12	7+6=13	8+6=14	9+6=15
1+7=8	2+7=9	3+7=10	4+7=11	5+7=12	6+7=13	7+7=14	8+7=15	9+7=16
1+8=9	2+8=10	3+8=11	4+8=12	5+8=13	6+8=14	7+8=15	8+8=16	9+8=17
1+9=10	2+9=11	3+9=12	4+9=13	5+9=14	6+9=15	7+9=16	8+9=17	9+9=18

The Times Tables

Multiplication is just repeated addition. The following examples demonstrate what that means.

Example 1: 4 x 3 means four three's or four rows of three objects each.

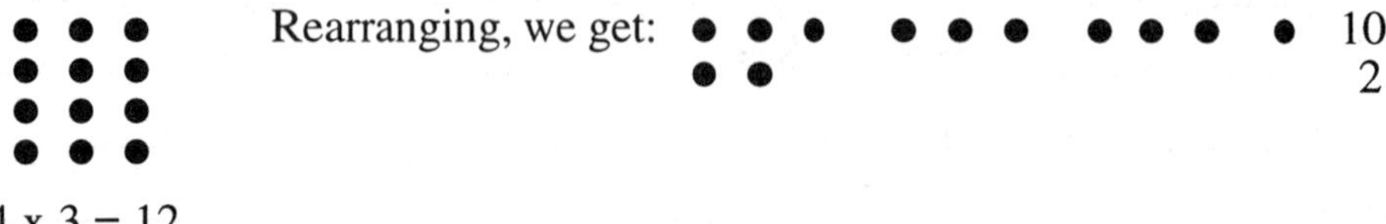

4 x 3 = 12

Example 2: 2 x 7 is two seven's or two rows of seven objects.

● ● ● ● ● ● ● Rearranging, we get ● ● ● ● ● ● ● ● ● ● 10
● ● ● ● ● ● ● ● ● ● ● 4

2 x 7 = 14

Exercise 2--Creating the Times Tables

Purpose of the Exercise

- To learn the times tables by working with objects

Materials Needed

- 100 beans

In each of the following, represent the multiplications with the appropriate number of rows with the same number of beans in each row as in examples 1 and 2. (The first number is the number of rows and the second number is the number of beans in each row.) After the rows are set up, rearrange the beans in rows of tens with the extra beans in the last row. From the rearrangement of beans, determine the product of the two numbers. Record the results below.

1. 1 x 1 = __________	2. 1 x 2 = __________	3. 1 x 3 = __________
4. 1 x 4 = __________	5. 1 x 5 = __________	6. 1 x 6 = __________
7. 1 x 7 = __________	8. 1 x 8 = __________	9. 1 x 9 = __________
10. 2 x 1 = __________	11. 2 x 2 = __________	12. 2 x 3 = __________
13. 2 x 4 = __________	14. 2 x 5 = __________	15. 2 x 6 = __________
16. 2 x 7 = __________	17. 2 x 8 = __________	18. 2 x 9 = __________
19. 3 x 1 = __________	20. 3 x 2 = __________	21. 3 x 3 = __________
22. 3 x 4 = __________	23. 3 x 5 = __________	24. 3 x 6 = __________
25. 3 x 7 = __________	26. 3 x 8 = __________	27. 3 x 9 = __________
28. 4 x 1 = __________	29. 4 x 2 = __________	30. 4 x 3 = __________
31. 4 x 4 = __________	32. 4 x 5 = __________	33. 4 x 6 = __________
34. 4 x 7 = __________	35. 4 x 8 = __________	36. 4 x 9 = __________
37. 5 x 1 = __________	38. 5 x 2 = __________	39. 5 x 3 = __________
40. 5 x 4 = __________	41. 5 x 5 = __________	42. 5 x 6 = __________

43. 5 x 7 = __________ 44. 5 x 8 = __________ 45. 5 x 9 = __________

46. 6 x 1 = __________ 47. 6 x 2 = __________ 48. 6 x 3 = __________

49. 6 x 4 = __________ 50. 6 x 5 = __________ 51. 6 x 6 = __________

52. 6 x 7 = __________ 53. 6 x 8 = __________ 54. 6 x 9 = __________

55. 7 x 1 = __________ 56. 7 x 2 = __________ 57. 7 x 3 = __________

58. 7 x 4 = __________ 59. 7 x 5 = __________ 60. 7 x 6 = __________

61. 7 x 7 = __________ 62. 7 x 8 = __________ 63. 7 x 9 = __________

64. 8 x 1 = __________ 65. 8 x 2 = __________ 66. 8 x 3 = __________

67. 8 x 4 = __________ 68. 8 x 5 = __________ 69. 8 x 6 = __________

70. 8 x 7 = __________ 71. 8 x 8 = __________ 72. 8 x 9 = __________

73. 9 x 1 = __________ 74. 9 x 2 = __________ 75. 9 x 3 = __________

76. 9 x 4 = __________ 77. 9 x 5 = __________ 78. 9 x 6 = __________

79. 9 x 7 = __________ 80. 9 x 8 = __________ 81. 9 x 9 = __________

None of the above problems involved multiplying by zero. Zero multiplied by any number is zero. For example, if you give $0 to 5 friends, this will cost you 0 x 5 dollars. 0 x 5 = 0.

0 x 0 = 0 1 x 0 = 0 2 x 0 = 0 3 x 0 = 0 4 x 0 = 0 5 x 0 = 0
6 x 0 = 0 7 x 0 = 0 8 x 0 = 0 9 x 0 = 0

Putting together the information from exercise 2 and the products involving 0, we get the following times table.

1x0=0	2x0=0	3x0=0	4x0=0	5x0=0	6x0=0	7x0=0	8x0=0	9x0=0
1x1=1	2x1=2	3x1=3	4x1=4	5x1=5	6x1=6	7x1=7	8x1=8	9x1=9
1x2=2	2x2=4	3x2=6	4x2=8	5x2=10	6x2=12	7x2=14	8x2=16	9x2=18
1x3=3	2x3=6	3x3=9	4x3=12	5x3=15	6x3=18	7x3=21	8x3=24	9x3=27
1x4=4	2x4=8	3x4=12	4x4=16	5x4=20	6x4=24	7x4=28	8x4=32	9x4=36
1x5=5	2x5=10	3x5=15	4x5=20	5x5=25	6x5=30	7x5=35	8x5=40	9x5=45
1x6=6	2x6=12	3x6=18	4x6=24	5x6=30	6x6=36	7x6=42	8x6=48	9x6=54
1x7=7	2x7=14	3x7=21	4x7=28	5x7=35	6x7=42	7x7=49	8x7=56	9x7=63
1x8=8	2x8=16	3x8=24	4x8=32	5x8=40	6x8=48	7x8=56	8x8=64	9x8=72
1x9=9	2x9=18	3x9=27	4x9=36	5x9=45	6x9=54	7x9=63	8x9=72	9x9=81

Appendix 3 *String Operations--the Calculator and Fractions*

When doing computations involving several fractions, sometimes we get decimal approximations instead of exact answers. The following examples show how to efficiently use a calculator to key in a string of computations involving fractions.

Example 1: Compute $\frac{4}{7}-\frac{4}{9}$ **by string operations on the calculator.**

First, rewrite the problem on paper on one line.

4/7 - 4/9

Type the problem into the calculator in the following way.

4 [÷] 7 [-] 4 [÷] 9 [=]

The calculator displays the answer, 0.126984127. The number of digits may vary from calculator to calculator.

Example 2: Use a calculator to approximate $3\frac{1}{7}-4\frac{3}{8}$.

Remembering that $3\frac{1}{7}$ is the same as $3+\frac{1}{7}$ and that $4\frac{3}{8}$ is the same as $4+\frac{3}{8}$, we can write the problem in the following way:

(3 + 1/7) - (4 + 3/8) and on the calculator key in:

[(] 3 [+] 1 [÷] 7 [)] [-] [(] 4 [+] 3 [÷] 8 [)] [=] The answer is about −1.232

Example 3: Find a decimal approximation for $\dfrac{\frac{4}{53}-\frac{29}{37}}{\frac{49}{56}-\frac{8}{11}}$.

First write in the implied parentheses around the entire numerator and entire denominator.

$$\frac{\left(\frac{4}{53}-\frac{29}{37}\right)}{\left(\frac{49}{56}-\frac{8}{11}\right)}$$

Rewrite the problem in the following manner.

(4/53 - 29/37) / (49/56 - 8/11)

Key in the following key strokes on the calculator.

[(] 4 [÷] 53 [-] 29 [÷] 37 [)] [÷] [(] 49 [÷] 56 [-] 8 [÷] 11 [)] [=]

This gives an answer of −4.794727965

Example 4: Compute a decimal approximation for $\dfrac{-7-\left(-\dfrac{14}{49}\right)}{\dfrac{32}{17}+(-29)}$.

Rewrite the problem in the following manner.

$\left(-7+\dfrac{14}{49}\right)\div\left(\dfrac{32}{17}-29\right)$ Now to write this to compute on the calculator.

On a calculator where one number is displayed at a time:

[(] 7 [+/-] [+] 14 [÷] 49 [)] [÷] [(] 32 [÷] 17 [-] 29 [)] [=]

On a calculator where the entire expression is displayed at once:

[(] [-] 7 [+] 14 [÷] 49 [)] [÷] [(] 32 [÷] 17 [-] 29 [)] [=]

This gives an answer of 0.2475983886

Exercise 1--String Operations With Fractions on the Calculator

Purpose of the Exercise

- To practice efficient use of the calculator for computing arithmetic with a scientific calculator

Materials Needed

- Calculator

Using string operations on your calculator, compute the approximate answers to each of the following problems. Round off the final answer to three decimal places. The first one is done as an example.

Fraction	The Problem Rewritten for the Scientific Calculator	The Rounded Answer
1. $\dfrac{\dfrac{7}{19}-\dfrac{14}{9}}{\dfrac{-25}{33}+\dfrac{-5}{9}}$	$\left(\dfrac{7}{19}-\dfrac{14}{9}\right)\div\left(\dfrac{-25}{33}+\dfrac{-5}{9}\right)=$ $\left(\dfrac{7}{19}-\dfrac{14}{9}\right)\div\left(\dfrac{-25}{33}-\dfrac{5}{9}\right)=$ (7÷19 - 14÷9) ÷ (25 [+/-] ÷ 33 - 5÷9) =	0.904
2. $\dfrac{\dfrac{84}{99}-\dfrac{-22}{81}}{-88+\dfrac{83}{276}}$		
3. $\dfrac{97-\dfrac{15}{83}}{-88-\dfrac{83}{276}}$		

4. $\dfrac{\frac{4}{5}-(-23)}{77}$		

5. $\dfrac{-92}{\frac{88}{7}+\frac{52}{9}}$		
6. $\dfrac{1}{\frac{1}{670}+\frac{1}{1290}+\frac{1}{795}}$		
7. $\dfrac{\frac{1}{670}+\frac{1}{1290}+\frac{1}{795}}{\frac{1}{324}}$		
8. $\dfrac{\frac{92}{63}-898}{-89+\frac{25}{86}}$		
9. $-4\left(7\frac{3}{4}-9\frac{2}{13}\right)$		
10. $\dfrac{-4\frac{2}{9}+3}{5\frac{3}{4}+1\frac{3}{5}}$		
11. $\dfrac{-\frac{74}{17}+\frac{23}{35}}{11+\frac{5}{3}}$		

Index